Generative Artificial Intelligence

This book comprehensively covers designing and implementing misinformation intervention strategies and risk communication approaches for generative artificial intelligence (AI) technologies.

Key Features:

- Showcases cutting-edge technology and offers insights into one of the most rapidly evolving fields in AI.
- Provides a comprehensive understanding of the underlying principles, algorithms, and applications of this cutting-edge technology, offering insights into both its theoretical foundations and practical implementations.
- Discusses generative AI models, unraveling the contrasts between machine learning, deep learning, and generative AI.
- Presents AI-based models for the detection and generation of fraudulent content.
- Covers topics such as speech generation, deepfake technology, pose synthesis for animation, and sports performance analysis.

This book is primarily written for senior undergraduates, graduate students, and academic researchers in the fields of electrical engineering, electronics and communications engineering, computer science and engineering, and information technology.

Generative Artificial Intelligence

Technology and Applications

Edited by
Deepika Varshney, Preeti Nagrath,
Srishti Vashishtha, and
Victor Hugo C. de Albuquerque

CRC Press
Taylor & Francis Group
Boca Raton London New York

CRC Press is an imprint of the
Taylor & Francis Group, an **informa** business

Designed cover image: Shutterstock

First edition published 2026
by CRC Press
2385 NW Executive Center Drive, Suite 320, Boca Raton FL 33431

and by CRC Press
4 Park Square, Milton Park, Abingdon, Oxon, OX14 4RN

CRC Press is an imprint of Taylor & Francis Group, LLC

ISBN: 9781032931647 (hbk)
ISBN: 9781032933986 (pbk)
ISBN: 9781003565703 (ebk)

DOI: 10.1201/9781003565703

Typeset in Times
by codeMantra

Contents

Preface

This book provides a comprehensive and application-driven perspective on Generative Artificial Intelligence (GenAI), emphasizing its transformative role in a broad spectrum of real-world scenarios. It not only introduces the foundational principles of explainable AI (XAI) but also demonstrates how explainability can be embedded within generative models to make them more transparent and trustworthy.

This book delves into state-of-the-art generative techniques, including GANs, VAEs, and transformer-based models, while also exploring interpretability methods such as SHAP, LIME, feature attribution, and counterfactual reasoning. These tools are essential in understanding and trusting the decisions made by complex GenAI systems. What sets this book apart is its diverse set of application areas where GenAI is making a significant impact. From image generation and computer vision to education technologies, plant disease diagnosis, marketing analytics, stock market forecasting, and social media content moderation and security, this book highlights how generative models are being tailored to solve domain-specific challenges with precision and creativity.

Bridging theory with practical use cases, this book includes case studies and real-world examples that showcase successful deployments of GenAI and explainable frameworks in various sectors. The content is designed to be accessible yet technically rigorous, making it ideal for graduate students, researchers, and professionals in computer science, electrical and electronics engineering, biomedical domains, and information technology. Overall, this book serves as a practical guide to harnessing the power of GenAI with explainability, equipping readers with both conceptual depth and hands-on strategies to implement cutting-edge solutions across a wide array of disciplines.

Biography

Dr. Deepika Varshney has been working as an Assistant Professor and serving academics and research in the Department of Computer Science and Engineering & Information Technology, Jaypee Institute of Information Technology, Noida, UP, since 2022. During her research career at Delhi Technological University, New Delhi, since 2019, she has worked on the cutting-edge issue on social web platforms, specifically fraudulent content detection on social media platforms, and has received commendable research award of 50,000 for publishing article in SCI-indexed journals in 2020, 2022, 2023, and 2024. She was also awarded with Senior Research Fellowship (SRF). She has published 20 research papers in SCI, SCOPUS, and international conferences. She has supervised both major and minor projects and is presently guiding two Ph.D. scholars. In addition, she has been actively involved in organizing short-term training programs, faculty development programs (FDPs), and various academic initiatives. She also serves as a reviewer for several prestigious SCI-indexed journals.

Dr. Preeti Nagrath is working as an Associate Professor at Bharati Vidyapeeth's College of Engineering. She has more than 20 years of academic experience. She holds a B.Tech, an M.Tech, and a Ph.D. in Computer Science and Engineering. Her areas of research include network security, delay tolerant networks, machine learning, and deep learning. She is a member of ISTE. She has more than 60 research papers in SCI-indexed journals, highly reputed journals, and international conferences. She has worked on a government-funded DST project on women security. She has chaired many sessions at international conferences. She is appointed as a reviewer for many conferences and journals. She has organized many faculty development programs, workshops, hackathons, and guest lectures. She has mentored teams in Smart India Hackathons. She has worked as an Associate Editor for the *Journal of Multidisciplinary Engineering Technologies* at Bharati Vidyapeeth's College of Engineering in New Delhi. She is currently teaching deep learning and machine learning, web engineering, network security, software testing, and object-oriented software engineering.

Dr. Srishti Vashishtha is currently working at Bharati Vidyapeeth's College of Engineering (BVCOE), GGSIPU, as an Assistant Professor in the Department of Computer Science and Engineering. She has received her Ph.D. degree in Sentiment Analysis and Social Media domain from Delhi Technological University in 2022. She has received a B.Tech degree in Information Technology from the Maharaja Surajmal Institute of Technology, GGSIPU, New Delhi, India, and an M.Tech degree in Computer Science from the University School of Information, Communication and Technology, GGSIPU. Her current areas of interest include machine learning, deep learning, data mining, natural language processing, sentiment analysis, speech emotion recognition, and fuzzy logic. She has about 10+ years of teaching experience in esteemed institutions. She qualified GATE and UGC-NET. She has published 26 research papers in peer-reviewed/Scopus/Web of Science-indexed international journals and international conferences. She, with a current h-index of 11, has published three Scopus-indexed and six SCIE journal papers. She has published a book entitled *Metaverse Technologies in Healthcare* and holds two Indian design patents. She is a reviewer for various SCIE and peer-reviewed journals. She has received DTU Research Excellence awards for commendable research (four times) in 2020, 2021, 2022, and 2023. She has received three Best Researcher awards from BVCOE in 2023 and 2024. She has received the Researcher of the Year award at the Universal Innovators Leadership Awards (UILA) organized by ICICC in 2024.

Victor Hugo C. de Albuquerque (Senior Member of IEEE) is currently a Professor and Senior Researcher at the Department of Teleinformatics Engineering (DETI)/Graduate Program in Teleinformatics Engineering (PPGETI) at the Federal University of Ceará (UFC), Brazil. He earned a Ph.D. in Mechanical Engineering from the Federal University of Paraíba (UFPB, 2010) and an M.Sc. in Teleinformatics Engineering from the PPGETI/UFC (UFC, 2007). He completed a BSE in Mechatronics Engineering at the Federal Center of Technological Education of Ceará (CEFETCE, 2006). He has experience in biomedical science and engineering, mainly in the research fields of applied computing, intelligent systems, visualization and interaction, with specific interest in pattern recognition, artificial intelligence, image processing and analysis, as well as automation with respect to biological signal/image processing, biomedical circuits, and human/brain–machine interaction, including augmented and virtual reality simulation modeling for animals and humans. He is a Full Member of the Brazilian Society of Biomedical Engineering (SBEB). He is an Editor-in-Chief of the *Journal of Biomedical and Biological Sciences*, as well as the *Journal of Artificial Intelligence and Systems* and the *Journal of Biological Sciences*, and an Associate Editor of the *IEEE Journal of Biomedical and Health Informatics*, *Computers in Biology and Medicine*, *Frontiers in Cardiovascular Medicine*, *Computational Physiology and Medicine*, *Applied Soft Computing*, *IEEE Access*, *Frontiers in Communications and Networks*, *Computational Intelligence and Neuroscience*, *Measurement*, and *IET Quantum Communication*. He has been a Lead Guest Editor of several highly reputed journals and a TPC member of many international conferences.

Contributors

Yashika Agarwal
Department of Computer Science & Engineering and information technology
Jaypee Institute of Information Technology
Noida, Uttar Pradesh, India

Kirti Aggarwal
Department of Computer Science and Engineering
Jaypee Institute of Information Technology
Noida, Uttar Pradesh, India

Ajaydeep
Department of Computer Science and Engineering & Information Technology
Jaypee Institute of Information Technology
Noida, Uttar Pradesh, India

Anuja Arora
Department of Computer Science and Engineering
Jaypee Institute of Information Technology
Noida, Uttar Pradesh, India

Deepika Bansal
Department of Information Techonology
Maharaja Agrasen Institute of Technology
Delhi, India

Shikha Bhalla
Department of Information Techonology
GTBIT, GGSIPU
Delhi, India

Deepika Bhatia
School of Engineering & Technology
Vivekananda Institute of Professional Studies – Technical Campus
Delhi, India

Saloni Bhushan
Department of Computer Science/IT
Chhatrapati Shivaji Maharaj University
Panvel, Maharashtra, India

Abhishek Chaudhary
Department of Computer Science and Engineering
Jaypee Institute of Information Technology, Noida

Poonam Chaudhary
Department of Computer Science and Engineering
The NorthCap University
Gurugram, Haryana, India

Arunita Chaukiyal
Department of Computer Science
Acharya Narendra Dev College, University of Delhi
Delhi, India

Hem Dutt Dabral
STMicroelectronics Pvt. Ltd.
Greater Noida, Uttar Pradesh, India

Harsh Dhanwani
Department of Computer Science and Engineering & Information Technology
Jaypee Institute of Information & Technology
Noida, Uttar Pradesh, India

Chhavi Dhiman
Department of Electronics and Communication Engineering
Delhi Technological University
New Delhi, India

Tina Dudeja
Department of Computer Science and Engineering
Maharaja Agrasen Institute of Technology
Delhi, India

Abhinav Garg
Department of Electronics and Communication Engineering
Delhi Technological University
New Delhi, India

Nitin Garg
Department of Computer Science and Engineering
Maharaja Agrasen Institute of Technology
Delhi, India

Dhanalekshmi Gopinathan
Department Computer Science and Engineering
Jaypee Institute of Information Technology
Noida, Uttar Pradesh, India

Punit Gupta
Department of Computer Science and Engineering
National college Ireland
Dublin, Ireland

Kirti Jain
Department of Computer Science and Enginnering and Information Technology
Jaypee Institute of Information Technology
Noida, Uttar Pradesh, India

Kunal Kartikeya
Jaypee Institute of Information Technology
Noida, Uttar Pradesh, India

Priyansh Kashyap
Department of Computer Science and Engineering
Bharati Vidyapeeth's College of Engineering
Delhi, India

Mansimran Kaur
Department of Computer Science and Engineering
Bharati Vidyapeeth's College of Engineering
Delhi, India

Ajay Kaushik
Department of School of Computing
College of Science and Engineering, University of Derby
Derby, United Kingdom

Surbhi Bhatia Khan
University of Salford
Salford, Greater Manchester, England

Sarthak Khanna
Department of Computer Science and Engineering
Bharati Vidyapeeth's College of Engineering
Delhi, India

Tejna Khosla
Department of Information Technology
Maharaja Agrasen Institute of Technology
Delhi, India

Anshu Khurana
Department of Computer Science and Engineering
Maharaja Agrasen Institute of Technology
Delhi, India

Utku Köse
Department of Computer Science and Engineering
University of North Dakota
Grand Forks, USA

Devansh Kukreja
Department of Computer Science and Engineering & Information Technology
Jaypee Institute of Information & Technology
Noida, Uttar Pradesh, India

Anil Kumar
Department of Applied Science
Bharati Vidyapeeth's College of Engineering
Delhi, India

Ashish Kumar
School of Computer Science Engineering & Technology
SCSET, Bennett University,
Greater Noida, Uttar Pradesh, India

Vijay Kumar
Department of Computer Science and Engineering
Bharati Vidyapeeth's College of Engineering
Delhi, India

Prakhar Madnani
Department of Computer Science and Engineering & Information Technology
Jaypee Institute of Information & Technology
Noida, Uttar Pradesh, India

Archit Malhotra
Department of Computer Science & Engineering and information technology
Jaypee Institute of Information & Technology
Noida, Uttar Pradesh, India

Preeti Nagrath
Department of Computer Science and Engineering
Bharati Vidyapeeth's College of Engineering
Delhi, India

Rachna Narula
Department of Computer Science and Engineering
Bharati Vidyapeeth's College of Engineering
Delhi, India

Kaustubh Pandey
Department of Computer Science and Engineering & Information Technology
Jaypee Institute of Information & Technology
Noida, Uttar Pradesh, India

Megha Rathi
Department of Computer Science and Engineering & Information Technology
Jaypee Institute of Information & Technology
Noida, Uttar Pradesh, India

Harsha Ratnani
Department of Information Technology
Jagannath International Management School
Delhi, India

Neetu Sardana
Department of Computer Science and Engineering & Information Technology
Jaypee Institute of Information & Technology
Noida, Uttar Pradesh, India

Aishwarya Shanker
Department of Computer Science & Engineering and information technology
Jaypee Institute of Information & Technology
Noida, Uttar Pradesh, India

Surabhi Shanker
Dept. of CSE, School of Engineering & Technology, Centre of Excellence-Cyber Security
K.R. Mangalam University
Gurugram, Haryana, India

Pratima Sharma
Computer Science Department
Roosevelt University
Illinois, USA

Prachi Shrivastava
School of Engineering & Technology
Vivekananda Institute of Professional Studies – Technical Campus
Delhi, India

Harikesh Pratap Singh
Department of Computer Science and Engineering & Information Technology
Jaypee Institute of Information & Technology
Noida, Uttar Pradesh, India

Adwitiya Sinha
Department of Natural and Applied Sciences
TERI School of Advanced Studies
Delhi, India

Alka Singhal
Department Computer Science and Engineering
Jaypee Institute of Information Technology
Noida, Uttar Pradesh, India

Jaya Srivastava
Department of Computer Science & Engineering and information technology
Jaypee Institute of Information Technology
Noida, Uttar Pradesh, India

Vikalp Srivastava
Jaypee Institute of Information Technology
Noida, Uttar Pradesh, India

Raghav Tiwari
School of Engineering & Technology
Vivekananda Institute of Professional Studies – Technical Campus
Delhi, India

Ajay Tripathi
Department of Information Technology
Jaipuria Institute of Management
Ghaziabad, Uttar Pradesh, India

Sonal Tuteja
Department of Computer Science & Engineering and information technology
Jaypee Institute of Information Technology
Noida, Uttar Pradesh, India

Nitish Uppal
Department of Computer Science and Engineering
Maharaja Agrasen Institute of Technology
Delhi, India

Ashwani Varshney
Dean of student welfare
Jaipuria Institute of Management
Ghaziabad, Uttar Pradesh, India

Deepika Varshney
Department of Computer Science and Engineering & Information Technology
Jaypee Institute of Information Technology
Noida, Uttar Pradesh, India

Preeti Varshney
Management Department
Institute of Professional Excellence & Management
Ghaziabad, Uttar Pradesh, India

Adithya Venkateswaran
School of Engineering & Technology
Vivekananda Institute of Professional Studies – Technical Campus
Delhi, India

Priyanka Verma
University of Galway
Galway, Ireland

Neha Yadav
Department Computer Science and Engineering
Jaypee Institute of Information Technology
Noida, Uttar Pradesh, India

1 Generative Artificial Intelligence

Foundations and Its Transformative Impact on Human Lives

Surabhi Shanker and Saloni Bhushan

1.1 INTRODUCTION

Generative Artificial Intelligence (AI) is a remarkable area of technology that doesn't just analyse and manipulate data—it also produces new content, including text, images, music and program codes, by using what is already known from datasets. Unlike traditional AI, which mainly focuses on classification or prediction, generative AI emphasizes the creation of new data and content, which uses advanced models to produce surprisingly creative and comprehensive original outcome, and it is a common feature of its work. As a result, this capability advances machine learning (ML) while uncovering new opportunities for people. Computer interaction and innovation are important topics. Because technology is changing so rapidly, generative AI has gained importance in our lives and is useful across different sectors. The technology is useful for creating content automatically in the media, and entertainment sectors are helping to transform design, healthcare and education. Firms are using this technology to make their processes more efficient and create personalized services for users to take part in experiences and even come up with new answers to tough problems. When we talk about generative AI, we can see how it is changing the way various industries function over time. Technologies that once belonged to the world of science fiction are now being used. We will use this chapter to examine the core ideas of generative AI and look into its major effect on the way we live. It is designed to give readers a clear idea of why these systems work, how they are important now and the many ways because of which the way we interact with technology is evolving. The chapter first discusses the opportunities and challenges of generative AI before moving on to further investigation of its contribution to innovation and transformation in our lives.

DOI: 10.1201/9781003565703-1

1.2 FOUNDATIONS OF GENERATIVE ARTIFICIAL INTELLIGENCE

1.2.1 Historical Development and Evolution

Historical Development and Evolution of Generative AI: The historical development and evolution of generative AI is an interesting journey, which has been evolving for several decades with continuous innovations and breakthroughs. In the early days, researchers tried rule-based systems and simple algorithms that simply produced random outputs or patterns. These first steps paved the way for a notion that computers could be taught to generate new content as well as to analyse data. Early neural network architectures appeared in the field in the 1980s and 1990s. While rudimentary by today's standards, these models started to look at how computers could learn patterns in the underlying data. Boltzmann machines and autoencoders were introduced as techniques to show machines could capture and reproduce aspects of the data they were fed. These were foundational approaches that hinted at the potential for more sophisticated models, which could produce more and more complex and meaningful outputs. Deep learning made the real revolution in generative AI happen in the 2010s. The period was one of great change in which researchers developed more advanced models that were able to work with enormous amounts of data and complex problems. Then in 2014, the introduction of Generative Adversarial Networks (GANs) by Ian Good fellow changed everything. A new framework presented by GANs, in which two neural networks, the generator (which produced the content) and the discriminator (which assessed its authenticity) worked together to generate impressive realistic outputs. This breakthrough proved that machines could not only be taught to replicate, but also to be creative, to generate outputs that were strikingly similar to human-generated content [1]. Following on, Variational Autoencoders (VAEs) were developed and, more recently, transformer-based models such as GPT and BERT. These models have expanded the realm of what natural language processing (NLP) can do; machines can now produce coherent, contextually appropriate text that can be surprisingly human. The progress of generative AI is closely linked to the progress of computational power, large datasets and training methodologies in general that have together turned the potential of AI from simple pattern recognition to creative expression. Thus, it can be said that the history of the development of generative AI is a story of gradual progress from early rule-based systems and rudimentary neural networks to modern advanced deep learning architectures. This journey has been made up of each step, which has led to the wonderful capabilities of the modern generative AI and has set the stage for its transformational effect on human lives [2].

1.2.2 Key Concepts and Definitions

Generative AI is best understood through a few key concepts and definitions that form its backbone. Generative AI is all about creativity at its heart, it is not only meant to recognize patterns in data, but also generate new, unique content based on what it has learned.

Generative AI itself is one fundamental concept. This is referring to systems that can generate completely novel outputs (like text, images or sounds) by learning from huge datasets. It is like a digital artist that looks at thousands of paintings and then paints its own paintings using those styles.

Latent Space is closely related. Think of this as a compressed blueprint or map of all the important features of the data. Like an architect using blueprints to design different buildings, the model uses this latent space to understand and recreate variations of the original data.

Generative vs. Discriminative Models is another important distinction. Generative models are like creative storytellers; they learn the whole story of how data is formed and can then generate new examples. On the other hand, discriminative models are more like sharp eyed inspectors, who look for boundaries in the data (e.g., to decide if an email is spam or not).

Training Data is then the raw material that goes into these models. The better the quality and variety of this data, the better the model can learn to mimic real-world patterns. A well-trained model is a chef who, after tasting many recipes, learns how to combine the ingredients to make something new but familiar.

The other important concept is the **Loss Function**, which tells us how far the model deviates from the expected output. It's really the model's performance score. In training, the system tries to minimize this loss and fine tunes itself to make better and more accurate results.

During this training process, the training is done using **backpropagation and Optimization**. The model figures out where it went wrong using backpropagation (calculating errors) and optimization algorithms (like gradient descent), which help the model to correct these mistakes by adjusting the model's internal parameters (just as a musician would refine a performance through practice).

Lastly, **Overfitting and Regularization** means that a model is overfitted when it has become too much adapted to its training data, i.e., noise and little details, and this hurts it on new data. This can be prevented with regularization techniques, which are methods to keep the model flexible and able to perform well on different data, not just the data it has seen before.

Understanding these key concepts and definitions helps you understand how the generative AI learns, creates and evolves. These foundational ideas explain how these systems go from recognizing patterns to generating content, and ultimately, how these systems can change our interaction with technology in creative and meaningful ways.

1.2.2.1 Models: Generative vs. Discriminative

Generative vs. discriminative models are really two ways of looking at and using data. Similar to writers of creative stories, **generative models** learn the entire picture of how data is formed, including the underlying patterns and relationships, and are able to generate new similar data. A generative model is something like imagining reading a bunch of fairy tales and then writing a new story in the same style. It doesn't just decide what category a new story might fall into; it actually generates a new story that fits with what it has learned.

Discriminative models, on the other hand, are more like sharp-eyed judges. The main thing they are supposed to do is to discern what category something belongs to by looking at the boundaries between categories. For example, they are good at classifying an email as spam or not spam or telling if an image is of a cat or a dog. Discriminative models learn to distinguish precisely on the features in the data instead of creating new content. Generative models essentially aim to understand the whole distribution of the data in order to recreate or generate new examples, while discriminative models focus on drawing a clear line between different types or classes of data. One creates and the other classifies, and they both play a very important role in ML.

1.2.3 The Core Technologies and Algorithms

i. Generative AI is based on core technologies and algorithms that make it possible for AI to generate new content that resembles human creativity. GANs, for example, pit two neural networks against each other—a generator that generates synthetic data and a discriminator that determines how realistic the data is. VAEs complement GANs by compressing data in a compact latent space, then reconstructing it and generating smooth, varied outputs. Transformer-based models like GPT and BERT leverage self-attention mechanisms to understand context and produce coherent, contextually rich text, transforming NLP. These core technologies, taken together, demonstrate the innovative spirit of generative AI and, at the same time, are the basis for its innovative impact on creative, industrial and scientific domains.

ii. **Generative Adversarial Networks (GANs):**

The technique chosen to design the generator is GANs. GANs are powered by a simple yet powerful idea: two neural networks, the generator and the discriminator, play a friendly competition. The generator's job is to generate new data (e.g., images or text) that mimic real examples, and the discriminator's job is to identify the generator's fake data from real data from the real world. As time goes by, both networks get better as they compete. At this point, the generator is getting better and better at generating realistic outputs, and the discriminator is getting better and better at spotting the slightest differences. GANs achieve this dynamic push and pull and can generate incredibly authentic content, like photorealistic images, artistic renderings or even new music, making them crucial for creative AI applications. The workflow of GAN is shown in Figure 1.1. In this figure, we show the adversarial framework consisting of a Generator and a Discriminator. The Generator generates synthetic data from random noise, and the Discriminator judges the authenticity of data by comparing real and synthetic data samples. They train simultaneously and improve each other via adversarial feedback.

Over time, researchers have proposed several improvements to the original GAN architecture to enhance stability, control, and output quality. One of the most notable advancements is the introduction of **StyleGAN** [3],

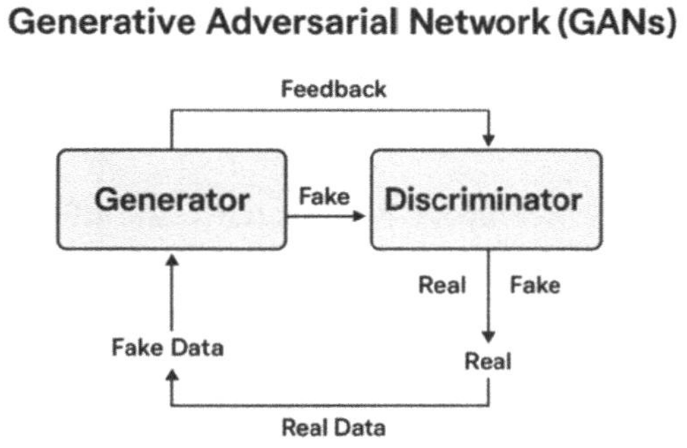

FIGURE 1.1 Architecture of a Generative Adversarial Network (GAN).

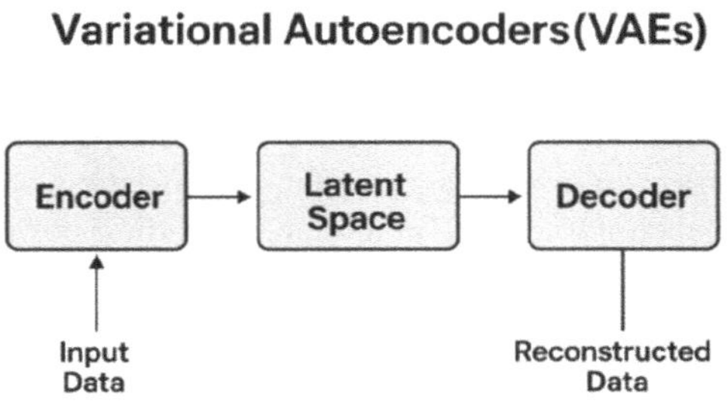

FIGURE 1.2 Architecture of a Variational Autoencoder (VAE).

a style-based GAN architecture. Unlike traditional GANs, StyleGAN enables fine-grained control over features at different levels of image synthesis by using a mapping network and adaptive instance normalization layers. This results in unprecedented photorealism and the ability to manipulate attributes such as facial expressions, hair style, and background independently.

1.2.3.1 Variational Autoencoders (VAEs)

There is another take on this called Variational Autoencoders, abbreviated as VAEs. It comes with two parts, which are encoder and decoder. An encoder compresses input data (e.g., an image) into a simpler, lower dimensional representation, often called latent space. The essence of the data is captured in this latent space in a compact form. Then, the compressed version of this is sent to the decoder, which tries to reconstruct the original data. What is special about VAEs is that they give a probabilistic twist—instead of learning a fixed code for each input, they learn to describe each input as a distribution. This allows us to generate new data by sampling from these learned distributions, which leads to smooth variations and creative outputs that are also used in image synthesis, anomaly detection, etc. Figure 1.2 shows the structure of a VAE: an encoder that maps the input data to a probabilistic latent space and a decoder that reconstructs the data from this latent representation. The model can generate new data by sampling from latent distributions learned.

Figure 1.2 shows the structure of a VAE, which is an encoder that projects the input data into a probabilistic latent space and a decoder that generates the data from this latent representation. The model can generate new data by sampling from latent distributions learned.

1.2.3.2 Other Transformer-Based Models (GPT and BERT)

A transformer-based model is a model that is based on a mechanism called self-attention and have revolutionized how the language data is handled. This mechanism permits the model to process all parts of a sentence or part of text at the same time, considering their relationships and context instead of sequentially processing the words. For example, GPT (Generative Pre-trained Transformer) models are trained on a huge amount of past text to predict the next word in a sequence conditioned on the entire past sequence of words. This makes GPT great for things like writing articles, writing dialogue or creative storytelling. For example, BERT (Bidirectional Encoder Representations from Transformers) is designed to get the full context of a word by considering words that come before and after it. The bidirectionality of this understanding makes BERT very effective for tasks that need deep understanding of text, like question answering and language translation. Models of both types have achieved new benchmarks in NLP and are providing powerful tools for many applications [4].

All these core technologies are key to pushing generative AI forward. GANs excel in creating realistic and innovative content, VAEs provide a robust framework for understanding and generating variations of data, and transformer-based models have transformed how we interact with and understand human language. Figure 1.3 illustrates the core architecture of transformer-based models

This figure illustrates a high-level view of transformer-based models, such as GPT and BERT, with a **bottom-to-top data flow**. The diagram captures the essential flow from input tokens through various **Transformer Blocks**, each consisting of

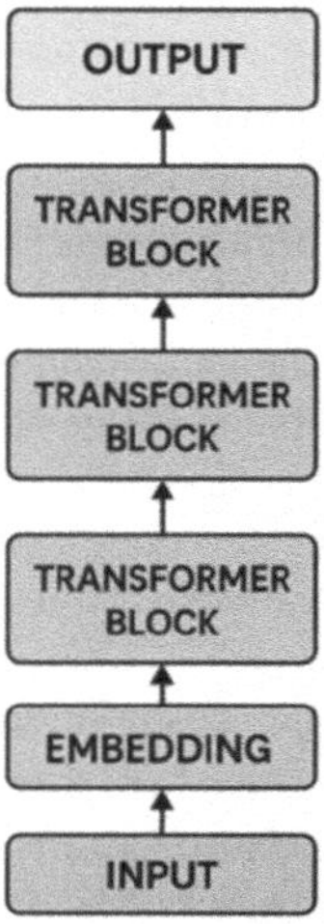

FIGURE 1.3 Architecture of Transformer-based models (e.g., GPT and BERT).

key operations like self-attention and normalization (implicitly represented by the layered structure). The **Input** layer at the base feeds data upward into a stack of transformer blocks, which progressively refine the representation of the input. The **Output** at the top represents the final contextualized embeddings or predictions, depending on the task. This bottom-up layout reflects how transformer-based models process input data through multiple layers to build a deep understanding of language, enabling powerful applications in generation (GPT) and comprehension (BERT).

1.2.4 Data, Training Methodologies and Computational Infrastructure

Generative AI is only possible because of the three pillars: data, training methodologies and computational infrastructure. Data is at the heart of this process; it is the vast and diverse collections of text, images, audio, etc. that form the learning material for the AI models. Because the quality and the representative power of the dataset determine how well a model will capture patterns, nuances and context from the real world. Basically, the richer the data, the more varied, the more robust the AI can be, the more creative its outputs will be. **Training methodologies** used to train these models to generate new content are equally important. Algorithms for training include adjusting the model's internal parameters to minimize errors with techniques like backpropagation and optimization methods (e.g., gradient descent). In training, models iterate through cycles, learning to reproduce the underlying structure of the data by comparing their outputs to actual examples and refining their predictions during many cycles. Moreover, techniques like regularization aid in preventing overfitting, i.e., ensuring that the model performs well on new (unseen) data and not just on the training set only. The **computational infrastructure**, the powerful hardware and software environments supporting both the data handling and the training methods, makes the training feasible. Generative AI models in modern days require massive computational power to perform elaborate calculations, which means they typically need the assistance of high-end GPUs (Graphics Processing Units), TPUs (Tensor Processing Units) and distributed computing systems. Specialized AI clusters and cloud computing platforms bring the scale and flexibility needed to train models on massive datasets in parallel, speeding up innovation. This is a symbiotic ecosystem where data drives learning, robust methodologies improve performance and high-powered infrastructure makes it all possible, together.

1.2.5 Building Blocks of Generative AI

Built upon a very strong foundation of interdependent parts that allow it to learn from data and create creative outputs, generative AI is being seen as the future of AI. High-quality data, diverse, labelled and representative datasets are at the heart of these systems: these are the raw materials for training. This data is then fed into preprocessing pipelines that consist of normalization, tokenization, augmentation and data formatting to make ML model ready. The next fundamental block is the design of the model architecture that tells AI how to interpret and transform the data. Deep neural networks, especially transformer-based architectures like GPT and BERT, are the most common generative systems used today, as they are able to capture complex data dependencies through things like self-attention [5,6].

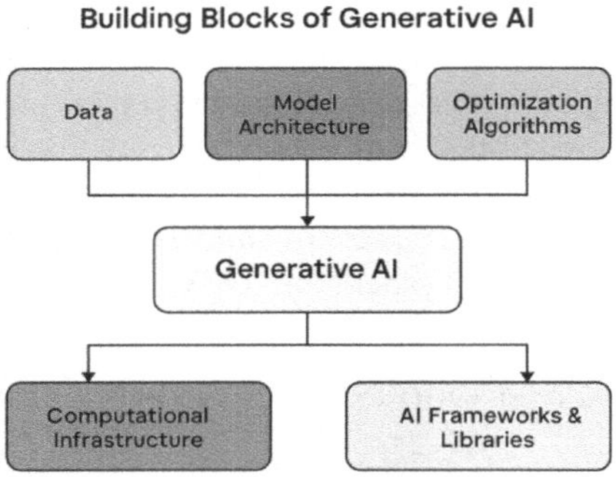

FIGURE 1.4 Building Blocks of Generative AI.

Powerful optimization algorithms, like gradient descent and backpropagation, are used to train these models by iteratively adjusting model parameters so as to reduce error according to a loss function. More sophisticated implementations of reinforcement learning can be used to achieve alignment of model behaviour with long-term goals or user feedback. An essential layer of computational infrastructure supports these models. Because large-scale models require significant computation, GPUs, TPUs and cloud-based high-performance computing environments are indispensable for efficient training and inference. AI frameworks and libraries, including TensorFlow, PyTorch and Hugging Face Transformers, finally accelerate development by supplying reusable components, pre-trained models and deployment utilities. Not only do these tools smooth up experimentation but they also make advanced AI capabilities accessible to a wider audience. Together, these building blocks provide the technological and methodological foundation upon which generative AI is built and which enables its growth, scalability and real-world impact. As illustrated in Figure 1.4, the following are the core components that power generative AI systems such as providing high-quality data, data preprocessing pipelines, deep learning model architectures (for example, transformers), optimization algorithms, computational infrastructure and supporting AI development frameworks.

1.3 TRANSFORMATIVE IMPACT ON HUMAN LIVES

The impact of generative AI on human lives is profound, far-reaching and multi-dimensional, changing not only how people relate to technology but also the very fabric of their day-to-day lives. Fundamentally, this is about expanding human creativity, efficiency and connectivity. Generative AI enables machines to make art, music, literature and even new business strategies, which serves as a catalyst for creative industries to push the limits of what's possible and give people new ways to explore and express ideas.

Generative AI is not only about creativity, it is changing the way we work and how we do business. Some ways AI is enhancing operational excellence and efficiency across sectors including automation of routine tasks, personalized customer service through advanced chatbots and targeted marketing campaigns. This shift not only benefits productivity but also allows human talent to do what it does best: higher level problem solving and strategic innovation [7].

It is equally transformative in healthcare. AI-driven tools are helping speed drug discovery, enabling personalized medicine that personalizes treatment to individual genetic profiles and improving diagnostic precision with advanced imaging and analysis. Not only do these breakthroughs promise faster, more effective treatments, but they also lead to earlier interventions and more accurate diagnoses, which ultimately improve patient outcomes and quality of life.

Generative AI is also changing education. Personalized learning experiences and intelligent tutoring systems have paved the way for education that is customized to the individual needs of each student in a more engaging and effective learning environment. This method closes learning gaps and gives learners an opportunity to reach their full potential, which is not possible in traditional education methods.

Finally, generative AI has wide-ranging societal and cultural impacts. *Democratization of creativity*: With powerful AI tools available to anyone, individuals from all walks of life can create compelling content, which shifts the cultural landscape to be more inclusive.

At the same time, the evolving workforce dynamics that are a result of automation and AI integration require new skills and adaptability, which demands a fundamental shift in how we approach work and career development. The sweeping change that generative AI will bring to human lives is transformational, affecting every aspect of our lives, from art and business to healthcare and education, ushering in new eras of innovation and personalization and new opportunities for everyone. Each sub-area [8] is explained in detail with a detailed explanation of how generative AI will transform human lives.

1.3.1 Creative Industries Impact

1.3.1.1 Arts and Design

Art and design have been revolutionized by generative AI, which is allowing machines to create works that were previously the sole domain of human creativity. Now artists employ AI algorithms, including GANs, to create new virtual patterns, abstract shapes and even realistic portraits that combine the traditional with the digital. These algorithms can play around with different styles and colour palettes, which give artists an infinite canvas to work with. And so designers are beginning to find ways to innovate in fields as far flung as fashion and architecture, where AI-generated designs can be refined and personalized further.

1.3.1.2 Music and Content Creation

Music and Content Creation (MC2) Generative AI is being a powerful collaborator in the realm of music and content creation. AI-powered tools can write original pieces of music, create other sounds with it and write lyrics inspired by a particular mood or style. Musicians and composers use these abilities to try out new genres or break out of creative ruts. Content creators are also using AI for more than just music: drafting articles, writing scripts and creating multimedia content is streamlined, and new avenues for storytelling are opening up.

1.3.1.3 Literature and Storytelling

Literature and Storytelling are amongst the most important mediums for the transmission of culture. Also, generative AI is also leaving its mark in the literature and storytelling by assisting writers to produce narratives, poems and entire novels. AI models can analyse huge amount of textual data and suggest plot twists, character arcs or even generate entire passages with a coherent style. It doesn't replace human authors, but rather provides an inspirational tool, offering creative prompts, helping refine language and experiment new narrative structures and ultimately expanding the boundaries of storytelling.

1.3.2 Impact on Business and Industry

1.3.2.1 Automation and Content Generation

Demand for creating high quality and quick content is increasing day by day in today's fast-paced digital world. Whether it's writing reports, updating social media or analysing market trends, there is simply a never ending demand for fresh, engaging content. But Generative AI is a game changer, automating content creation and changing the way companies work.

1.3.2.2 Automating Report Generation Allows More Time for Strategy

The process of creating business reports has traditionally been a time-consuming affair requiring data collection, analysis and careful formatting. Now, with generative AI, this process is so much faster and efficient. With AI-powered tools, you can extract insights from large datasets, summarize key findings and generate well-structured reports in minutes. It helps business professionals spend more time on interpretation of results and strategic decisions instead of hours of compiling information [8]. For example, financial firms use AI to create earnings reports, marketing teams use AI to track campaign performance and create detailed analysis reports and so on. AI makes sure businesses are agile and data driven by handling these repetitive yet important tasks.

1.3.2.3 Engaging with Audiences in Real Time: Social Media Updates

Social media is managed by consistent, relevant and timely content. Audiences need to be engaged with fresh posts, tweets and updates, sometimes multiple times a day and brands need to do that. This is simplified with generative AI, which automatically writes social media posts, captions and even picks the right images or hashtags for your audience. There are some AI-driven tools that can crawl trending topics and come up with content ideas that are suited to the brand's identity, so companies never have to worry about coming up with new ideas. Also, AI responses to customer queries or comments keep the engagement going, but without a social media manager online 24/7.

1.3.2.4 Market Analysis and Content Personalization

To succeed at business, you need to understand market trends and customer preferences but trawling through mountains of data can be daunting. AI-powered analytics tools can **scan news articles, competitor updates, customer reviews and social media discussions to identify emerging trends and insights.** These insights allow businesses to create data-backed content that resonates with their target audience.

Moreover, AI can **personalize content for different customer segments**, ensuring that advertisements, emails and product recommendations are more relevant. For example, e-commerce platforms use AI to generate personalized product descriptions, while streaming services like Netflix or Spotify use AI to recommend content based on user behaviour [9].

1.3.2.5 Ensuring Consistency, Speed and Cost Efficiency

One of the biggest advantages of using AI for content generation is the **consistency and speed** it brings. Unlike human writers or marketers who may have creative blocks or require breaks, AI can generate **error-free, well-structured and brand-aligned content at any scale**—whether it's one report or a thousand product descriptions.

Additionally, **automating content creation significantly reduces costs.** Instead of hiring large teams to manually draft reports, social media posts or marketing materials, businesses can leverage AI to handle these tasks efficiently, allowing employees to focus on higher-level strategic work.

1.3.2.6 The Human-AI Collaboration: AI as an Assistant, Not a Replacement

While AI can automate many aspects of content creation, it still lacks human creativity, emotional intelligence and the ability to craft deeply nuanced messages. The best results come from a **collaborative approach**, where AI handles the groundwork, and humans refine and enhance the output.

For instance, AI can generate the first draft of a blog post, but a human writer can add **storytelling elements, emotions and personal experiences** to make the content more engaging. Similarly, AI-powered design tools can suggest visuals, but human designers add the finishing touches to ensure a unique brand identity [10].

1.3.2.7 The Future of AI in Content Generation

Generative AI is **not here to replace human creativity, but to amplify it**. By taking over repetitive and data-intensive tasks, AI **frees up professionals to focus on strategy, creativity and innovation**. As businesses further embrace AI-driven automation, the success lies in the balance between AI efficiency and human creativity, so that content is both high quality and deeply impactful. AI will continue to get smarter in the next few years and will provide more personalization, deeper audience insights and seamless content integration with other digital platforms. The opportunity for businesses willing to adapt to AI-driven automation is huge in an increasingly competitive market.

1.3.2.8 Chatbots, Virtual Assistants, Customer Service

Nowadays, in the fast growing digital world, customers expect quick, accurate and convenient responses to their questions regarding a product, issue solving or personalized recommendations. The traditional models of customer service, which essentially rely on human agents, are often unable to cope with the increased demand for 24/7 support. When it comes to AI-powered chatbots and virtual assistants, they are smart, responsive and always there to help.

With these AI-driven systems, customer service is being revolutionized by providing real-time inquiry handling, shorter wait times, increased efficiency and instant, personalized customer assistance. So let's see how they are doing it [11].

1.3.2.9 Always Available: 24/7 Customer Support Without Delays

Waiting is one of the biggest frustrations for customers: waiting on hold when calling or waiting hours (or even days) for a response to an email. The problem of this is eliminated by AI-powered chatbots, which provide instant support all the time which means 24×7. A customer comes to an e-commerce website late at night with questions about a product's return policy. Rather than having to wait until business hours to chat with someone, they can chat with an AI assistant that will immediately provide them with the information they need.

1.3.2.9.1 Smarter Conversations: Understanding and Responding Like a Human

Early chatbots were frustrating because they relied on rigid, pre-programmed responses, often misunderstanding customer queries. Today's AI-driven virtual assistants are powered by **NLP and ML**, which enable them to:

- **Understand natural human language** (even with typos or slang)
- **Detect customer sentiment** (e.g., recognizing frustration or urgency)
- **Provide more personalized responses** based on previous interactions

For example, if a customer messages an airline chatbot asking, *"Can I bring my dog in the cabin?"*—a traditional bot might struggle. But a modern AI assistant can **interpret the intent**, check airline policies and provide a detailed answer while also offering additional options like pet-friendly seating or travel tips.

- Reducing Workload for Human Agents: AI Handles Routine and Humans Solve Complex Issues.

Not all customer inquiries require human intervention. In fact, **AI can handle up to 80% of routine questions** like:

- Order status updates
- Password resets
- Refund and return policies
- Basic troubleshooting

By managing these repetitive tasks, AI allows human agents to **focus on more complex, high-value interactions**—like handling disputes, offering emotional support or solving unique problems that require critical thinking.

A great example is in banking. Instead of having agents answer simple balance inquiries, an AI chatbot can instantly provide account details, while human representatives focus on more complicated cases like fraud investigations or financial advisory [12].

1.3.2.9.2 Personalized Assistance: AI Remembers and Adapts

One of AI's biggest advantages is its ability to **learn from past interactions** and tailor responses based on customer history. As a result, personal and relevant conversations occur. If a returning customer reaches out to a chatbot on an online clothing site, the AI might:

- Greet the person by their name.
- Suggest styles to you based on what you have bought in the past.
- Deliver order tracking details for each customer's recent purchase.

Personalization at this level helps customers feel appreciated and improves their experience, which raises loyalty to the brand and keeps them coming back.

1.3.2.10 Multichannel Support: Seamless Assistance across Platforms

Today's customers interact with brands through multiple channels—websites, social media, messaging apps and even voice assistants like Alexa or Siri. AI-powered customer service integrates seamlessly across these platforms, ensuring customers receive consistent support wherever they are [13].

Say a customer talks with a chatbot on a company's website and then decides to continue the chat on WhatsApp, without needing to explain their issue again. Using omnichannel, consumers have a smoother and more convenient experience.

1.3.2.11 Cost Savings and Scalability: Doing More with Less

Hiring and training human agents for large-scale customer support can be expensive. AI-powered chatbots **reduce operational costs** by handling thousands of conversations simultaneously, eliminating the need for massive support teams.

For startups and small businesses, this is a game changer. Instead of hiring a large customer service team, they can deploy AI assistants to **manage inquiries efficiently at a fraction of the cost**. As the business grows, AI can scale effortlessly to handle increased demand without additional overhead.

1.3.2.11.1 AI and Humans Working Together: The Best of Both Worlds

Despite their efficiency, AI chatbots and virtual assistants **aren't perfect replacements for human empathy, creativity, or deep problem-solving**. So, the best models in customer service use both automated AI and skilled human staff.

For example:

- AI can address common concerns and send the rest to people for further handling.

- During live chats, virtual assistants can share suggestions, show the history of each customer and recommend appropriate answers for agents.
- AI helps companies look at customer feedback and conversations, which enables them to make their service strategies better over a period.

Because of this cooperation, customers get their problems solved quickly, accurately and with empathy.

1.3.2.12 The Future of AI-Powered Customer Service

In the end, AI appears to be an important part of customer service in the future. AI has changed customer service by making it quicker, more intelligent and available to more people. AI-powered customer service is available all the time, offers personal interactions and supports users across all channels. This is making users happy and allowing businesses to cut costs and expand their operations.

As AI develops further, virtual assistants will be able to interact with people by voice, understand their feelings and tailor experiences to each person. Still, the human touch is something that will always be needed. AI is improving the way businesses and support teams can serve their customers, not replacing humans in the process.

1.3.2.13 Personalized Marketing and Advertising

Generative AI is changing the way marketing and advertising work by allowing more personalized promotions. Analysing consumer data like what people look at online, what they buy and their social media activity, AI helps brands make personalized offers. It helps to increase customer interaction and the number of conversions for businesses, and it also helps them spend their advertising funds wisely by reaching the best audience.

1.3.3 Impact on Healthcare

1.3.3.1 Drug Discovery and Personalized Medicine

Think of a day when new medicines can be found in weeks and each treatment is made for an individual based on their genes, so it works best and causes fewer problems. Because of Generative AI, we are getting closer to this future at a fast pace. For many years, drug discovery has taken a lot of time, cost a lot of money and had unpredictable results. A drug usually takes 10–15 years and a lot of money to bring to market, and most possible treatments do not work in trials. AI is now making a difference—it helps researchers work more efficiently, predict how drugs might react when combined and makes personalized medicine more available.

1.3.3.2 How AI Is Transforming Drug Discovery

The main purpose of drug discovery is to discover molecules that can work on biological targets (such as proteins or enzymes) to treat various diseases. Before, scientists used trial-and-error testing, which required testing thousands of compounds in laboratories. AI is making it possible to do this process much faster and with better results.

Here's how:

- **Simulating Molecular Interactions:** AI models can be used to study how billions of molecules behave in a digital environment and guess how they may interact with our bodies. This drastically reduces the need for physical experiments.
- **Generating New Drug Candidates**: Instead of searching for existing molecules, AI can create entirely new ones, designing drugs that are optimized for effectiveness and safety.
- **Predicting Side Effects Early**: AI can analyse vast datasets of chemical structures and biological responses, allowing researchers to anticipate potential side effects before clinical trials even begin.

In short, AI-powered drug discovery allows researchers to **test thousands of potential drugs in a fraction of the time**, making the process faster, cheaper and more precise.

1.3.3.3 Personalized Medicine: Tailoring Treatments to Individuals

Not all patients respond to the same medication in the same way. A treatment that works wonders for one person might be ineffective—or even harmful—for another. This is where **personalized medicine** comes in, and AI is making it more powerful than ever.

Using ML, AI can:

- **Analyse a person's genetic data** to determine how their body will respond to certain medications.
- **Identify biomarkers** (biological signals) that indicate how likely a person is to develop a specific disease.
- **Design custom treatments** that are optimized for an individual's unique genetic and biological makeup.

For example, in cancer treatment, AI-powered personalized medicine can identify which chemotherapy drugs will be most effective for a patient, minimizing harmful side effects and maximizing survival rates. Because of this approach, medicine is changing, as it now offers more targeted and effective solutions instead of the general approach it used in the past.

1.3.3.4 AI in Action: Real-World Breakthroughs

Generative AI is now playing a major role in drug development. Let me give you some exciting examples:

AI-Designed Drugs in Clinical Trials: In Silico Medicine and BenevolentAI have used AI to develop drugs for clinical trials, and their candidates have shown good results in treating fibrosis and Parkinson's.

AI-Powered COVID-19 Treatments: AI was used in the COVID-19 pandemic to look at drugs that were already available and recommend new combinations that could treat the virus. Because of this, researchers could identify successful treatments more quickly than with old methods.

Personalized Cancer Treatments: AI is helping healthcare professionals choose the right treatment for each patient based on their genetic results, which leads to more people surviving and less harm from side effects.

They prove that AI is helping to make drugs more effective, lowering costs and ensuring everyone can get the medicine they need.

1.3.3.5 Challenges and the Road Ahead

Problems and the Future AI is transforming drug discovery, but it also has its own set of issues.

Getting Medical Data: For AI to work properly, it needs access to a lot of clear medical data, but this information is often spread and hard to obtain because of privacy rules.

Regulatory Hurdles: It is much regulated to develop drugs and AI-generated drugs are no exception; they must still be tested thoroughly before approval.

Trust and Adoption: Both doctors and patients need to trust the suggestions made by AI, which requires that AI models are clear, easy to understand and fair.

These problems have not stopped AI from making great progress in medicine. With AI models getting better, drug discoveries will quicken, treatments will become more effective and healthcare will be more personalized than ever.

1.3.3.6 A New Age of Medical Care

Generative AI is changing the way we think about and manage diseases, not only making drug discovery faster. AI is speeding up the discovery of new drugs and helping to customize treatments for each person, which benefits healthcare safety, effectiveness and accessibility.

We might soon live in a world where:

- Cancer, Alzheimer's and rare genetic disorders are now treated with highly targeted treatments designed by AI.
- Doctors give patients drugs according to their DNA, making sure they get the most benefit with the least risk.
- It takes only a few months to move from discovering a drug to treating patients, not many years.

Because of AI's influence on drug discovery and personalized medicine, we now have medicine that is more intelligent, faster and adapted to every individual—and this leads to hope and healthier outcomes for millions.

1.3.3.7 Diagnostic Imaging and Analysis

AI helps doctors examine and interpret complex images from X-rays, MRIs and CT scans in healthcare diagnostics. Some algorithms are able to spot early signs of cancer or neurological problems that a person might miss. By making diagnosis more accurate, this better supports patients by giving doctors the chance to treat them early and also reduces the stress on medical staff from analysing many images.

1.3.4 Impact on Education

1.3.4.1 Personalized Learning Experiences

Generative AI is providing students with learning that matches their abilities and the way they learn. AI-powered software can evaluate a person's skills and knowledge as they go, sharing unique lessons, interactive tasks and helpful advice. When teachers use this approach, students stay interested and handle challenges better, which supports better educational results.

1.3.4.2 Intelligent Tutoring Systems

AI-driven intelligent tutoring systems imitate the experience of working with a tutor by immediately helping students with difficult topics. They make use of advanced NLP to analyse what students ask and provide detailed answers, so they can better understand complex ideas. AI tutors provide a helpful learning buddy, which improves understanding and helps students remember more information when human tutors are scarce.

1.3.5 Societal and Cultural Impacts

1.3.5.1 Democratization of Creativity

Generative AI has greatly democratized creativity. Because AI tools are now easier to get, anyone can make art, music, literature and design without being formally trained. Thanks to the democratization, more creative people are included and encouraged to share their original ideas.

1.3.5.2 Changes in Workforce Dynamics

Changes in the Way People Work More and more AI is being used, which is changing the way workforces operate. Even though automation gets rid of routine work, it also results in the creation of jobs and industries that call for advanced technical and creative abilities. Because of this shift, workers must enhance their skills to fit the changing job market. In addition, AI being used in many industries is transforming the way organizations are built and how work is done, encouraging greater agility and innovation.

All these sectors highlight that generative AI goes beyond being a technology and is now shaping all kinds of human activities, including creative jobs, business improvements, medical discoveries, education and social change.

1.4 DEALING WITH CHALLENGES AND ETHICAL MATTERS

Generative AI has a lot of potential, yet it brings with it some problems and ethical issues. We should look at some significant issues:

1.4.1 Can AI Really Avoid Bias?

AI learns using the data it has been trained with, and if that data has biases (which is common), AI might unknowingly support stereotypes. As an example, if an AI model is trained mostly based on Western media, it may not give all cultures the fairness they deserve. You can do this by causing issues in hiring, making content and providing healthcare. To make AI more fair and inclusive, researchers are working on ways to remove biases from training data and improve how AI makes decisions [14].

1.4.2 Who Owns AI-Generated Content?

If an AI creates a piece of artwork, a song or even writes a book, who gets credit? The person who trained the AI? The company that built it? Or does AI-generated content even count as "original"? These are big legal questions, and copyright laws haven't fully caught up with AI's capabilities. Some artists and writers worry that AI is using their work without permission, while businesses are looking for clear rules on how they can use AI-generated content legally.

1.4.3 The Rise of Misinformation and Deepfakes

Generative AI can create incredibly realistic fake images, videos and articles. While this is great for entertainment and creative industries, it also means AI can be misused to spread fake news, impersonate people or manipulate public opinion. Deepfake videos, for example, can make it look like someone said something they never did, which could be dangerous in politics, journalism, or cybersecurity. AI companies are working on detection tools, but it's a constant battle between creators and bad actors [15].

1.4.4 The Need for Regulations and Laws

Governments around the world are still figuring out how to regulate AI. Should AI-generated misinformation be illegal? Who is responsible if an AI system causes harm? These are tough questions, and different countries are taking different approaches. The EU, for example, is pushing for strict AI laws, while the US is focusing more on innovation-friendly guidelines. Finding the right balance between encouraging AI development and ensuring ethical responsibility is crucial [16].

Generative AI has the power to change industries, improve creativity and make life easier—but only if we handle it responsibly. By addressing issues like bias, ownership, misinformation and legal frameworks, we can ensure that AI remains a force for good rather than a source of harm. As AI continues to evolve, it's up to researchers, policymakers and everyday users to shape its future in a fair and ethical way.

1.5 FUTURE TRENDS AND EMERGING RESEARCH DIRECTIONS

As technology advances, the landscape of ML and sustainable resource management is evolving rapidly. Below are some key trends and research directions that will shape the future.

1.5.1 Advancements in Model Architectures

ML models are constantly improving in terms of efficiency, accuracy and scalability. Recent breakthroughs in deep learning, such as transformer-based architectures (like GPT and BERT), have demonstrated remarkable capabilities in processing complex datasets. In the future, we can expect:

- **More efficient and lightweight models** that require fewer computational resources, making AI accessible for a broader range of applications.
- **Self-learning and adaptive models** that can update themselves without extensive retraining, allowing them to remain relevant in dynamic environments.
- **Hybrid AI models** that combine multiple architectures, such as neural-symbolic AI, which merges deep learning with symbolic reasoning to improve interpretability and decision-making [17,18].

These advancements will enable smarter, more flexible AI systems that can be applied to a variety of fields, from resource optimization to real-time environmental monitoring.

1.5.2 Integration with Emerging Technologies (e.g., Quantum Computing, AR/VR)

AI is not evolving in isolation—its integration with other groundbreaking technologies is opening new possibilities. Some key areas of convergence include:

- **Quantum Computing and AI**: Quantum computing has the potential to revolutionize ML by solving problems that are currently computationally impossible. For example, it could optimize resource allocation models or enhance the speed of AI-driven climate simulations.
- **Augmented Reality (AR) and Virtual Reality (VR)**: AI-powered AR/VR applications are enhancing training, simulations and data visualization. In sustainable resource management, AR/VR can be used to create digital twins of cities or natural ecosystems, helping policymakers and researchers test different scenarios before implementing real-world solutions.
- **IoT and Edge AI**: The combination of Internet of Things (IoT) devices with AI at the edge (on-device processing) is making real-time resource monitoring and management more efficient. For example, AI-powered sensors can analyse soil health or air quality without relying on cloud computing [19,20].

By leveraging these emerging technologies, AI-driven solutions will become more powerful and effective across multiple domains.

1.5.3 Long-Term Societal Implications

As AI continues to grow, it is crucial to consider its broader impact on society. Some long-term implications include:

- **Ethical and Fair AI**: Bias in AI models remains a challenge. Researchers must focus on developing more inclusive and transparent algorithms to ensure fair decision-making.
- **Job Market Transformations**: AI-driven automation is changing the nature of work. While new job opportunities will emerge, there will also be a need for upskilling and reskilling programs to prepare the workforce for an AI-driven economy.
- **Sustainability & Climate Impact**: AI itself consumes energy, particularly large-scale models that require extensive computational resources. Finding ways to make AI models more energy-efficient will be a critical research focus [21–23].

By addressing these societal implications proactively, we can ensure that AI contributes positively to humanity's future.

1.5.4 Research Challenges and Opportunities

Despite the rapid progress, many challenges still need to be addressed in AI research:

- **Data Scarcity and Quality**: Many AI models rely on large datasets, but in some cases (e.g., rare environmental events), data is limited. Developing AI models that can work with small, noisy, or synthetic datasets is an ongoing challenge.
- **Interpretability and Trust**: AI decision-making is often seen as a "black box." Improving model transparency and explainability will help build trust among users and policymakers [24].
- **Security and Privacy**: AI models are vulnerable to adversarial attacks, where subtle manipulations can lead to incorrect predictions. Strengthening AI security and ensuring user privacy will be critical for future adoption [25].
- **Scalability and Real-World Deployment:** Many AI inventions stay in labs as calculation cost is high or there is no infrastructure. Scaling and deploying AI solutions in the real world, especially in developing regions [26], is both a challenge and an opportunity.

These areas each offer intriguing possibilities for work at the intersection of researchers, industries and policymakers to advance just what AI can do.

Through these future trends and challenges, the next generation of AI technology will be not only more powerful but also more ethical, inclusive and beneficial to society.

1.6 CONCLUSION

1.6.1 Summary of Key Insights

AI generates the new and the novel as a transformative force, remaking industries, reimagining human creativity and reinventing human interactions with technology. In this discussion, we walked through the basics of generative AI: its history, the core technologies, GANs, VAEs and Transformer-based models. Through these advancements, AI has not only been able to analyse data but also create new and meaningful content, filling the space between man's imagination and machine intelligence.

Generative AI effects many domains from creative industries to businesses, healthcare and education. AI-powered tools have democratized content creation in creative fields, and artists, musicians and writers have never been able to experiment and innovate like they can now. AI is being used by businesses to automate workflows, improve customer service and personalize the marketing strategies. In healthcare, meanwhile, AI-driven drug discovery and diagnostic imaging have yielded medical advances, and in education, AI has revolutionized learning experiences to become more personalized.

While these are benefits, there are also challenges facing generative AI—bias in AI models, intellectual property disputes, deepfakes producing misinformation and regulatory concerns. Ensuring that AI is a force for good requires ethical AI development, fair model training and responsible deployment of models. The future of generative AI will be defined by further evolution in AI architectures, coupling with emerging technologies, such as quantum computing and AR/VR, and tackling society implications.

1.6.2 Reflections on the Transformative Potential of Generative AI

In generative AI, we don't just see some new technology, we see a new paradigm in how we interact with machines and how machines can amplify our humanity. Enabling AI to create (instead of only classifying or predicting) opens new frontiers of creativity, problem solving and automation. AI has the ability to produce realistic images, music, text and even synthetic biological structures, which can help drive breakthroughs in science, art and industry.

Generative AI is adaptable and that is its transformative power. It's revolutionizing workflows, enabling us to do things faster and more efficiently, and it's pushing the boundaries of what machines can do. This power also brings great responsibility, but AI-created content has ethical implications. It can disrupt job markets, and misinformation must be actively addressed to make sure AI benefits are equitably distributed.

However, as AI progresses, its connection with other contemporary innovative technologies like quantum computing for speedier model training, AR/VR for more profound user experiences and IoT for real-time information analysis is going to open up more potential. The future of AI is that of humans and machines working together, with each enhancing the strengths of the other rather than replacing the other.

1.6.3 Final Thoughts on Responsible Innovation

Generative AI is no exception to the rule with great power comes great responsibility. At the foot of this technological revolution, it is our responsibility to innovate responsibly. Development and deployment of AI should be prioritized with a focus on fairness, inclusivity and ethical considerations.

Responsible AI innovation includes:

Transparency and Explainability: AI models should be interpretable such that the users and stakeholders understand how the decision is being made.

Bias Mitigation and Fairness: AI systems should not be biased towards certain groups of people and should be based on training on a variety of datasets.

Regulation and Governance: Policymakers and the industry need to create legal and ethical frameworks to thwart abuse and guarantee that AI works for the public benefit.

Sustainability: Since the large AI models have high energy consumption, energy efficient architectures should be created.

Generative AI is promising, but its future path is up to us. The potential of AI can be best realized when all these stakeholders work in concert—researchers, businesses, policymakers and society. Instead of replacing human intelligence, AI should be seen as another tool that helps us boost our creativity, increase our efficiency and solve problems.

Ultimately, generative AI will succeed or fail less based on how sophisticated its systems are, but rather on how they affect people. But, developed responsibly, it can improve human life in ways we are only starting to grasp.

REFERENCES

1. Bengio, Y., Goodfellow, I., & Courville, A. (2016). *Deep Learning.* MIT Press, Cambridge, MA.
2. Brown, T., Mann, B., Ryder, N., Subbiah, M., Kaplan, J., Dhariwal, P., … & Amodei, D. (2020). Language models are few-shot learners. *Advances in Neural Information Processing Systems, 33*, 1877–1901.
3. Dosovitskiy, A., Beyer, L., Kolesnikov, A., Weissenborn, D., Zhai, X., Unterthiner, T., … & Houlsby, N. (2020). An image is worth 16×16 words: Transformers for image recognition at scale. *arXiv preprint arXiv:2010.11929.*
4. Chollet, F. (2017). *Deep Learning with Python.* Manning Publications, Shelter Island, NY.
5. Devlin, J., Chang, M. W., Lee, K., & Toutanova, K. (2019). BERT: Pre-training of deep bidirectional transformers for language understanding. In: *Proceedings of the 2019 Conference of the North American Chapter of the Association for Computational Linguistics: Human Language Technologies*, vol. 1, pp. 4171–4186.
6. Karras, T., Laine, S., & Aila, T. (2019). A style-based generator architecture for generative adversarial networks. In: *Proceedings of the IEEE/CVF Conference on Computer Vision and Pattern Recognition*, pp. 4401–4410.

7. Kingma, D. P., & Welling, M. (2013). Auto-encoding variational Bayes. *arXiv preprint arXiv:1312.6114.*
8. LeCun, Y., Bengio, Y., & Hinton, G. (2015). Deep learning. *Nature, 521*(7553), 436–444.
9. Li, J., Monroe, W., & Jurafsky, D. (2016). A simple, fast diverse decoding algorithm for neural generation. *arXiv preprint arXiv:1611.08562.*
10. He, K., Zhang, X., Ren, S., & Sun, J. (2016). Deep residual learning for image recognition. In: *Proceedings of the IEEE Conference on Computer Vision and Pattern Recognition*, pp. 770–778.
11. Open AI. (2023). GPT-4 technical report. *arXiv preprint arXiv:2303.08774.*
12. Radford, A., Wu, J., Child, R., Luan, D., Amodei, D., & Sutskever, I. (2019). Language models are unsupervised multitask learners. *OpenAI Blog, 1*(8).
13. Vats, P., Phogat, A.K., Shanker, S., Upadhyay, G.M., Upreti, K. (2025). Exploring the nexus of deepfakes and VFX technology: Unveiling insights, challenges, and innovations. In: *Intelligent Solutions for Smart Adaptation in Digital Era. InCITe 2024. Lecture Notes in Electrical Engineering*, vol. 1278. Springer, Singapore. https://doi.org/10.1007/978-981-97-8193-5_32
14. Ramesh, A., Pavlov, M., Goh, G., Gray, S., Voss, C., Radford, A., ... & Sutskever, I. (2021). Zero-shot text-to-image generation. *arXiv preprint arXiv:2102.12092.*
15. Hinton, G. E., Osindero, S., & Teh, Y. W. (2006). A fast learning algorithm for deep belief nets. *Neural Computation, 18*(7), 1527–1554.
16. Rumelhart, D. E., Hinton, G. E., & Williams, R. J. (1986). Learning representations by back-propagating errors. *Nature, 323*(6088), 533–536.
17. Hochreiter, S., & Schmidhuber, J. (1997). Long short-term memory. *Neural Computation, 9*(8), 1735–1780.
18. Vaswani, A., Shazeer, N., Parmar, N., Uszkoreit, J., Jones, L., Gomez, A. N., ... & Polosukhin, I. (2017). Attention is all you need. *Advances in Neural Information Processing Systems, 30*, 5998–6008.
19. Zhang, H., Goodfellow, I., Metaxas, D., & Odena, A. (2019). Self-attention generative adversarial networks. In: *International Conference on Machine Learning*, pp. 7354–7363.
20. Bommasani, R., Hudson, D.A., Adeli, E., Altman, R., Arora, S., von Arx, S., et al. (2021). On the opportunities and risks of foundation models. *arXiv preprint arXiv:2108.07258.*
21. Rombach, R., Blattmann, A., Lorenz, D., Esser, P., & Ommer, B. (2022). High-resolution image synthesis with latent diffusion models. In: *Proceedings of the IEEE/CVF Conference on Computer Vision and Pattern Recognition (CVPR)*, pp. 10684–10695.
22. Chen, M., Tworek, J., Jun, H., Yuan, Q., Pinto, H., Kaplan, J., et al. (2021). Evaluating large language models trained on code. *arXiv preprint arXiv:2107.03374.*
23. Goh, G., Andrei, A., & Olah, C. (2021). Multimodal neurons in artificial neural networks. *Distill, 6*(3), e30. https://distill.pub/2021/multimodal-neurons
24. Kirillov, A., Mintun, E., Ravi, N., Mao, H., Rolland, C., Gustafson, L., et al. (2023). Segment anything. *arXiv preprint arXiv:2304.02643.*
25. Bubeck, S., Chandrasekaran, V., Eldan, R., Gehrke, J., Horvitz, E., Kamar, E., et al. (2023). Sparks of artificial general intelligence: Early experiments with GPT-4. *arXiv preprint arXiv:2303.12712.*
26. Open AI. (2023). DALLE 3 Technical Report. https://openai.com/dall-e-3

2 Generative AI Models
Unraveling How Does Generative AI Work

Saloni Bhushan and Surabhi Shanker

2.1 INTRODUCTION

The way machines were doing tasks that required human intelligence has been redesigned by Artificial Intelligence (AI). Generative models are one of the diverse domains of AI that are getting very common due to their ability to design content like text, visuals, audio, etc. It is done by studying and learning patterns in available datasets. In traditional AI models, we usually do the classification and prediction. However, generative AI models can design completely new, realistic data, which makes them very useful for creative and technical purposes. This chapter introduces the fundamentals of Machine Learning (ML), Deep Learning (DL), and Generative AI and explains the differences between the three, as well as the working mechanism of generative models. It also looks into the important model architectures, training methods, real-world applications, ethical considerations, and the future of Generative AI.

2.1.1 Overview of Artificial Intelligence

Artificial intelligence is a broader domain which focuses on developing machines that can manage jobs that need human thinking skills. AI is created to develop technology that can learn, respond to updates and handle tasks such as problem solving, decision-making, processing language, and pattern finding. AI is being used in healthcare, self-driving cars, creative areas, and finance, and its development has moved fast [1].

Based on how much AI can do and how much it resembles human thinking different types of AI are defined. The two main categories of AI are Narrow AI (Weak AI) and General AI (Strong AI). Specific AI (Narrow AI or Weak AI) refers to systems that are designed and taught to handle particular tasks or challenges. Some of these systems are limited to a small area of use such as voice assistants (for example, Siri and Alexa), recommendation systems, and image recognition software. They are capable of many impressive things, but they do not actually understand or think, they are just very good at their specific jobs. General AI (Strong AI) is a hypothetical kind of AI capable of performing any mental task that a human can. General AI would be able to handle more situations than narrow AI. It would be able to reason, solve

 DOI: 10.1201/9781003565703-2

problems, learn from what it experiences and grasp difficult abstract concepts like a human being. There has not yet been a general AI, and it is still a major goal for AI researchers [2].

2.1.2 The Evolution of AI

There are several milestones of development of AI:

1. **1950s–1960s**: Early research focused on symbolic AI, where intelligence was represented by rules and logic-based approaches. Alan Turing's seminal work in 1950 with the Turing Test was pivotal in defining AI as the imitation of human thought processes.
2. **1970s–1980s**: AI saw the rise of expert systems and the application of knowledge-based approaches. Researchers also explored neural networks during this period. Although the computational limitations of the time prevented significant breakthroughs.
3. **1990s–2000s**: The evolution of ML algorithms, particularly its use in the field of speech recognition and data mining, led to practical applications in commercial products.
4. **2010s–Present**: The rapid growth in the size of data, enhanced computational power and advancements in DL techniques (e.g., neural networks with multiple layers) led to next generation applications such as autonomous vehicles, AI-driven health diagnostics, and advanced language models like GPT.

This chapter aims to give a complete overview of Generative AI and show how it is different from ML and DL. Although AI, ML, and DL have made big waves in many industries, generative AI is a separate and fast-changing area with its own special abilities and uses [3].

We want to discover how generative AI is able to generate brand new, artificial data. This chapter also looks into different architectures that back its models, such as Generative Adversarial Networks (GANs), Variational Autoencoders (VAEs), and Diffusion Models. They allow AI to make new things like images, text, and music, which changes our ideas about creativity and automation [4]. Also, this section aims to explain what sets generative AI apart from regular AI methods, mainly when considering supervised learning, unsupervised learning, reinforcement learning in ML and DL's neural networks. While ML usually aims at classification or regression (predicting based on existing data), generative AI looks to form data that looks similar to real examples but is not copied from any sample.

The training methods, the way models are built and uses of generative AI are also greatly different. For example, traditional AI is mainly concerned with pattern recognition and making decisions, whereas generative AI works on producing data that can be useful for many creative purposes such as making simulated images, producing natural-sounding audio, or creating text that sounds like it was written by a person [5].

2.2 THE BASICS OF AI INVOLVE MACHINE LEARNING AND DEEP LEARNING

AI is built upon the ideas of ML and DL, in which computers improve by analyzing data. So, ML algorithms are able to identify patterns for prediction, but DL goes a step further by working with multi-layered neural networks on complex and unstructured data [6].

2.2.1 The Practice of Machine Learning (ML)

ML is a field of AI that works on building algorithms and models that let machines learn and make forecasts or conclusions on their own. These systems learn from the patterns in data and get better with time, instead of having someone write the code for every task.

Supervised learning, unsupervised learning, and reinforcement learning are the main types of ML.

- **Supervised Learning**

  Supervised Learning is called Supervised because the training process uses labeled (identified) data. Some data is put in and the correct output is shown. The purpose is to find a way to map the input variables (X) to an output label (Y).

Some examples are:

- Spam detection
- Predicting loan eligibility (e.g., credit scoring)
- It can also be used to identify objects in images (image classification).

The common Algorithms of Supervised learning are Linear Regression, Logistic Regression, Decision Trees, Support Vector Machines (SVM), Neural Networks, etc. (Figure 2.1)

- **Unsupervised Learning**: With Unsupervised Learning, the data is not labeled and the aim is to spot hidden patterns in the data. Examples of unsupervised learning are:
  - Grouping customers together based on how they buy (customer segmentation)
  - Anomaly detection (fraud detection in financial transactions)
  - Market basket analysis (identifying frequently bought items together)

The Market basket analysis helps in finding out frequently bought items together (Frequently occurring products) (Figure 2.2).

- **Reinforcement Learning**: The model learns by doing things in the environment and seeing if its actions result in rewards or penalties. Reinforcement learning has an agent learning by interacting with an environment to get the

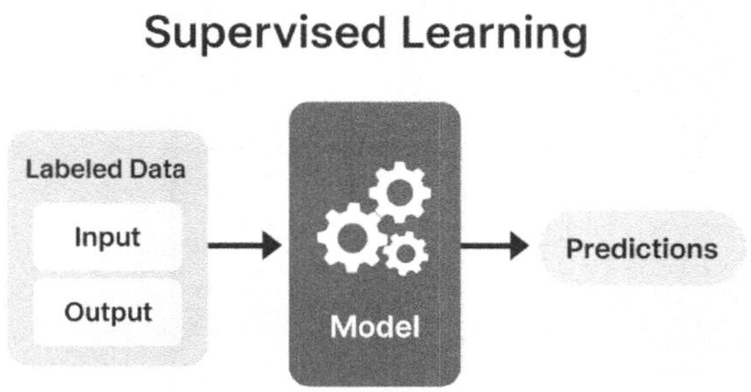

FIGURE 2.1 Supervised learning.

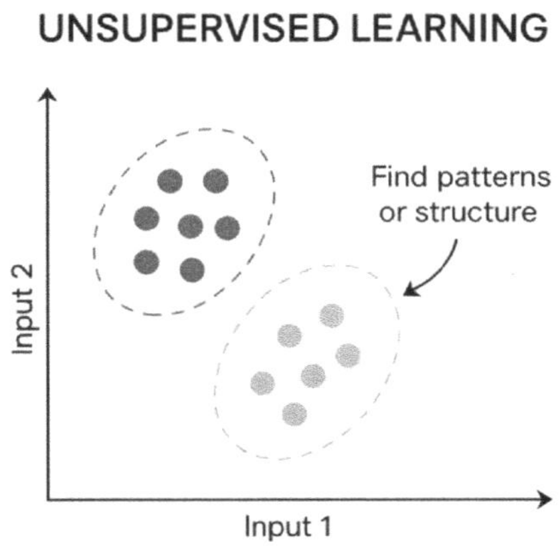

FIGURE 2.2 Unsupervised learning.

most rewards. The agent picks up new information through the process of trying and getting either positive feedback (a reward) or negative feedback (a penalty). Examples are:

- Robotics (self-learning robots)
- Game AI (AlphaGo, Chess-playing AI)
- Autonomous vehicles (self-driving cars)

What makes up RL are:

- The **agent** is the learner or person making the decisions.
- The **environment** is the outside world the agent deals with.
- **Actions** are the different things the agent might do.
- **Feedback** is the way rewards let you know for your decisions (Figure 2.3).

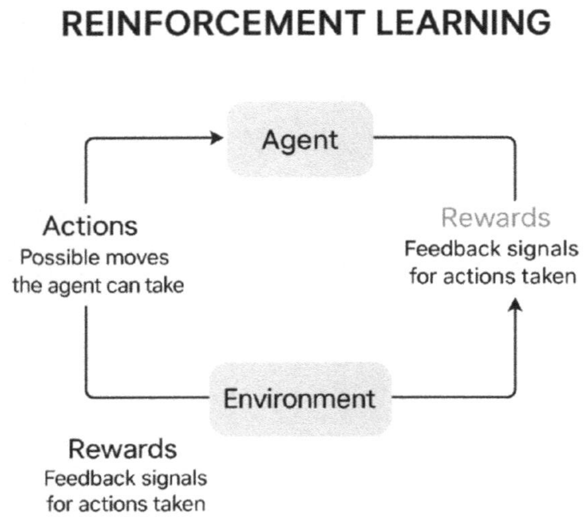

FIGURE 2.3 Reinforcement learning.

Basic Principles of ML:

a. **Data-Driven Approach**: ML models rely on large datasets to identify patterns and improve accuracy.
b. **Generalization**: The ability of a model to perform well on unseen data, not just the training data.
c. **Feature Engineering**: Selecting and transforming input variables to improve model performance.
d. **Model Training**: The model is optimized using training data and adjusting parameters to minimize errors.
e. **Evaluation**: It is the process in which you are going to measure the performance of your model through different metrics like accuracy, precision, recall, and F1-score.

2.2.2 Deep Learning (DL)

Deep Learning (DL) is the branch of ML that employs neural networks with many layers to find patterns and representations in vast sets of data. DL is based on the way the human brain works and can discover patterns in raw data by itself which is why it is very good at image recognition, natural language processing, and speech recognition [7]. DL models are particularly very powerful in handling vast amount of unstructured data such as text, audio and image (Figure 2.4).

Few popular DL architectures are listed below:

- **Convolutional Neural Networks (CNNs)**: Image and video recognition mainly rely on Convolutional Neural Networks (CNNs).

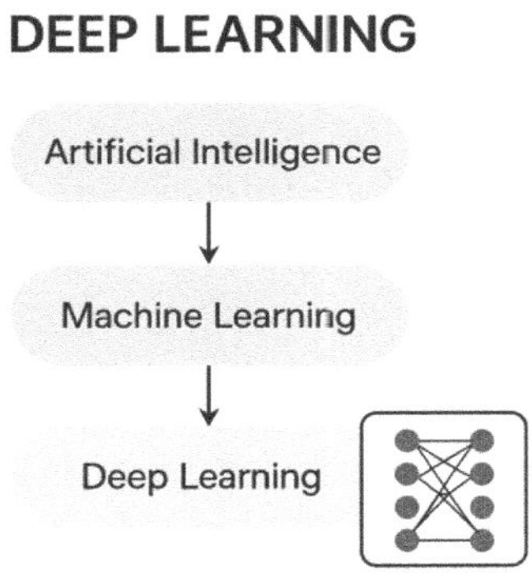

FIGURE 2.4 Deep learning.

- **Recurrent Neural Networks (RNNs)**: Recurrent Neural Networks (RNNs) are well suited for working with data that has a sequence like time series or language.
- **Transformers**: The use of Transformers has completely changed how NLP tasks such as machine translation and text generation are performed.

2.3 UNDERSTANDING GENERATIVE AI

Generative AI is an important branch of AI. It focuses on making new content as text, images, audio, or video, instead of only analyzing and predicting existing data. Instead of looking for specific patterns, generative models in AI study the overall data structure and produce outputs that look real [8].

Some of the DL methods these models depend on are GANs, VAEs, and Diffusion Models for producing high quality data. In artistic fields, entertainment, healthcare, and software development, generative AI has transformed the way we do things, allowing us to do things faster and better, like deepfake technology, text generation (e.g., GPT), and AI creativity [9].

Generative AI tries to learn the form (distribution) of the input data and then creates new samples that are similar to what is found in reality. Few examples of generative AI models are:

- **Generative Adversarial Networks (GANs):** In GANs, two neural networks (a generator and a discriminator) try to outdo one another to make the generated data better.
- **Variational Autoencoders (VAEs):** VAEs are a type of neural network that learns how data is distributed and creates new instances by sampling from the learned latent space.
- **Diffusion Models:** A more recent approach that generates data by simulating a gradual process of denoising, often used for image generation tasks.

State-of-the-art applications, such as deepfakes, AI-generated artwork, automated text generation (like GPT models) and AI-generated music, are made possible by generative AI.

2.3.1 Purpose of Generative AI

Generative AI aims to:

- Create human-like text, images, and audio.
- Enhance creativity in fields like art, design, and music.
- Assist with data augmentation for training ML models.
- Simulate real-world scenarios for research and gaming.
- Automate repetitive creative tasks (e.g., code generation and writing assistance).

Generative AI examples include OpenAI's GPT for generating text, DALL•E for creating images and DeepMind's AlphaFold for predicting protein structure [10].

2.3.1.1 Comparison between Generative AI and Discriminative AI

Generative AI and Discriminative AI represent two distinct paradigms within artificial intelligence, each serving different purposes in data modeling. Generative AI focuses on learning the underlying distribution of data to create new, realistic examples, such as generating synthetic images, text, or audio. It captures both the features and the joint probability of inputs and outputs, making it suitable for tasks that require creativity, simulation, or data augmentation. In contrast, Discriminative AI concentrates on learning the boundary between classes by modeling the conditional probability of outputs given inputs, making it highly effective for classification, regression, and prediction tasks. While generative models are more computationally intensive and complex to train, discriminative models are often simpler, faster, and more accurate in distinguishing between categories. Together, these two approaches complement each other and highlight the evolving landscape of AI applications. The key distinctions between Generative AI and Discriminative AI are summarized in Table 2.1.

2.3.2 Key Types of Generative AI Models

Generative AI models have to learn from data and then create new data. It uses a variety of architectures to do this. Examples of common architectures are GANs, VAEs, and Diffusion Models. A competitive approach is used by GANs and VAEs that deal with data by encoding and decoding it probabilistically, and Diffusion Models make high-quality outputs by gradually reducing noise. All of these models have different specialties such as generating images, text, helping with drugs and developing creative AI tools [11].

TABLE 2.1
Comparison of Generative AI with Discriminative AI

Aspect	Generative AI	Discriminative AI
Objective	Generates new data similar to the training data	Classifies or differentiates between data points
Example task	Creating a realistic image of a non-existent person	Determining whether an image contains a cat or a dog
Output	Synthesized content (text, images, audio, etc.)	Probabilities or labels for classification
Model type	Generative Adversarial Networks (GANs), Variational Autoencoders (VAEs), Transformer-based models (GPT, DALL·E)	Logistic Regression, Decision Trees, Random Forest, CNNs for image classification
Use cases	Art generation, text generation, synthetic data creation	Spam detection, sentiment analysis, medical diagnosis

2.3.2.1 Generative Adversarial Networks (GANs)

Generative Adversarial Networks (GANs) are a class of Generative AI models. It consists of two neural networks known as generator and a discriminator. These are two opposing processes which compete with each other, the generator (G) is able to create data that looks very much like the real thing. First, it uses random noise and improves more and more as it learns. It might, for example, make a fake picture of a human face. The discriminator (D) is a type of classifier. It can distinguish between actual data and made-up (generated) data. The output of the generator is presented to the discriminator, which uses it to adjust its results. An example is that it can tell if a picture was taken by a person or an AI. They are learned at the same time with each network having a different task: the generator aims to make fake data that looks real, and the discriminator tries to tell which data is real and which is fake. This type of training also pushes the generator to improve, so it produces more realistic results [12].

2.3.2.1.1 How GANs Work

GANs training process consists of several key steps performed iteratively in a training process. First, the generator makes fake data from random noise and makes outputs that look like the real data it was trained with. The discriminator then evaluates generated data, which is given both real and fake samples, and learns to tell them apart. In the adversarial learning phase, the generator tries to learn its parameters so as to fool the discriminator. On the other hand, the discriminator is doing this every step of the way to become better and better at detecting fake data. Through iterative training, this competition cycle between two networks keeps happening, and the outputs of the Generator keep getting better and better. The process eventually produces data indistinguishable from real data, which is the goal of the GAN [13].

2.3.2.1.2 Applications of GANs

Applications of GANs are listed below:

- **Image Generation**: Deepfake technology and AI-generated artwork (e.g., DALL·E).
- **Text-to-Image Synthesis**: Generating images based on text descriptions (e.g., MidJourney).
- **Data Augmentation**: Creating synthetic training data for ML.
- **Super-Resolution**: Enhancement of image resolution beyond original quality.
- **Video and Animation**: Generating realistic animations from minimal inputs.

2.3.2.2 Variational Autoencoders (VAEs)

VAEs are a kind of generative models that use probabilistic inference to make new data. Instead of GANs, VAEs take input data and reduce it to a smaller representation called the latent space before generating new samples from it. VAEs consist of two important parts called the encoder and the decoder. The input data is turned into a representation with fewer dimensions by the encoder.

Here, what is learned is a probability distribution (mean and variance) instead of just mapping inputs to a single point in the latent space. It allows the computer to blend the samples together smoothly. In contrast, the decoder gets a sample from the latent space and gives a result that is very similar to the original input. It is able to make new samples by transforming the latent representations, which helps it to produce new data according to the distribution it learned [14].

2.3.2.2.1 Key Idea: Latent Space Representation

Unlike traditional autoencoders, which map data to a fixed vector, VAEs learn a distribution over the latent space. This ensures smooth sampling and controlled data generation.

2.3.2.2.2 Role of VAEs in Data Generation

- **Synthetic Data Generation:** Creating realistic but new samples (e.g., faces, handwriting).
- **Anomaly Detection**: Identifying outliers by comparing reconstructed vs. real data.
- **Dimensionality Reduction**: Extracting meaningful compressed features from high-dimensional data.
- **Interpolation and Morphing**: Blending between different data points (e.g., transitioning between faces).

2.3.2.3 Diffusion Models

Diffusion models are a class of generative AI models that create high-quality synthetic data, such as images, by gradually refining noise into meaningful structures. These models have gained popularity due to their ability to generate realistic images, often surpassing even GANs in certain applications like Stable Diffusion and DALL·E 2. Inspired by non-equilibrium thermodynamics, diffusion models approach data generation as a gradual denoising process. They operate in two main phases: the forward process and the reverse process. In the forward process, Gaussian noise is gradually added to an image (or any data) over multiple steps until it becomes pure noise, creating a smooth transition from structured data to complete randomness. In the reverse process, the model learns to reverse the noise addition, step by step, to recover the original data. This is accomplished through a deep neural network that predicts and removes noise iteratively. As a result, a meaningful image is eventually reconstructed from pure noise, showcasing the model's ability to generate highly realistic data [15].

2.3.2.3.1 Key Features of Diffusion Models

- **High-Quality Image Generation**: Produces detailed and realistic images.
- **Stable Training**: Unlike GANs, diffusion models do not suffer from mode collapse.
- **Latent Space Interpolation**: Allows smooth transitions between different data points.
- **Scalability**: Can generate high-resolution images with fine details.

Some of the applications of Diffusion Models are listed below:

- **Image Generation and Editing**: Used in tools like Stable Diffusion and DALL·E 2.
- **Super-Resolution**: Enhancing low-resolution images (e.g., AI upscalers).
- **Inpainting**: Filling in missing parts of images seamlessly.
- **3D Object Generation**: Creating 3D assets for games and virtual environments.
- **Music and Text Generation**: Adapting diffusion models for structured data like text and audio.

2.4 THE WORKING MECHANISM OF GENERATIVE AI

Generative AI models learn patterns from large datasets and utilize probabilistic techniques to generate new, realistic content. The process typically unfolds in three key stages: training, inference, and fine-tuning. The model begins by analyzing a dataset to learn how to generate realistic data during the Training Phase. The primary phase is Data Collection and Preprocessing, where large datasets, such as text, images, or audio, are gathered and cleaned. The cleaned data is then transformed into a format that the model can understand like tokenization for text or normalization for images. In the next step, Model is trained. Usually, AI model tunes its parameters using the DL architectures like GANs, VAEs, Transformers, or Diffusion Models to minimize the errors and improving the quality of generation. Loss Function and Optimization entails the model measuring closeness of generated outputs to real data by using a loss function. Optimizers like Gradient Descent are used to adjust weights to make performance better [16].

After the model is trained, Inference Phase comes into the picture where the model generates new data. It starts by listening to an Input Prompt, such as a random noise vector (used by GANs, VAEs, and Diffusion Models) or a text prompt (for models such as GPT or DALL•E). Then, this input is processed by model using its learned Latent Space. Mapping the input to innovative, similar outputs represents meaningful variations in the training data from the learned Latent Space. Then, Decoding and Output

Generation happens in which the AI reconstructs data from the latent space using post-processing methods like filtering or fine-tuning to produce the end result [17].

Lastly, in the Fine-Tuning and Feedback Loop, users can use techniques like transfer learning or reinforcement learning (RLHF in ChatGPT, for example) to improve the model's outputs. The AI's generation quality is continuously trained including user feedback. Take Text-to-Image Models such as DALL•E or Stable Diffusion as an example; a user prompt could be 'A futuristic cityscape at sunset'. This text is modeled into a numerical representation, a diffusion model begins with random noise and slowly improves it into a sensible image by learned patterns, resulting in a high-quality AI-generated image of a hi-tech futuristic city.

2.4.1 Data Generation Process

Generative AI models create new data, such as images, text, or audio. This is done by learning patterns from large datasets and using probability-based transformations. While the specific process varies by model type, it generally follows several key steps. It starts with Input Representation (Encoding), where the model receives an input, which may be in the form of random noise (for GANs, VAEs, and Diffusion Models), a prompt or seed text (for Transformers like GPT), or a piece of sound or melody (for audio-based AI like WaveNet). If using a latent space, the model encodes the input into a compressed representation that captures meaningful variations. The next step is Model Processing and Learning Patterns, where the AI processes the input through deep neural networks that have been trained on large datasets. Each model type has its own approach: GANs employ a generator and discriminator to generate outputs and iterate their ability to tell real from fake data; VAEs use an encoder to encode inputs into a probabilistic latent space and decoder to reconstruct the new samples; Transformers (GPT, BERT, etc.) utilize self-attention to understand the context of the text and predict the next word or phrase; Diffusion Models work by sequentially de-noising a random noise into a meaningful structure.

The next step after processing the input is Sampling from the Model (Decoding) in which it samples from the learned distribution to produce new data. The sampling process is important because it results in variability of the output while preserving some realistic properties, i.e., text predictions, image generation, or feature construction. The next step is Refining and Post-processing the output. It may need to be fine-tuned for quality improvement at initial output. For images, it can involve filtering and smoothing, grammar correction for text and noise reduction for audio. Finally, in Final Output Generation, the model produces a final output refined version of the generated data. For example, a text model like GPT can generate human-like paragraphs, an image model like Stable Diffusion generates high-quality AI-generated images, and an audio model like WaveNet generates realistic speech or music [18].

2.4.2 Training Generative Models

To study the pattern of data and generate meaningful results, generative models require to perform training neural networks. It consists of specifying loss functions, using optimization algorithms, and iterating through training cycles until the model outputs high-quality results. The first step in training is the preparation of Dataset, where a large amount of dataset (e.g., images, text, audio) is collected and processed. Data is normalized (e.g., scaling images to [0,1] or tokenizing text) and then split into training and validation sets. Second, the Model Architecture is Selected, which depends on the task at hand. For example, GANs are typically used for high-quality image and video generation, VAEs for structured latent space and data interpolation, Diffusion Models for generating highly detailed images, and Transformers (such as GPT and BERT) for text, code, and language generation [19].

After choosing the architecture, Defining the Loss Function follows. Loss functions guide the model's learning by measuring the difference between generated outputs and real data. For GANs, adversarial loss is used, where the generator aims to produce fake data that appears real, and the discriminator learns to distinguish real from fake data. The formulas for GANs' loss functions are as follows:

Generator Loss: Encourages fake data to appear real.

$$L_G = -E\left[\log \log D\left(G(z)\right)\right]$$

Discriminator Loss: Helps distinguish real from fake data.

$$L_D = -E\left[\log \log D(x)\right] - E\left[\log\left(1 - D\left(G(z)\right)\right)\right]$$

In VAEs, the loss function combines reconstruction loss (the difference between the original and generated data) with KL divergence loss to regularize the latent space. The formula for VAE loss is:

$$L = \text{Re construction Loss} + \beta.\text{KL Divergence}$$

Diffusion models use noise prediction loss to teach the model how to remove noise step by step. Transformers like GPT and BERT use cross-entropy loss to predict the next token in text.

Finally, in the Optimization and Training Process, models update their weights using optimization techniques like Gradient Descent, Adam, or RMSprop. Training proceeds through multiple epochs, where the model iterates over the dataset repeatedly, continuously improving its ability to generate high-quality data [20].

Ways to make your performance better:

There are several ways to boost how well generative models perform during their training. Batch Normalization helps to make training more stable and faster by normalizing the inputs to each layer, which ensures the network's distributions stay consistent. Dropout is a way to help a neural network avoid learning too much by randomly removing some units during training, so it works well with new cases. Data Augmentation helps training by adding more diverse examples to the dataset, such as by rotating, flipping, or cropping images, which improves the model's robustness [21].

Evaluation and Fine-Tuning should be done after the training phase. The model is checked on separate validation data to confirm it is creating quality outputs. The data type determines which metrics are used. FID (Fréchet Inception Distance) is usually applied to check the quality and diversity of images, while BLEU (Bilingual Evaluation Understudy) measures how accurate the generated text matches reference translations. After evaluation, it is decided if the model can be adjusted with training or transfer learning to make its output better and suitable for new or specific roles.

2.4.3 Challenges in Training Generative Models

There are several challenges in training generative models that impact their stability, efficiency, and output quality. Each model type—GANs, VAEs, Diffusion Models, and Transformers—has its own set of concerns. One of the major issues in GANs is mode collapse, in which the generator repeatedly produces limited types of outputs instead of capturing the full diversity of the dataset. This happens when the generator finds patterns, which consistently fool the discriminator. To overcome such problems, techniques such as mini-batch discrimination (where the discriminator examines batches instead of individual samples), Wasserstein GANs (WGANs, which use Earth Mover's distance), and label smoothing can improve diversity and stability [22].

Another frequent challenge is the vanishing gradient problem, especially common in GANs, Diffusion Models, and Transformers. When gradients become too small to meaningfully update weights, the generator may stop learning. This imbalance often arises when one part of the network, like the discriminator, learns faster than the other. Solutions include using Leaky ReLU instead of standard ReLU, training the discriminator more frequently, and applying gradient penalty as in WGAN-GP to ensure training remains stable.

Overfitting is another issue, especially in VAEs and large models like Transformers. These models may memorize training data instead of generalizing, particularly if the latent space is not properly regularized in VAEs. Mitigating this requires expanding the dataset and using regularization techniques such as KL divergence in VAEs, dropout, and data augmentation. Early stopping is also effective—it halts training when validation loss begins to increase [23].

Slow convergence is a practical problem for Diffusion Models and large-scale Transformers like GPT. Diffusion models often need a big number of denoising steps, making them computationally intensive. Similarly, large Transformers require massive datasets and prolonged training. Efficiency can be improved by reducing the number of diffusion steps, leveraging pre-trained models through transfer learning, and fine-tuning hyperparameters like batch size and learning rate.

In VAEs, a specific challenge called posterior collapse may arise, where the model effectively ignores its latent space and outputs become overly similar to the training data. Solutions include adjusting the β term in β-VAEs to balance reconstruction and regularization, or using hierarchical VAEs to enforce structure in the latent space [24].

Another major barrier is compute and resource constraints. Training deep generative models demands significant GPU/TPU resources and access to vast datasets. Additionally, inference time can be slow, making real-time deployment difficult. Strategies such as quantization, model distillation, pruning, and using lower-precision formats (e.g., FP16 instead of FP32) help in reducing computational demands. For scalable and faster training, Cloud-based infrastructure can also be used.

One of the significant considerations is ethical and bias concerns, which must also be addressed. Sometimes, there may be societal biases present in training data, and Generative AI systems can inadvertently reproduce or even amplify those biases, thereby generating misleading or harmful content. Such issues can be solved by curating diverse, representative datasets, deploying robust content filtering systems, and involving human-in-the-loop oversight to facilitate responsible deployment and ethical use of these technologies [25].

2.5 COMPARISON BETWEEN GENERATIVE AI AND TRADITIONAL MACHINE LEARNING AND DEEP LEARNING

Generative AI, traditional ML, and DL share some common features, but they are different in purpose, architecture, and capabilities. This section outlines their dissimilarities in aspects like learning paradigms, model architectures, and applications.

2.5.1 Key Differences: Generative AI vs. ML vs. DL

Although Machine Learning (ML), Deep Learning (DL), and Generative AI share common foundations, they significantly differ in scope, methodology, and applications. ML focuses on building models which can learn from data that can be used to make predictions or classifications. While DL, a subset of ML, works with complex, high-dimensional data such as images, speech, and natural language using deep neural networks. Generative AI, on the other hand, goes beyond prediction and classification by creating entirely new data that resembles the training distribution, such as generating realistic images, text, or audio. These distinctions highlight how Generative AI builds upon the principles of ML and DL but expands their potential into creative and data-synthesis domains. The detailed comparison is presented in Table 2.2.

2.5.2 What Makes Generative AI Different?

Unlike traditional AI, which can just predict, Generative AI creates entirely new data. It is not limited to mapping inputs to labels (classification) or values (regression), rather it learns underlying distributions to generate realistic outputs. For example:

- **ML/DL Task**: Provided a dataset of cat images, classify whether an image contains a cat.
- **Generative AI Task**: Generate entirely new, realistic cat images that are not given in the dataset (Table 2.3).

TABLE 2.2
Comparative analysis Between Generative AI vs. ML vs. DL

Feature	Traditional Machine Learning	Deep Learning	Generative AI
Definition	Algorithms that learn from data to make predictions	A subset of ML using neural networks with multiple layers to process data	A type of DL focused on generating new content (images, text, audio, etc.)
Input and output	Input data → Features extracted manually → Prediction	Input data → Neural network processes raw data → Classification/regression	Input data (text, images, etc.) → Model learns patterns → Generates new data
Data dependency	Works well with structured/tabular data	Works with both structured and unstructured data	Requires large-scale datasets for high-quality generation
Feature engineering	Requires manual feature selection	Learns feature representations automatically	Learns data distributions for creative content generation
Main goal	Predict/classify based on existing data	Extract hierarchical features and learn deep patterns	Create new, realistic data based on learned patterns
Example models	Decision Trees, SVM, k-NN, Naïve Bayes	CNNs, RNNs, Transformers	GANs, VAEs, Diffusion Models, Large Language Models (GPT)
Computational cost	Lower	Higher	Very high (requires GPUs/TPUs)

2.5.3 Use Cases of Generative AI vs. ML vs. DL

TABLE 2.3
Applications of Generative AI vs. ML vs. DL

Application	Machine Learning	Deep Learning	Generative AI
Image recognition	SVM, k-NN	CNNs (ResNet, VGG)	Not applicable
Speech recognition	HMMs, GMMs	LSTMs, Transformers	Not applicable
Chatbots	Rule-based models	Sequence models (RNNs)	GPT, ChatGPT
AI art generation	Not possible	Limited	DALL·E, Stable Diffusion
Drug discovery	Statistical models	Deep neural networks	GANs for molecule generation
Game AI	Decision trees, Reinforcement learning	AlphaGo (Deep RL)	Generative AI for NPC dialogues

2.5.4 Generative AI and Discriminative Models

In AI, Generative AI and discriminative models have different roles to play. Generative models work on learning the underlying data distribution to create new samples, such as images, text, or audio. On the other hand, discriminative models classify or predict labels based on input data. For instance, a generative model like GANs can generate realistic human faces, while a discriminative model like CNN can do classification of an image whether it contains a cat or a dog. The primary distinction is in their objectives.

The generative AI is designed for content creation. Whereas discriminative AI is optimized for decision-making and classification tasks. Furthermore, the training strategies used for generative models are more sophisticated than discriminative models and often include adversarial learning or probabilistic techniques, while discriminative models define decision boundaries in data [26].

Key Takeaways:

a. Generative AI wants to create and discriminative AI wants to classify.
b. Learning the underlying data distribution is called generative models and discriminative models search for decision boundaries in a given dataset.
c. Discriminative models tend to be faster and more efficient, while generative models require large datasets and involve complex training processes.
d. Hybrid models (e.g., Semi-Supervised Learning) use a combination of both approaches for enhanced learning.

2.5.5 Key Differences: Generative AI vs. Traditional Deep Learning

The key distinctions between Generative AI and Traditional Deep Learning are summarized in Table 2.4.

Example:

- **Deep Learning Model**: A speech-to-text model predicts spoken words from an audio file (Table 2.4).
- **Generative AI Model**: A text-to-speech (TTS) model generates human-like speech from text.

TABLE 2.4
Generative AI vs. Traditional Deep Learning

Feature	Deep Learning (Predictive)	Generative AI
Primary goal	Learn patterns to make predictions (classification, regression)	Learn data distributions to generate new content
Model type	CNNs, RNNs, Transformers (for prediction tasks)	GANs, VAEs, Diffusion Models, Transformers (for generation)
Output type	Labels, numerical values, probabilities	New images, text, speech, videos, etc.
Training objective	Minimize error between predicted and actual labels	Learn the probability distribution of data to generate realistic samples
Feature engineering	Extracts meaningful features for decision-making	Learns how to create new features from patterns in the data

2.5.6 Specific Features of Generative AI

Generative AI is distinct from traditional predictive models in that it focuses on data generation rather than prediction. For instance, instead of identifying whether an image contains a cat, a generative model is capable of creating entirely new images of cats. This creative capability comes with more complex training requirements. Generative models often rely on sophisticated techniques such as adversarial training (as seen in GANs) or probabilistic methods (used in VAEs and Diffusion Models), which demand substantial computational resources and fine-tuning. Interestingly, generative AI can perform well even with limited labeled data, as it leverages self-supervised learning and can extract patterns from unlabeled datasets—unlike many traditional DL models that depend on large amounts of annotated data [27].

This ability makes generative AI especially powerful in creative domains, enabling innovations in AI-generated art (like DALL·E and Stable Diffusion), natural language generation (such as ChatGPT), and even AI-composed music (e.g., Jukebox AI).

2.5.7 Where Generative AI and Deep Learning Overlap

Generative AI and Deep Learning are closely connected, as many generative models are built upon deep learning architectures. The relationship and differences between Generative AI and Deep learning are summarized in Table 2.5.

Example:

- **Deep Learning (CNNs)**: Identify pneumonia in an X-ray.
- **Generative AI (GANs)**: Generate synthetic X-rays to train a model with limited real data (Table 2.5).

2.6 APPLICATIONS OF GENERATIVE AI

Generative AI has rapidly transformed numerous industries with its ability to create new content that mirrors human-like creativity and problem-solving capabilities. In healthcare, these systems assist in drug discovery by generating novel molecular structures, while also creating synthetic medical images that help train diagnostic systems without compromising patient privacy. Generative AI has been

TABLE 2.5
Generative AI and Deep Learning Overlap

Task	Deep Learning Model	Generative AI Model
Image classification	ResNet, EfficientNet (CNNs)	Not applicable
Object detection	YOLO, Faster R-CNN	Not applicable
Text generation	Not applicable	GPT-4, BERT (pre-training)
Speech recognition	DeepSpeech, Wav2Vec	Not applicable
Image generation	Not applicable	GANs, VAEs, Diffusion Models
AI art creation	Not applicable	DALL·E, Midjourney

taken up by the art and media sector to create everything from artwork and music to literature and film scripts. It democratized creative tools and made creative tools accessible to professionals and amateurs. These technologies are used by marketing departments to create personalized content for the sake of creating customized advertisements, product descriptions, and customer communications that are relevant to certain audience segments. Code generation tools have revolutionized the way software development tasks are performed by generating functional code from natural language descriptions, drastically accelerating development cycles and making programming more accessible to non-experts.

Generative AI enables generation of optimized designs in the manufacturing and product design domain by techniques such as generative design, which explores thousands of possible configurations to satisfy specified performance criteria. These systems are used in the education sector to create personalized learning materials, practice questions, and educational content, which adjust to the needs of each student. For customer service operations, sophisticated AI chatbots and virtual assistants are used to have customers engage in natural conversations, resolve issues, and provide information without human intervention.

Generative AI can simulate complex systems, formulate hypothesis, and predict outcomes across disciplines, from physics to climate science, and scientific research can benefit from this. These technologies are used by financial institutions in risk assessment, fraud detection, and market prediction by creating scenarios and finding patterns in huge datasets. Generative techniques are used in urban planning and architecture to design buildings and city layouts to be as efficient as possible for sustainability, traffic flow, and human well-being. Perhaps most transformative, generative AI continues to advance accessibility through tools that generate real-time translations, convert text to speech for the visually impaired, and create sign language animations from text, making digital content and services available to previously underserved populations.

Generative AI has a wide range of applications across various industries. In image and video generation, models like GANs create realistic images, deepfakes, and AI-generated art. In text generation, language models such as GPT produce human-like text for chatbots, content creation, and summarization. Music and audio synthesis uses AI to generate original music compositions and realistic voice cloning. Generative models in healthcare are used for medical image enhancement and personalized treatment recommendation. It can be used in drug discovery for molecule generation for drug development. This proves that the Generative AI is transformative in automating creativity, increasing productivity and propelling innovation in multiple domains.

2.7 ETHICAL CONSIDERATIONS AND CHALLENGES

Generative AI (GenAI), which can generate text, images, music, and even code, is a major step forward in AI, but it also comes with a whole new set of ethical issues beyond those seen in traditional ML or DL models. Misinformation and disinformation present one of the major challenges. ChatGPT, DALL•E, Midjourney, and ElevenLabs Voice AI are GenAI tools that can generate content that looks like human language, art style, or even speech patterns with remarkable precision. Deepfakes synthetic media has surged as a result. It can be used to pretend to be public figures, spread false political narratives, or commit fraud. For instance, AI-created political ads and fake statements of a candidate can go viral during elections and thus confuse voters and distort public debate. An additional issue of urgent ethical concern is intellectual property (IP) and creative ownership.

The generative AI models are trained on large datasets scraped from the internet, which may include copyrighted texts, images, and audio. GitHub Copilot uses OpenAI Codex to suggest code. Debates sparked after it would generate code snippets that were almost identical to open source repositories without the proper licensing references. Also, backlash has been raised against AI-generated artworks, which mimic the style of living artists, claiming that their aesthetic identity is cloned and commodified without any consent. As a result, there are critical questions to be answered: Can AI-generated content be copyrighted? Should the original human creators of the data used for training get credit (or compensation)? Core challenges also arise from bias and fairness.

Generative models often inadvertently embed skewed demographics and cultural norms in training datasets sometimes. For example, in language models such as GPT-3, if not carefully fine-tuned, it produces stereotypical responses that associate ethnicities with crime or genders with specific job roles. When generating human faces, darker skin tones or misgendered individuals may be underrepresented with visual models like Stable Diffusion. These kinds of responses can further discriminate and marginalize underrepresented communities. Reinforcement learning from human feedback (RLHF), bias audits, and inclusive dataset curation are all mitigation strategies, but they are used differently by different platforms. GenAI's privacy implications are just as serious concerns. Large language models have been prompted to reproduce personal data, including email addresses, financial details, or confidential conversations, which appeared in their training data, on many occasions. This breaches user privacy and can contravene data protection laws like the General Data Protection Regulation (GDPR) in Europe or India's Digital Personal Data Protection Act (DPDPA).

The technical ability to 'forget' or delete data from a trained model is still limited. It makes it difficult to enforce data erasure rights. It is especially problematic in contexts of sensitive data such as medical records, legal documents, or corporate secrets, which may have been inadvertently used for training. The lack of explainability or, as it is sometimes known, the black box problem, presents another major challenge. Generative models operate through billions of parameters and dense neural layers unlike traditional rule-based systems. It is difficult to understand or trace why a given image was created or how a given sentence was generated because of its complexity. The lack of transparency in these cases causes concern regarding accountability in safety critical applications such as healthcare diagnostics, financial forecasting, or autonomous vehicles. This is similar to a case where a generative model misinterprets a patient's symptoms leading to a hazardous course of action being recommended (and whether such a responsibility is to be held by the developer, the deploying institution, or the algorithm itself). In addition, GenAI is a dual

TABLE 2.6
Ethical Challenges in Generative AI and Possible Solutions

Ethical Challenge	Description	Example	Mitigation Strategy
Misinformation/ deepfakes	Fake content mimicking real people or events	AI-generated political speeches or false celebrity videos	Watermarking, digital signatures, fact-checking AI, platform regulation
Copyright infringement	Use of protected content in training datasets	AI art mimicking known artists; code copied by GitHub Copilot	Consent-based datasets, copyright reform, citation mechanisms
Bias and fairness	Discrimination or stereotyping in outputs	Job ads showing men for tech roles, women for caregivers	Bias audits, inclusive training datasets, RLHF (Reinforcement Learning from Human Feedback)
Privacy breaches	Exposure of sensitive or personal data from training corpora	Model outputs containing personal emails or addresses	Dataset anonymization, differential privacy, privacy laws (e.g., GDPR, DPDPA)
Lack of explainability (black box)	Inability to trace how the model reached an output	Diagnostic AI suggests wrong treatment without explainable logic	Use of interpretable models, AI audits, model documentation
Dual use (abuse potential)	Technology used for both helpful and harmful purposes	AI used for scams, phishing, cheating	Usage licenses, access controls, ethics by design

use major dilemma. Tools that can be applied for wrong things can also be applied for innovation and creativity. Now, open source GenAI platforms such as LLaMA 2 or Stable Diffusion provide great power that can be used both for good and to inflict harm in education. Students can use text generators for academic dishonesty. Scammers might be able to automate phishing emails or use text-to-speech engines to impersonate voices. To address this, many platforms have begun to incorporate content filtering, watermarking, and ethical usage policies, and these are still developing and the implementation is inconsistent across tools and jurisdictions.

So ethical concerns in Generative AI are not limited to theoretical debates. The consequences are real world and the systems are real world. To tackle these, we need a multifaceted approach that includes robust legal frameworks, governance at platform level, transparent algorithms, and public awareness. Building powerful models should not only be the goal, but to expand the goal of building them responsibly, so that their outputs are consistent with societal values, human rights, and the public good. Table 2.6 shows Ethical Challenges in Generative AI and Possible Solutions.

2.8 THE FUTURE OF GENERATIVE AI

Generative AI is the future and its potential capabilities are poised to transform industries, change the way we interact with computers, and drive unprecedented creativity. As these models get increasingly intelligent, adaptable, and personalized, they are facilitating AI systems that are not only reactive but also proactive collaborators in human workflows. Human-centered creativity is one of the most transformative applications of Generative AI. Tools like ChatGPT, MidJourney, and Runway ML are already being used by writers, designers, and filmmakers for brainstorming, drafting, and even creating high-quality content. In the future, AI could be a co-author of books, a co-composer of music, or co-inventor of things in research labs, pushing the boundaries of what individuals and small teams can do.

Generative AI is expected to be applied in healthcare for synthetic data generation, which is needed for medical training and simulation without violating patient privacy. AI models may help to create personalized treatment plans. It may be used to generate synthetic medical images of rare conditions or even in drug discovery by predicting molecular structures. With far-reaching implications in biomedical science, DeepMind's AlphaFold is already proving how AI can solve complex protein folding problems, which is a great achievement. In education, the future has AI-powered personalized learning systems, which generate lesson plans, questions, and feedback dynamically for each student based on the student's learning style and pace. These systems could accommodate real-time adaptation to improve learners experience, bridging the standardized curricula gap to individual learning needs, especially in underserved regions. Generative AI is growing rapidly in the software development domain, from code completion to autonomous coding agents. They can build whole applications from high-level specifications in natural language instructions. This could democratize software creation allowing even non-programmers to see their digital ideas come alive. But this powerful progression also brings new responsibilities. Future developments would demand stronger ethical frameworks, regulatory oversight, and transparency mechanisms. Researchers are now increasingly calling for explainable AI, and efforts are being taken to ensure that AI outputs are traceable, verifiable, and aligned with human values.

Technologically, AI systems that understand and generate across text, image, audio, and video will become more prominent. Such models are called multimodal models. This will enable truly immersive experiences such as AI-powered customer service avatars, virtual teachers, or digital twins in industrial simulations.

The democratization of Generative AI tools is expected to shift power from big tech companies to individuals, small businesses, and local communities in future. Open-source models and decentralized architectures would empower people to build, customize, and deploy AI solutions tailored to their specific requirements, languages, and cultures.

The future of Generative AI is beyond smarter machines. It is about augmenting human potential. However, challenges related to ethics, equity, and governance are going to exist, and the continued collaboration between researchers, policymakers, and the users will decide how responsibly and inclusively this future unfolds.

2.9 CONCLUSION

Generative AI represents a substantial rise ahead in AI, providing the facility to design new content rather than just analyzing existing data. It is based on the ML and DL foundations. It uses state-of-the-art models like GANs, VAEs, and Diffusion Models to produce images, text, music, and more. These advanced capabilities are now transforming nearly all industries, from entertainment and healthcare to research, by allowing for things such as AI-generated art, realistic deepfake videos, and drug discovery. But these advances come with many ethical challenges, including bias, misinformation, and the potential for abuse. To deal with these concerns, we need responsible AI development, transparency, and regulation to make sure that Generative AI is used for positive, ethical purposes.

Moreover, the evolution of Generative AI is in the direction of learning systems that are more efficient, interpretable, and self-supervised and that can be seamlessly integrated into human creativity and decision-making. With the technology advancing every day, Generative AI will soon be the key to changing the way of the next generation of AI applications by breaking the frontiers of what machines can create and do.

REFERENCES

1. Russell, S., & Norvig, P. (2020). *Artificial Intelligence: A Modern Approach* (4th ed.). Pearson.
2. Reed, S., & De Freitas, N. (2016). *Learning to Generate Images with Perceptual Similarity Metrics*. NeurIPS.
3. Goodfellow, I., Pouget-Abadie, J., Mirza, M., Xu, B., Warde-Farley, D., Ozair, S., Courville, A., & Bengio, Y. (2014). *Generative Adversarial Networks*. NeurIPS.
4. Kingma, D. P., & Welling, M. (2013). Auto-Encoding Variational Bayes. *arXiv:1312.6114.*
5. Sohl-Dickstein, J., Weiss, E., Maheswaranathan, N., & Ganguli, S. (2015). *Deep Unsupervised Learning using Nonequilibrium Thermodynamics*. ICML.
6. Brown, T., Mann, B., Ryder, N., Subbiah, M., Kaplan, J., Dhariwal, P., … & Amodei, D. (2020). Language Models are Few-Shot Learners. NeurIPS.
7. Ramesh, A., Pavlov, M., Goh, G., Gray, S., Voss, C., Radford, A., Chen, M., & Sutskever, I. (2021). Zero-Shot Text-to-Image Generation. *arXiv:2102.12092.*
8. Open AI. (2023). GPT-4 Technical Report. Retrieved from https://openai.com/research
9. DeepMind. (2021). AlphaFold: AI for Protein Folding. *Science*, 373(6557), 1147–1152.
10. MIT Technology Review. (2023). The Future of Generative AI. Retrieved from https://www.technologyreview.com/
11. Karpathy, A., & Fei-Fei, L. (2016). Stanford CS23. In: DeLiang Wang (Ed.) *Convolutional Neural Networks for Visual Recognition*. Elsevier.
12. Ng, A. (2017). *Deep Learning Specialization*. Coursera.
13. Vaswani, A., Shazeer, N., Parmar, N., Uszkoreit, J., Jones, L., Gomez, A. N., Kaiser, Ł., & Polosukhin, I. (2017). *Attention is All You Need*. NeurIPS.
14. Devlin, J., Chang, M. W., Lee, K., & Toutanova, K. (2019). *BERT: Pre-Training of Deep Bidirectional Transformers for Language Understanding*. NAACL-HLT.
15. Heusel, M., Ramsauer, H., Unterthiner, T., Nessler, B., & Hochreiter, S. (2017). *GANs Trained by a Two Time-Scale Update Rule Converge to a Local Nash Equilibrium*. NeurIPS (introduces FID).
16. Ho, J., Jain, A., & Abbeel, P. (2020). *Denoising Diffusion Probabilistic Models*. NeurIPS.

17. Dhariwal, P., & Nichol, A. (2021). Diffusion Models Beat GANs on Image Synthesis. *arXiv:2105.05233.*
18. van den Oord, A., Vinyals, O., & Kavukcuoglu, K. (2017). *Neural Discrete Representation Learning.* NeurIPS (VQ-VAE).
19. Karras, T., Laine, S., & Aila, T. (2019). *A Style-Based Generator Architecture for Generative Adversarial Networks.* CVPR (StyleGAN).
20. Oord, A. v. d., Dieleman, S., Zen, H., Simonyan, K., Vinyals, O., Graves, A., ... & Kavukcuoglu, K. (2016). *WaveNet: A Generative Model for Raw Audio.* SSW.
21. Zhang, H., Goodfellow, I., Metaxas, D., & Odena, A. (2019). *Self-Attention Generative Adversarial Networks.* ICML.
22. Karras, T., Aittala, M., Hellsten, J., Laine, S., Lehtinen, J., & Aila, T. (2020). *Training Generative Adversarial Networks with Limited Data.* NeurIPS.
23. Rombach, R., Blattmann, A., Lorenz, D., Esser, P., & Ommer, B. (2022). *High-Resolution Image Synthesis with Latent Diffusion Models.* CVPR.
24. Chen, T., Kornblith, S., Norouzi, M., & Hinton, G. (2020). *A Simple Framework for Contrastive Learning of Visual Representations.* ICML (SimCLR).
25. Crawford, K., & Paglen, T. (2021). *Excavating AI: The Politics of Training Sets for Machine Learning.* The AI Now Institute. Retrieved from https://excavating.ai
26. Ng, A. (2022). *Generative vs. Discriminative Models: A Comparison.* DeepLearning. AI Blog. Retrieved from https://www.deeplearning.ai
27. Rombach, R., Blattmann, A., Lorenz, D., Esser, P., & Ommer, B. (2022). *High-Resolution Image Synthesis with Latent Diffusion Models.* IEEE/CVF Conference on Computer Vision and Pattern Recognition (CVPR), 10684–10695. https://arxiv.org/abs/2112.10752

3 How GenAI Is Replacing Humans: Myth or Real

Ethical Considerations in AI: Bias and Fairness in Generative Models

Deepika Bhatia, Raghav Tiwari, Adithya Venkateswaran, and Prachi Shrivastava

3.1 INTRODUCTION

Generative Artificial Intelligence (GenAI) is a giant leap for artificial intelligence—a paradigm shift in the capability of artificial intelligence. GenAI is different from regular AI, which mostly understands data and makes predictions because it goes into creation—starting new content such as text, images, audio, and video. Its creative potential has sparked wide interest and speculation placing GenAI on the brink of being a revolutionary force that can reshape industries, reconfigure workflows, and test our definition of human creativity. The historical perspective presented in the chapter offers a valuable framework for understanding the present impact and future implications that arise from GenAI.

GenAI has been arising as an echo of those long-standing patterns of tech development and effects on work. From the Industrial Revolution mechanization of work to digital age automation, tech eras have drastically transformed social organization and labor processes at every turn. GenAI is next in this evolution, a step beyond mere automation, capabilities that challenge existing boundaries and intrude increasingly into domains. This was once thought to be reserved for human creativity. Its uses range across many areas, from fun and wellness to learning and investigation, showing its ability to both improve and change old industries.

In addition to its practical uses, GenAI prompts important ethical and societal questions. Since these systems learn from historical datasets, they can in fact learn and perpetuate the biases and inequalities of those same histories. Prominent issues in fairness, transparency, and accountability—particularly within the domains of justice and finance—machine learning have now brought several critical concerns to our attention. The development of technical innovations under a strong regulatory framework with continuous societal dialogue would address these challenges ensuring AI systems to be equitable, trustworthy, and aligned with human values.

DOI: 10.1201/9781003565703-3

TABLE 3.1
Timeline: Evolution of Work and AI Integration

Time	Description	Impact	Challenges
Pre-18th Century	Work dominated by manual labor, agriculture, and craftsmanship	Localized production	Skill-intensive and inefficient.
18th–19th Century (Industrial Revolution)	Mechanization revolutionized production	Rise of factory systems, urbanization, and mass production	Exploitation of workers, environmental degradation, and widening economic inequalities
Early 20th Century	Advancements in electrification and mechanized tools	Increased efficiency, gradual reduction of manual labor	Job displacement for traditional crafts, unequal access to technology
End of the 20th Century (Digital Revolution)	Computers and automation emerge, transforming industries	Outsourcing and global connectivity	Technological unemployment, digital divide, cybersecurity concerns
21st Century	AI integrates into daily life and industries, driving intelligent automation	Creation of new industries, personalized services, and global innovation	Job displacement in traditional sectors: ethical issues, privacy, and misuse

This chapter initiates a multifaceted inquiry into GenAI by providing its historical context. Analyzing the effects of past technological revolutions lets us identify the prospects and challenges that GenAI brings along—furthermore, Table 3.1 illustrates the evolution of Work and AI Integration. Case studies from multiple fields—marketing, manufacturing, healthcare, etc.—demonstrate GenAI's transformation potential while bringing to light paramount issues concerning job displacement, the loss of unique human skills, and all sorts of ethical questions around its deployment.

The way we work has always been shaped by our ability to adapt and innovate. Back in the pre-industrial days, work was all about physical labor and craftsmanship. People relied on manual skills for farming, weaving, and trade—things that kept ancient and medieval economies running. Work was local, with artisans creating goods tailored to their communities. It was previously an art that was typically tiring with much reliance on human and animal energy. This emerging technology is rapidly being adopted around the globe; however, the true consequences of its application remain cloudy. The ethical implications of AI get hairy—big on fairness, accountability, and transparency, and how this affects human well-being. AI models learn from vast amounts of data, including historical biases, meaning they could unintentionally reinforce social inequalities. In fields like hiring, law enforcement, and banking, algorithms can unintentionally favor certain groups, deepening systemic discrimination. Fixing these biases is critical to making AI fair and trustworthy, and strong regulations are needed to ensure accountability.

This chapter explores how GenAI is transforming work, looking at the past, analyzing where we are now, and imagining the future. The goal is to spark discussions and help shape a world where human creativity and AI come together to build a more just and thriving society.

3.2 HISTORY OF THE EVOLUTION OF WORK: FROM HUMAN LABOR TO AI AUTOMATION

During the 18th and 19th centuries, the Industrial Revolution diverted economies from agriculture. This supported the existence of factories and cities, establishing a strong demarcation between home and work. Machines are essential to the process of production. There was a structural change in society as the workers shifted from being independent craftspeople and tradesmen to members of organized hierarchical labor systems [1]. Though these advancements boosted productivity and generated wealth, they also introduced challenges such as harsh working conditions and labor exploitation, deepening the divide between the rich and the poor [2]. The 20th century saw a deep change as technology progressed, mainly in electricity, electronics, and computing, leading to automation. This major shift allowed machines to take on repetitive work, which cut down on the need for human labor in both production and administrative fields [3]. In this digital time, not just productivity improved but also helped connect the world, which gives power to firms to export jobs and reach global markets. While this tech growth made new jobs in areas such as computer coding and IT management, it also made some jobs obsolete [4,5].

What started as a thought in the 20th century has now become one of the most important aspects of creativity. AI can be defined as technologies using which computers and machines perform human functions including learning, comprehension, problem-solving, decision-making, inventiveness, and self-direction [6].

The idea was once limited to science fiction and academic debate on the influence of technology, but now AI is a significant part of human life. It is very important in industries like manufacturing, healthcare and supply chains, where its ability to do work is more than human powers and thus output and productivity get better [7].

3.3 WHAT IS GENAI?—AN INTRODUCTION

GenAI is a more advanced type of AI that does not use existing data for analysis or operations but creates new content. Traditional AI identifies patterns and makes predictions within the scope of its programming and available datasets. GenAI produces text, images, audio, and video well comparable to actual human output [1].

What sets GenAI apart is creativity. Conventional AI is good at solving specific, predefined problems, but it does not create new possibilities the way GenAI can. For example, rather than just recognizing objects in an image, a GenAI model can create a realistic image of something that does not exist in real life [3].

A GenAI system is organized into three major layers: the model layer, the connection layer, and the application layer. It gets integrated into information systems, making boundaries between itself on one hand and external entities such as users, organizations, and data sources—either public or enterprise-based, as highlighted by Samtani et al. (2023). Within these boundaries, factors like creativity and variation influence all parts of the system [3].

- **Model Layer**: This layer holds ready-to-use GenAI components like a Deep Generative Model meant for application systems. Whereas broad models such as GPT-4 use wide-ranging data for training, specialized models—like CodeBERT—get trained on domain-specific data to perform particular tasks.
- **Application Layer**: In this layer, the models are taken into environments that have an impact on users and organizations so that practical use cases can be developed within different domains.
- **Connection Layer**: It acts as a middleman making sure that the model talks well with application layers helping the easy joining with outside systems and data sources [3].

Transformers, proposed by Vaswani et al., drastically changed the architecture of neural networks. Their central contribution, attention mechanisms, allows models to attend input elements long-range dependencies in sequences. Self-attention and multi-head attention form the basic building blocks that enable Transformers to learn compound relationships between input items. Positional encoding is added so that word order can be learned by Transformers in any text sequence. This very sturdy architecture has contributed to enormous advancement in natural language processing, resulting in models like GPT and BERT for applications involving text generation and understanding [4].

Generative adversarial network (GAN) is a machine learning setup and widely used setup for tackling GenAI. A novelty in this GAN setup was that it did not greatly rely on labeled train data. Also its arch is very different from traditional deep neural nets [4]. At its core, GenAI blurs the line between human and machine creativity, raising both excitement and curiosity about its potential applications and implications in various fields.

3.4 IS GENAI REPLACING HUMAN BEINGS? - MYTH OR REALITY

Due to the rapid evolution of GenAI, there are common fears of an imminent mass flip out of the human workforce from different industries. Today, GenAI can create content such as articles, stories, and scripts. It can summarize long texts, create fascinating images and videos, explain coding syntax, and even make fully functioning websites. However, the idea of GenAI replacing the human workforce is an exaggeration. While it is true that it will disrupt certain industries and job roles, a straightforward answer to this question is not possible.

With the introduction of GenAI models such as ChatGPT, a 21% decline was recorded in jobs about coding and writing as compared to jobs that required simple skills. Image-creation jobs also saw a 17% decline post the advent of image-generating technologies [5].

A study was conducted to examine GenAI's role in the work environment from a managerial view point. It revealed a substantial augmentation in employee engagement and performance. However, additional time is required for managers to develop confidence in the effectiveness of such models owing to the deteriorating effect on GenAI's overall positive impact [6]. The benefits of AI models and tools for business have widely been recognized by media outlets. However, it is imperative to keep in mind the shortcomings of such models, which can lead to plagiarism, offensive content, and false information. Organizations thus have to be mindful when adopting AI models to mitigate these problems [8].

AI can serve as a coach to assist humans in navigating through various phases of life such as weight loss [9], physical activities [10], and even when developing new skills [11]. AI coaches help people guide their time on self-care. The GenAI models may serve as members of hybrid innovation teams, the wide solution spaces that lead to much better innovation performances [12]. Because these models are generative in nature, they create new ideas [13] and help humans better understand some problems and sets of solutions.

AI models were first used by humans for automation. It has rapidly taken a more scenario-based approach, where humans and AIs are counterparts of a hybrid team that works collaboratively to get work done [14]. The discussion of whether AI will replace human jobs is an extremely complicated issue that is unfortunately dealt with in much too simplistic a manner. Whereas AI, particularly GenAI, has shown better abilities in content creation, the effect on the workforce is much more nuanced than outright replacement. Research consistently indicates that GenAI is more likely to augment human capabilities than to eliminate them. By working with AI, humans can enhance problem-solving and solution creation, resulting in more effective and efficient outcomes. Additionally, the content created by AI often includes complex, real-world problems requiring human judgment and ethical analysis. This highlights the ongoing necessity of human oversight and leadership, despite advanced AI technologies. Thus, while AI has come a long way, it is still far from achieving the full replacement of humans as a self-sufficient workforce.

3.5 ETHICAL IMPLICATIONS OF GenAI

Fast growth and development in AI-based technologies are revolutionizing the landscape of contemporary industries. While these developments occur, the ethical concerns and societal implications of artificial intelligence have emerged as topics of important interest among industries, researchers, and the public at large. Nonetheless, participating in discussions about the ethical challenges of such AI-based technologies involves a clear perception of how such technologies are viewed and adopted globally [15]. To understand ethics, it is imperative to grasp the culture and vice versa. Terminology like "fairness" and "privacy" can have various meanings across different cultures because each society has different ethical vocabularies [16,17]. To understand AI ethics, there is a need to critically evaluate the connection between technology and society across the world. This has led to the following findings:

- **Variations across Regions are Profound**: Social science research indicates that the impact of AI is influenced by the cultural setting. Just as importantly, the cultural and social contexts of a specific region are to be considered significant factors in the perception of AI and AI-driven technologies [15].
- **Potential to Intensify Social Inequalities**: Studies conducted in the US reveal that AI and AI-driven technologies can often lead to entrenched discrimination, thereby aggravating social inequalities, particularly affecting historically marginalized groups. The findings have shown that such a pattern exists on a global scale among low- and middle-income countries who are far more likely to be vulnerable to the negative social impacts of AI. This can have wide-ranging geopolitical repercussions [15].
- **Increased Efforts are Imperative**: Despite the need for a worldwide wide comprehension of AI and AI-driven technologies, limited research in the area has led to significant gaps in understanding the social impacts of AI. Therefore, thorough on-ground research is crucial to understanding and identifying areas where AI may be exacerbating social inequalities and developing mitigation measures for its negative effects [15].

The use of AI-driven technologies in workplaces has the potential to either enhance or diminish meaningful work. However, how these changes along with their ethical implications may happen remain largely unexplored [18]. The incorporation of AI in decision-making has presented itself with ethical dilemmas. The crypticity of these algorithms raises concerns about transparency, accountability, and bias and thus demands scrutiny. The machine learning models trained on biased historical datasets can amplify social biases. Rigorous examination of training data and algorithmic design are therefore necessary to ensure fairness and mitigate discrimination [19].

Conversational AI systems can also harbor domain-specific ethical issues. Figure 3.1 shows that out of 100 conversational AI chatbots, 37% were female-gendered. It was

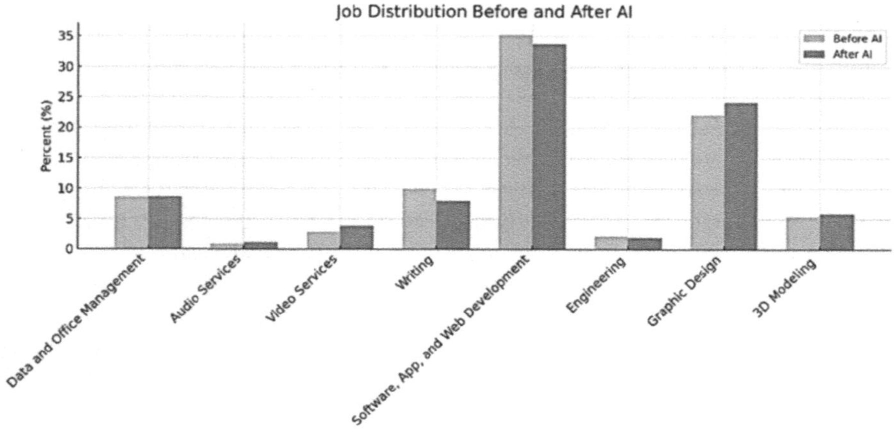

FIGURE 3.1 Job posts for pre- and post-periods of ChatGPT [5].

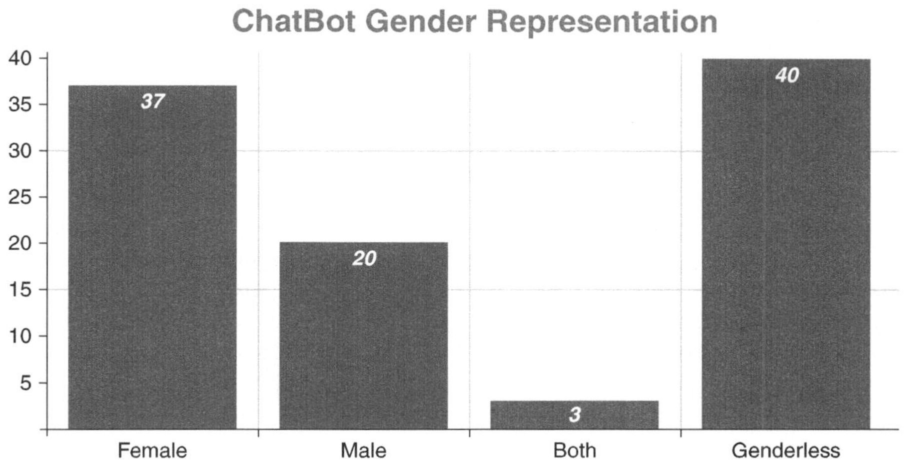

FIGURE 3.2 Gender Representation in Chatbots, 2022 [21].

also found that 62.5% were female by default, highlighting the trend which results from women being the face of glitches stemming from flaws in AI [20] (Figure 3.2).

An experiment designed to explore the difficulties in differentiating between AI-generated content and human media raised concerns about the societal and ethical repercussions in the context of the advancement of AI-generated content. Therefore, future discussions and strict preventive measures are to be given priority [22]. To ensure ethical AI and AI-driven technologies, their development and deployment should be guided by ethical standards. Ethical decision-making frameworks are required to aid in deterring prejudice, bias, and negative social effects. Such frameworks are of utmost importance to help build trust with the users and understand the larger implications of AI [23].

Ethical decision-making frameworks for AI development [23]:

- **Principle-Based**: Broad ethical principles are implemented to govern AI research and development. Transparency, fairness, and accountability are a few of the seven key requirements for an AI system to be acknowledged as reliable and trustworthy [24].
- **Evaluating the Outcomes of the Solutions**: Such frameworks insist that the developers conduct a thorough assessment of the impact of their AI solutions, particularly for those who are the most vulnerable. This may aid the developers to avoid or lessen the gravity of their outcomes.
- **Human-Centered Approaches**: Protection of an individual's privacy, free speech, and equal treatment are considered to develop technologies that are in harmony with social norms [25].

As AI gradually marks its presence in various fields and industries, ethical and just technologies have become paramount. Social inequalities and differences need to be prevented by addressing bias. Transparency, fairness, accountability, and human rights should be prioritized for AI to be acknowledged as ethical, thereby promising user protection, trust, and conscientious innovation.

3.5.1 Bias and Fairness in Generative AI

The rapid advancements of GenAI through deep learning and neural networks have revolutionized fields like computer vision, natural language processing, and content creation. However, the power of GenAI also brings forth pressing ethical concerns regarding bias, transparency, accountability, and potential societal repercussions. As GenAI adoption accelerates, proactively addressing these ethical challenges becomes crucial.

Bias, a tendency or prejudice toward or against something or someone, manifests in machine learning and GenAI as algorithm bias or AI bias. This occurs when an algorithm systematically produces prejudiced results due to flawed assumptions in its training process [26]. Essentially, algorithmic bias leads to consistent errors in the output, deviating from the true or expected outcome [27]. Bias can manifest in various forms. Table 3.2 illustrates how data, algorithms, and user interactions can contribute to different model biases concerning GenAI.

AI bias, a silent yet insidious threat, can exacerbate societal inequalities. In healthcare, algorithms trained on biased data may prioritize white patients over equally ill Black patients, perpetuating systemic racism. Gender bias, too, can lead to misdiagnosis, as seen in the case of misdiagnosis of heart attacks in women, due to the models being trained predominantly on male datasets [29]. The specter of bias looms large even in the realm of creative AI. Popular models like StableDiffusion, despite their impressive capabilities, often generate images that reinforce harmful stereotypes and underrepresented marginalized groups [30]. This not only reflects societal prejudices but also risks amplifying them, demanding urgent action to ensure that AI should be developed and deployed ethically.

The concept of fairness in the case of AI has emerged to ensure fair and unbiased AI. This principle guarantees that AI systems treat all individuals and groups equitably, regardless of protected characteristics such as race, gender, age, or religion [31]. We can further classify fairness into group fairness and the other one is individual fairness. Group fairness focuses on statistical parity and calibration across protected groups, aiming for similar outcomes for variety of groups. On the other hand, Individual fairness emphasizes similar treatment for similar individuals, regardless of group membership. This ensures that AI systems treat everyone fairly, preventing discrimination and promoting inclusivity [32].

Three primary techniques are employed for mitigating bias. Preprocessing involves transforming training data to eliminate underlying discrimination. In-processing modifies algorithms during training to prevent bias. Post-processing, applied after training, reassigns labels to ensure fairness without altering the model's core structure. Each technique offers a unique approach to mitigating bias, tailored to specific constraints and the nature of the data and model [28].

Preparing data is another crucial method to reduce bias in AI models. Techniques like Random Undersampling (RUS) remove majority class samples, Arbitrary Oversampling

TABLE 3.2
Classification of Biases in GenAI Algorithms [28]

Bias Type	Relevance to AI	Example
Measurement Bias	Misleading features in training data lead to biased outcomes	Using arrest records as a proxy for "risk" in AI reflects the over-policing of minority groups
Omitted Variable Bias	Missing critical variables cause incomplete or incorrect predictions	Ignoring market competition in predicting subscription cancellations
Representation Bias	Skewed sampling causes AI to underrepresent diverse populations	Image generation models underrepresent non-Western cultures
Aggregation Bias	Generalizing population-level data leads to inaccurate individual predictions	Text models assume stereotypes about individuals based on aggregate trends
Algorithmic Bias	Bias is introduced by model architecture, optimization, or design choices	Language models reinforce stereotypes through biased token probabilities
User Interaction Bias	User interface or interaction patterns skew the outcomes	Presentation bias in generative search results favoring top-ranked outputs
Popularity Bias	Overemphasis on popular content due to biased exposure metrics	Generative models amplify viral topics regardless of quality
Emergent Bias	Outputs fail to adapt to evolving societal norms or cultural shifts	AI continues using outdated terminology that is no longer appropriate
Historical Bias	Past inequalities in data perpetuate biased outputs	Text models generating male-dominated CEO examples due to historical trends
Temporal Bias	Outdated training data results in irrelevant or incorrect outputs	Generative models fail to reflect recent events or slang in their outputs

(ROS) copy minority lesson tests, and Engineered Minority Oversampling Method (Destroyed) produce modern minority course tests. These strategies point to a more adjusted dataset, progressing to demonstrate execution [26].

In conclusion, as AI proceeds to progress, tending to predisposition is foremost. Prioritizing decency, straightforwardness, and responsibility in AI advancement and sending is basic to relieve potential hurt and guarantee that AI benefits society even-handedly.

3.6 FUTURE OF GENERATIVE AI

As technology advances, its potential for industrial revolutionization and human creativity grows exponentially. The future of this powerful tool is brimming with possibilities, shaped by emerging trends and challenges. According to a study by Wang et al., researchers from around the globe are actively making efforts to contribute to the field of GenAI by improving the accuracy and efficiency of these models [33]. From its humble beginnings as a tool for simple tasks like answering questions and translating languages, GenAI has rapidly evolved. Today, it can outperform human professionals in various fields, including writing, coding, and design [34].

Personalization will be a key part of GenAI moving forward. Rather than one-size-fits-all models, there will be tailored systems for individual users. For example, in education, AI is already revolutionizing the field by giving personal feedback and making the learning experience better. Now educators can give more targeted feedback by combining ChatGPT with their advanced techniques to better help students [35].

In business, GenAI drives personalized customer experiences. Using tools like deep learning, smart data, and IoT, firms can make special deals and build better bonds with customers [36].

In healthcare, GenAI might help find new ways in personal medicine by making drugs suited to a person's genetic makeup. Also, AI-based diagnostic tools could mimic patient results, helping doctors to make exact and well-informed choices [37]. As AI continues to evolve, personalized experiences will become the norm, shaping the future of various industries.

GenAI also has the ability to turn into self-improving systems [38], which may eventually lead to AGI. This is a type of strong AI that can emulate human adaptability and competence across multiple domains [39]. Reinforcement learning, a separate method concerned with trial-and-error optimization, acts as a supplement to GenAI [40]. In unison, these technologies would turn around AI completely and bring in unprecedented levels of creativity and adaptability. Their interaction could create an acceleration in the development of generalized smart systems. The future for GenAI is dazzling and complex all at once. Navigating between the opportunities and challenges will play out how influential GenAI technologies become based on how responsibly developed and integrated instances are within societies—balancing innovation with ethical foresight.

3.7 CONCLUSION

This industry is going through a very big change, much like an industrial revolution, but this time with AI at the center. Advanced and Flexible GenAI has redefined many sectors, changed how work gets done, and created great content. GenAI keeps pushing new limits from automating regular jobs to even creating highly complex work results; it delivers truly amazing new things.

Despite its strides, the societal and ethical concerns are what will drive the large adoption of AI. This, therefore, calls for thoughtful oversight and responsible implementation in matters about reinforcing biases, widening social inequalities, and dismantling established industries. In fact, without proper governance in place, the risks could overshadow the benefits potential that AI holds; hence, a balanced and fair integration calls for ethical considerations. To leverage on this technology and avoid its attendant drawbacks equity, transparency and accountability principles should be laid emphasis. This will entail creating strong regulatory frameworks, ethical guidelines, and inclusive practices to reduce harm as well as promote constructive human–AI interactions. In adapting to new work dynamics with a structured approach, this shall be the key for fair treatment while upholding integrity in its usage.

REFERENCES

1. Martineau, K. (2024). *What is Generative AI?* IBM Research. https://research.ibm.com/blog/what-is-generative-AI
2. McKinsey & Company. (2024, April 2). *What is Generative AI?* https://www.mckinsey.com/featured-insights/mckinsey-explainers/what-is-generative-ai
3. Banh, L., & Strobel, G. (2023). Generative artificial intelligence. *Electronic Markets, 33*(1), 11–26. https://doi.org/10.1007/s12525-023-00680-1
4. Sengar, S. S., Hasan, A. B., Kumar, S., & Carroll, F. (2025). Generative artificial intelligence: A systematic review and applications. *Multimedia Tools and Applications, 84*, 23661–23700. https://doi.org/10.1007/s11042-024-20016-1
5. Demirci, O., Hannane, J., & Zhu, X. (2023, October 15). *Who Is AI Replacing? The Impact of Generative AI on Online Freelancing Platforms.* Available at SSRN: https://ssrn.com/abstract=4602944 or https://dx.doi.org/10.2139/ssrn.4602944
6. Manresa, A., Sammour, A., Mas-Machuca, M., Chen, W., & Botchie, D. (2024). Humanizing GenAI at work: Bridging the gap between technological innovation and employee engagement. *Journal of Managerial Psychology*, Vol. ahead-of-print No. ahead-of-print. https://doi.org/10.1108/JMP-05-2024-0356
7. Schuckart, A. (2024). Introduction to work with GenAI. In *Proceedings of the 28th European Conference on Pattern Languages of Programs (EuroPLoP '23)* (p. 8). Association for Computing Machinery. https://doi.org/10.1145/3628034.3628042
8. Dwivedi, Y. K. et al. (2023). Opinion paper: "So what if ChatGPT wrote it?" Multidisciplinary perspectives on opportunities, challenges and implications of generative conversational AI for research, practice, and policy. *International Journal of Information Management, 69*, 102642. https://doi.org/10.1016/j.ijinfomgt.2023.102642
9. Stein, N., & Brooks, K., (2017). A fully automated conversational artificial intelligence for weight loss: Longitudinal observational study among overweight and obese adults. *JMIR Diabetes*, *2*(2), e28. https://doi.org/10.2196/diabetes.8590
10. Terblanche, N., & Cilliers, D. (2020). Factors that influence users' adoption of being coached by an artificial intelligence coach. *Philosophy of Coaching: An International Journal*, *5*(1), 61–70.
11. Wlasak, W., Zwanenburg, S. P., & Paton, C. (2023). Supporting autonomous motivation for physical activity with chatbots during the COVID-19 pandemic: Factorial experiment. *JMIR Formative Research*, *7*(1), e38500.
12. Bouschery, S. G., Blazevic, V., & Piller, F. T. (2023). Augmenting human innovation teams with artificial intelligence: Exploring transformer-based language models. *Journal of Product Innovation Management*, *40*(2), 139–153.
13. Stevenson, C., Smal, I., Baas, M., Grasman, R., & van der Maas, H. (2022). Putting GPT-3's creativity to the (alternative uses) test. *arXiv preprint arXiv:2206.08932.*
14. Vassilakopoulou, P., Haug, A., Salvesen, L. M., & Pappas, I. O. (2023). Developing human/AI interactions for chat-based customer services: Lessons learned from the Norwegian government. *European Journal of Information Systems*, *32*(1), 10–22.
15. Hagerty, A., & Rubinov, I. (2019). Global AI ethics: A review of the social impacts and ethical implications of artificial intelligence. *ArXiv, abs/1907.07892*, https://doi.org/10.48550/arXiv.1907.07892
16. Strathern, M. (1990). The *Gender of the Gift: Problems with Women and Problems with* Society in Melanesia, No. 6. Univ of California Press; Linklater, A. (1998). *The Transformation of Political Community: Ethical Foundations of the Post-Westphalian Era.* Univ of South Carolina Press; Lebow, R. N. (2012). *The Politics and Ethics of Identity: In Search of Ourselves.* Cambridge University Press.

17. Agre, P. E., & Marc, R., eds. (1998). *Technology and Privacy: The New Landscape.* MIT Press; Bélanger, F., & Robert E. C. (2011). Privacy in the digital age: A review of information privacy research in information systems. *MIS Quarterly, 35*(4), 1017–1042; Hartman, L. P. (2001). Technology and ethics: Privacy in the workplace. *Business and Society Review, 106*(1), 1–27.
18. Bankins, S., & Formosa, P. (2023). The ethical implications of artificial intelligence (AI) for meaningful work. *Journal of Business Ethics, 185,* 725–740. https://link.springer.com/article/10.1007/s10551-023-05339-7
19. Femi, O., Olukunle, O. A., Akoh, A., Temitayo, O. A., Oluwatoyin, A. F., & Benjamin, S. A. (2024). Reviewing the ethical implications of AI in decision making processes. *International Journal of Management & Entrepreneurship Research, 6*(2). https://doi.org/10.51594/ijmer.v6i2.773
20. Adewumi, T. P., Liwicki, F.S., & Liwicki, M. (2022). State-of-the-art in open-domain conversational AI: A survey. *Information, 13,* 298.
21. Nestor Maslej, Loredana Fattorini, Erik Brynjolfsson, John Etchemendy, Katrina Ligett, Terah Lyons, James Manyika, Helen Ngo, Juan Carlos Niebles, Vanessa Parli, Yoav Shoham, Russell Wald, Jack Clark, and Raymond Perrault, "The AI Index 2023 Annual Report," AI Index Steering Committee, Institute for Human-Centered AI, Stanford University, Stanford, CA, April 2023, pp. 145, https://hai.stanford.edu/assets/files/hai_ai-index-report_2023.pdf
22. Partadiredja, R. A., Serrano, C. E., & Ljubenkov, D. (2020). AI or human: The socio-ethical Implications of AI-generated media content. *2020 13th CMI Conference on Cybersecurity and Privacy (CMI) - Digital Transformation - Potentials and Challenges (51275),* Copenhagen, Denmark, pp. 1–6, https://doi.org/10.1109/CMI51275.2020.9322673
23. Modi, T. B. (2023). Artificial Intelligence Ethics and Fairness: A study to address bias and fairness issues in AI systems, and the ethical implications of AI applications. *Revista Review Index Journal of Multidisciplinary, 3*(2), 24–35. https://doi.org/10.31305/rrijm2023.v03.n02.004
24. European Commission. (2019). Ethics Guidelines for Trustworthy AI. https://ec.europa.eu/digital-single-market/en/news/ethics-guidelines-trustworthy-ai
25. Romeo Casabona, C. M. (2018). Human rights and ethical considerations in the application of artificial intelligence to health care. In *Artificial Intelligence in Health Care* (pp. 73–84). Springer, Cham.
26. Kartal, E. (2022). A comprehensive study on bias in artificial intelligence systems: Biased or unbiased AI, that's the question! *International Journal of Intelligent Information Technologies, 18.* 1–23. https://doi.org/10.4018/IJIIT.309582
27. Žliobaitė, I. (2017). Measuring discrimination in algorithmic decision making. *Data Mining and Knowledge Discovery, 31*(4), 1060–1089.
28. Mehrabi, N., Morstatter, F., Saxena, N., Lerman, K., & Galstyan, A. (2019). A survey on bias and fairness in machine learning. *arXiv.Org.* https://arxiv.org/abs/1908.09635
29. Norori, N., Hu, Q., Aellen, F. M., Faraci, F. D., & Tzovara, A. (2021). Addressing bias in big data and AI for health care: A call for open science. *Patterns, 2*(10), 100347. https://doi.org/10.1016/j.patter.2021.100347
30. Nicoletti, L., & Bass, D. (2023). *Generative AI Takes Stereotypes and Bias from Bad to Worse.* Bloomberg.Com. https://www.bloomberg.com/graphics/2023-generative-ai-bias/
31. Dwork, C., Hardt, M., Pitassi, T., Reingold, O., & Zemel, R. (2012). Fairness through awareness. In: *Proceedings of the 3rd Innovations in Theoretical Computer Science Conference* Association for Computing Machinery, New York, NY, USA. (pp. 214–226).
32. Binns, R. (2019). On the apparent conflict between individual and group fairness. *arXiv.Org.* https://arxiv.org/abs/1912.06883

33. Wang, N., Li, S., Wang, C., & Zhao, L. (2024). Current status and emerging trends of generative artificial intelligence technology: A bibliometric analysis. *Journal of Internet Technology*, *25*(3). https://jit.ndhu.edu.tw/article/view/3065
34. Morgan, J. P. (n.d.). The Rise of Generative AI. J.P. Morgan Research. https://www.jpmorgan.com/insights/global-research/artificial-intelligence/generative-ai
35. Kuzminykh, I., Nawaz, T., Shenzhang, S., Ghita, B., Raphael, J., & Xiao, H. (2024). Personalised feedback framework for online education programmes using generative AI. *arXiv.Org*. https://arxiv.org/abs/2410.11904
36. Abrokwah-Larbi, K. (2023). The role of generative artificial intelligence (GAI) in customer personalization (CP) development in SMEs: A theoretical framework and research propositions. *Industrial Artificial Intelligence*, 1(1). https://doi.org/10.1007/s44244-023-00012-4
37. Gulia, K., Hamdan, I. A., Datta, N., Gupta, Y., Kumar, P., Yadav, A., Mitten, S., & Kumar, R. (2024). Machine learning models for personalized healthcare on marketable generative-AI with ethical implications. *World Journal of Advanced Research and Reviews*, 23(03), 707–720.
38. Khlaisamniang, P., Khomduean, P., Saetan, K., & Wonglapsuwan, S. (2023). Generative AI for self-healing systems In: *2023 18th International Joint Symposium on Artificial Intelligence and Natural Language Processing (iSAI-NLP)*, Bangkok, Thailand (pp. 1–6). https://doi.org/10.1109/iSAI-NLP60301.2023.10354608
39. Goertzel, B. (2014). Artificial general intelligence: Concept, state of the art, and future prospects. *Journal of Artificial General Intelligence, 5*(1), 1–48. https://doi.org/10.2478/jagi-2014-0001
40. Kaelbling, L. P., Littman, M. L., & Moore, A. W. (1996). Reinforcement learning: A survey. *Journal of Artificial Intelligence Research, 4*, 237–285. https://doi.org/10.1613/jair.301

4 Ethical Frontiers in AI Development

Navigating Innovation with Integrity in the Age of GenAI

Vijay Kumar, Rachna Narula, Sarthak Khanna, Anil Kumar, Poonam Chaudhary, and Surbhi Bhatia Khan

4.1 WHAT IS GENERATIVE AI AND ITS IMPACT ON SOCIETY

Generative AI, better known as GenAI, represents a transformative force in society, revolutionizing various areas from art to healthcare. At its core, it involves the creation of new content, whether images, text, or even music, by algorithms trained on vast datasets. In recent years, generative AI has turned out to be an essential tool for different applications and services. Nonetheless, this rapid advancement also raises an important question: is GenAI replacing humans or is this notion merely a delusion? To investigate this question, it is important to dive into the ethical considerations regarding AI development, particularly focusing on bias and fairness in generative models.

Generative AI holds gigantic potential to reshape society in significant ways. From producing similar pictures to composing music vague from human manifestations, the applications of GenAI are boundless. Within the domain of craftsmanship, for example, it enables makers to investigate modern wildernesses, pushing the boundaries of creative energy and inventiveness. In addition, in areas such as healthcare and instruction, generative models help in information examination, medication revelation, and personalized learning, expanding human capabilities and improving efficiency.

The rise of generative AI has in a general sense changed different segments of society, as outlined in Table 4.1. Later considerations show that 67% of organizations have as of now executed a few shape of GenAI in their operations, highlighting its developing noteworthiness within the cutting edge innovative scene (Figure 4.1).

However, with great power comes great responsibility. The pervasive influence of GenAI raises ethical concerns regarding its impact on employment, privacy, and societal norms. The fear of automation replacing human jobs looms large, necessitating a careful examination of how to ensure a smooth transition in the workforce. Additionally, the creation of very realistic fake material threatens privacy and security, resulting in debates about regulation and accountability. So, we have to tackle these issues in that technological age.

 DOI: 10.1201/9781003565703-4

TABLE 4.1
Impact of Generative AI Across Different Sectors

Sector	Primary Applications	Impact Level (1–5)	Key Challenges
Healthcare	Medical imaging, Drug discovery	4.5	Data privacy
Education	Personalized learning, Content creation	4.0	Plagiarism concerns
Finance	Risk assessment, Fraud detection	4.8	Algorithmic bias
Creative industries	Art generation, Music composition	3.9	Copyright issues
Manufacturing	Design optimization, Quality control	4.2	Integration costs

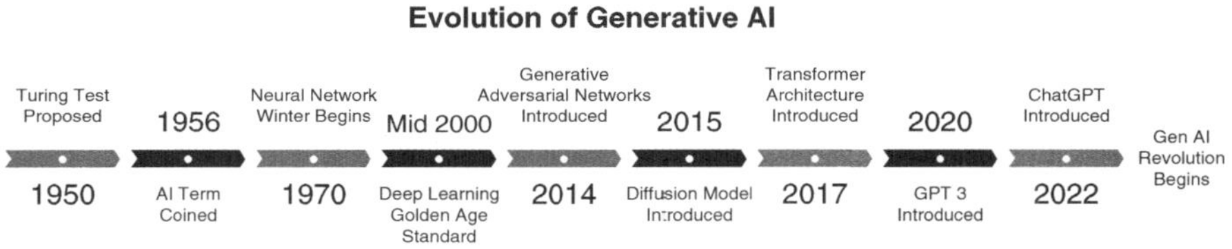

FIGURE 4.1 Evolution of generative AI applications.

4.2 THE USE OF GENERATIVE MODELS: ARE THEY MYTHS?

Generative models shine through their trigger imagination to create real-world content and disrupt the industry. With all these changes taking place, the question now is whether the rise of generative model is hype or success? Whether the very rise of such generative models, which are purported to be real-world content creators, sparks innovation and creativity and would in the future bring changes to industries. To start answering this question, we will look at the complexity of AI construction and its strengths and challenges. This technology shows how much AI can help humans use their creativity to change many things like entertainment, business, and education.

Figure 4.2 shows that the abilities of generative models have grown substantially over the past decade [1]. Table 4.2 presents a comparative analysis of different generative model architectures and their applications.

Even though excitement about generative models is on the rise, concern around the potential harm from their use is imminent. Many people around the globe are concerned about the ethical issues related to the text that was written by AI, including bias, fairness, and truthfulness. Misinformation is becoming more widespread because of the rise of deepfake technology. Another growing concern is that accidents have led to the development of generative AI [2].

In addition, generative models taking the place of human creativity and skill leads to tensions surrounding work displacement and social change. Though AI certainly enhances human ability and work, the idea of automation encroaching on human sphere makes us think about future of work and the important retraining and adaptations necessary in an AI world. It is essential as we progress into the future, to be

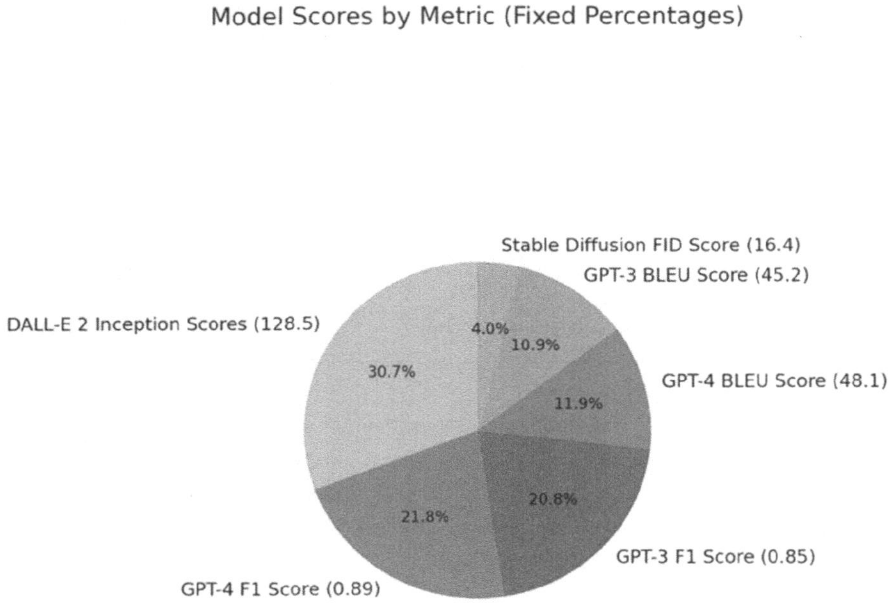

FIGURE 4.2 Performance metrics of leading generative models.

TABLE 4.2
Comparison of Generative Model Architectures

Architecture	Key Features	Use Cases	Performance Score
GANs	Adversarial training	Image generation	8.5/10
Transformers	Self-attention mechanism	Text generation	9.2/10
VAEs	Probabilistic encoding	Data compression	7.8/10
Diffusion models	Iterative denoising	Image synthesis	9.0/10

vigilant, develop responsible AI, and be transparent, accountable, and cautious with the ethical considerations. When all stakeholders come together, including technologists, academic researchers, policymakers, civil society, industry, and media, we will moderate the possible harmfulness of generative AI while harnessing the benefits [3].

4.3 ETHICAL CONSIDERATIONS IN AI DEVELOPMENT

As AI continues to grow, the ethical provision has become central to the dedicated development and deployment of AI, as per an expert. AI ethical considerations include technical, societal, political, economic, and legal dimensions.

AI development faces many ethical challenges, but bias is the major one. Since AI models are typically trained on historical data to predict future outcomes, they may inadvertently perpetuate existing inequalities and discriminate against disadvantaged communities. To reduce bias in AI, care is taken when making datasets; it is made sure that the algorithms used are transparent; these systems are continuously monitored and audited [4].

AI development must not be a black box that produces results. As AI systems increasingly influence decision-making processes across various domains, including healthcare, finance, and criminal justice, ensuring transparency and explainability is essential for building trust and accountability. Customers and partners need to be able to understand how AI systems make decisions and evaluate their reliability and fairness. In order to gain transparency and explainability, in many cases, we simply need to build AI approaches and designs that allow for human oversight and comprehension [5].

Privacy and information security touch on another fundamental ethical consideration in the development of AI. AI structures depend on unending totalities of singular information to be tried and run, raising critical concerns about information protection and security. Protecting security needs robust information, and the executives use frameworks based on their choices on rules like information minimization, anonymization, and client assent. And giving data to the individuals who ought to possess their very own information is essential for protecting security rights and working with moral AI.

Furthermore, as AI frameworks attended to be increasingly self-ruling and capable of making choices without direct human intercession, inquiries regarding responsibility and liability gotten to be more prominent. Also, establishing clear rules and components for responsibility and oversight is important to navigate ethical complexities associated with autonomous decision-making in AI. This involves defining aspects and obligations, establishing accountability mechanisms, and ensuring appropriate reparation mechanisms in the area of AI-related harms [6].

4.4 UNDERSTANDING BIAS AND FAIRNESS IN GENERATIVE MODELS

Models are most capable of producing outputs that are consistent and accurate; however, they are still subject to data bias. Design inequality includes gender, race, and economic classes, among other types of inequality. These prejudices can give rise to social inequalities, as well as ideas or actions of the bias in the design context (Figure 4.3).

The most common way to introduce deviations in generative models is by using biased dataset. If the dataset used in the model is in any sense biased toward a specific group of individuals, then there is a high chance that the generated models will mirror those deviations. Additionally, predispositions too can be presented amid the plan and execution of the generative demonstrate design. The optimization destinations utilized amid preparation may prioritize certain traits or characteristics over others, thereby advancing compounding inclination within the produced content [7].

Addressing predisposition and advancing decency in generative models requires a multifaceted approach. To begin with, it is basic to fundamentally assess the preparing information to identify and moderate predispositions some time recently preparing the demonstration. This may include procedures such as information enlargement, information adjustment, and cautious curation of preparing datasets to guarantee

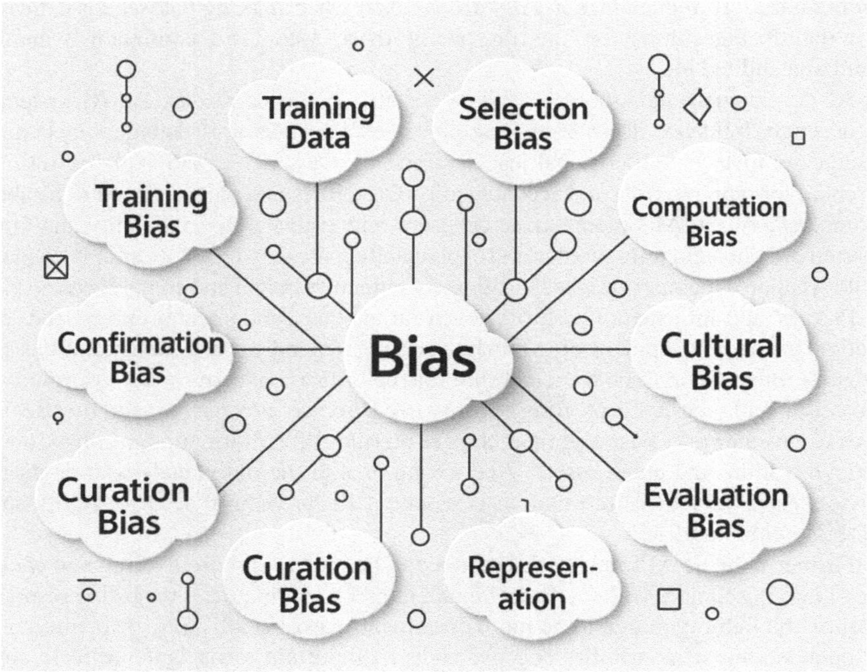

FIGURE 4.3 Common types of bias in GenAI.

representativeness and diversity. Furthermore, analysts and designers must embrace inclinationmindful techniques throughout the complete lifecycle of the generative demonstration, from planning and preparing to assessment and arrangement [8]. Besides, advancing differences and inclusivity within the generative demonstrates that investigating community is basic for addressing predisposition and cultivating decency.

By empowering intrigue collaboration and joining differing points of view into the investigation and improvement in preparation, we will create more impartial and comprehensive generative models that superiorly reflect the lavishness and differing qualities of human encounters (Table 4.3).

4.5 CHALLENGES OF RESPONSIBLE AI DEVELOPMENT

Dependable AI advancement faces various challenges that must be addressed to guarantee that AI advances are created and sent in a way that adjusts with moral standards and societal values. A few of the key challenges include:

1. AI frameworks prepared on one-sided datasets may sustain existing societal imbalances and lead to biased results. To control bias and promote fairness, it is important to control data and algorithmic transparency as well as to continuously monitor and audit AI systems [9].

TABLE 4.3
Common Bias Types in AI Systems

Bias Type	Definition	Example	Impact on AI
Training data	Bias from unrepresentative or skewed data used to train models	Facial recognition dataset lacking diversity	Model performs poorly on underrepresented groups
Training bias	Bias introduced during model training from data preprocessing or design	Oversampling certain classes while training	Unfair performance across different categories
Selection bias	Bias from how data is selected or sampled for training or testing	Only collecting user feedback from active users	Skewed model understanding of actual user base
Algorithmic bias	Bias arising from the design of the algorithm itself	A scoring system favoring certain profiles due to algorithmic structure	Systemic unfairness in predictions or decisions
Evaluation Bias	Bias from how a model's performance is measured or validated	Using metrics that favor accuracy over fairness	Misleading performance metrics; overlooked harm
Representation bias	Bias in how groups or scenarios are portrayed or included in datasets	Underrepresentation of disabled individuals in data	Reinforcement of societal stereotypes or exclusion
Confirmation bias	Tendency to focus on data that confirms pre-existing beliefs	Cherry-picking results to show success of an AI model	Misleading conclusions and lack of scientific objectivity

2. A lot of the AI dealings—particularly with the profound learning models—function as black boxes, making it difficult to comprehend how choices are made. Developing software and code mechanisms that will be readily interpretable and comprehensible has a substantial role to play in enabling accountable human oversight and understanding the ways in which AI systems work.
3. AI systems rely on large troves of personal data to train and perform efficiently, causing concerns about the protection of data as well as the security of individuals. Building data governance frameworks includes mechanism like data minimization, anonymization, consent, etc., to protect privacy rights and monitor behavior of AI applications [10].
4. Artificial intelligence systems are accused of wrongdoing, but who is to be blamed for that? How can AI be held accountable? Legal and ethical challenges arise to ensure accountability, which involves a challenging but essential step.

4.6 ADDRESSING BIAS AND FAIRNESS IN GENERATIVE MODELS

By counting a rich cluster of cases from different bunches, preferences can be moderated, generating content material that shows the multifaceted nature of human experiences. The inclusion of fairness constraints and regularization methods in the

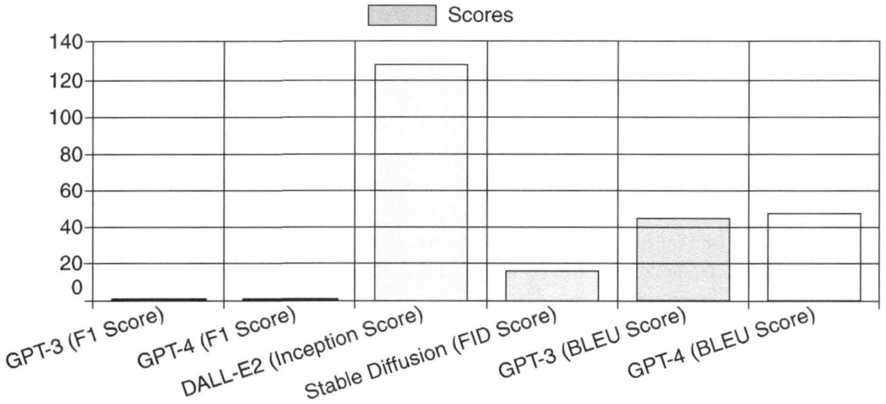

FIGURE 4.4 Bias mitigation pipeline in generative model.

training objective progressively promotes fairness in generative models. By specifically punishing unfair outcomes or encouraging diversity in the created substance, these practices help to alleviate biases and develop more attractive display yields (Figure 4.4).

Furthermore, the development of evaluation metrics, which help to design unbiased generative models, is crucial. These measurements allow the performance of models to be assessed and compared in order to identify and measure biases in generated content and to undertake actions aimed at improving fairness [11]. Locking in with assorted communities and seeking feedback from partners is additionally crucial. By consolidating a wide run of viewpoints into the advancement preparation, generative models can superiorly address inclinations and guarantee that the created substance reverberates with the values and needs of society [12,13].

Predisposition in generative models can be successfully receiving these multifaceted techniques and implanting standards of reasonableness and inclusivity into the advancement process. This clears the way for the creation of substance that really reflects the differing qualities of human encounters and contributes emphatically to societal well-being [14].

4.7 LEGAL AND REGULATORY FRAMEWORKS FOR ETHICAL AI DEVELOPMENT

Lawful and administrative systems play a pivotal part in guaranteeing that AI improvement follows to moral standards and shields against potential dangers and hurts. Here are seven such frameworks:

1. **General Information Security Direction (GDPR)**: The GDPR, actualized by the European Union, controls the preparation of individual information and forces strict necessities on the collection, capacity, and utilization of

information. It incorporates arrangements for straightforwardness, information minimization, and client assent, which are fundamental for moral AI development [15].

2. **Ethical AI Rules and Standards**: Different organizations and industry bunches have created moral AI rules and standards to advance capable AI improvement. Illustrations incorporate the OECD AI Standards, IEEE Morally Adjusted Plan, and the Asilomar AI Standards. These systems give direction on moral contemplations such as reasonableness, straightforwardness, responsibility, and human rights [16].
3. **Algorithmic Responsibility Act**: Proposed within the Joined together States, the Algorithmic Responsibility Act points to address the potential inclinations and oppressive results of AI algorithms by requiring companies to evaluate and relieve the impacts of their calculations on decency, straightforwardness, and accountability [17,18].
4. **EU Proposition for Fake Insights Act**: The European Union's proposed Manufactured Insights Act looks for to control the improvement and arrangement of AI frameworks, especially those regarding high risk. It incorporates prerequisites for straightforwardness, human oversight, and chance evaluation, pointing to guarantee that AI innovations are created and utilized responsibly [19].
5. **AI Affects Evaluation Systems**: A few nations and organizations have created AI impact appraisal systems to assess the potential social, financial, and moral impacts of AI innovations. These systems offer assistance in distinguishing and addressing dangers and challenges related with AI advancement, directing policymakers and engineers in making informed decisions [20].
6. **Sector-Specific Guidelines**: In addition to general AI regulations, some sectors may have specific regulations governing the use of AI technologies. As an example, the healthcare section should have rules appropriate to the utilization of AI in restorative determination and treatment, whereas the budgetary section may have controls relating to AI-supported algorithmic exchanging and hazard management.
7. **Global Cooperation and Actions**: Advancement of AI happens through benchmarks; step endeavors are fundamental for ensuring consistency and interoperability over unique locales. The global effort organizations such as ISO and ITU are also working to develop universal metrics interim (as we'll see) to facilitate global progress toward moral and mindful AI development.

Through the practical implementation of these legal and regulatory frameworks, policymakers and stakeholders can promote ethical AI development, mitigate risks, and ensure that AI technologies are developed and deployed in ways that align with societal values and norms (Table 4.4).

TABLE 4.4
Global Legal and Regulatory Frameworks for Ethical AI Development

Framework	Jurisdiction	Key Focus Area	Primary Objectives	Implementation Status
GDPR	European Union	Data Protection	Personal data privacy, consent, transparency	Fully Implemented
Algorithmic Accountability Act	United States	Algorithmic Bias	Bias assessment, algorithmic impact evaluation	Proposed
EU AI Act	European Union	High-Risk AI Systems	Risk classification, compliance requirements	In Advanced Stages
IEEE Ethically Aligned Design	International	Ethical Design Principles	Human rights, accountability, transparency	Recommended Guidelines
OECD AI Principles	Global	Responsible AI Development	Trustworthy AI, human-centric approach	Adopted by 42 Countries
China's AI Regulations	China	National AI Governance	State control, ethical development	Partially Implemented

4.8 CASE STUDIES: ETHICAL DILEMMAS IN GEN AI APPLICATIONS

4.8.1 Case Study 1: Deepfake Technology in Politics

Deepfake technology presents a real ethical issue in the world of law executive. What if a candidate creates a fake video committing a criminal or corrupt act, just days before an important election? Though totally fake, the video goes viral on social media, creating questions about the candidate's judgment and possibly influencing voters [21].

This presents complicated moral questions. In a single hand, the use of deepfake innovation to circulate incorrect data undermines constituent preparation and subverts the law-based standards of straightforwardness, which are useful for pluralism and responsibility. Controlled substance voters are naturally prone to make careless decisions and debase trust in political teaching (Figure 4.5).

On the other hand, restricting the use of deepfake innovation raises concerns over the issues of censorship and opportunity of expression. Modulating the need to combat trickery with the necessity for free discussion is a daunting moral dilemma for policymakers and tech organizations the alike.

4.8.2 Case Study 2: Bias in Facial Recognition Systems

Systems of facial recognition have become more and more ubiquitous in a number of different areas ranging from law enforcement to commercial applications.

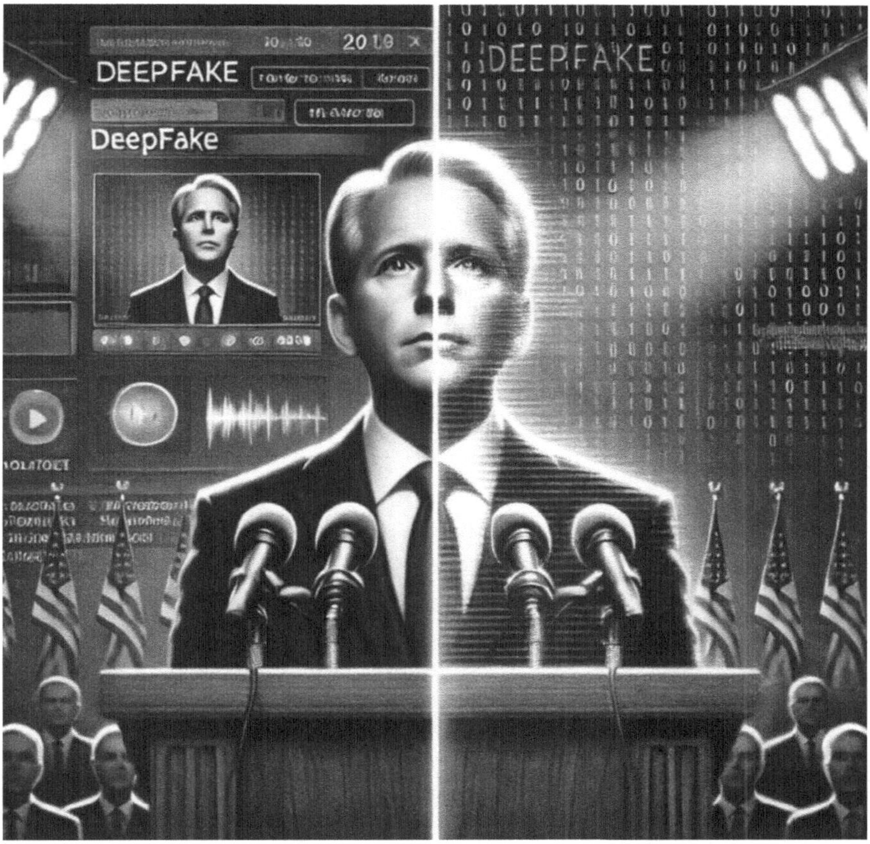

FIGURE 4.5 Deepfake technology in politics.

In any event, these frameworks have been known to show slants, particularly about race and sex. For illustration, consider research indicating that facial acknowledgment frameworks are less viable when recognizing individuals with more obscure skin tones, driving to uneven misidentifications and false positives among particular demographic groups [22,23].

This bias in facial recognition technology is why there are ethical considerations that are real, as it can reinforce systemic inequalities and produce unfair outcomes. Biased facial acknowledgment frameworks could target minority communities for law enforcement purposes, further exacerbating racial profiling and discriminatory issues (Figure 4.6).

Although deformatted, the outline needs to be set up for trading thing in facial acknowledgment frameworks amid upgrade and assessment. Furthermore, the use of facial recognition technology must be accompanied by transparency of use and accountability mechanisms to ensure that facial recognition technology is not developed or used irresponsibly and that its social impact is carefully considered [24].

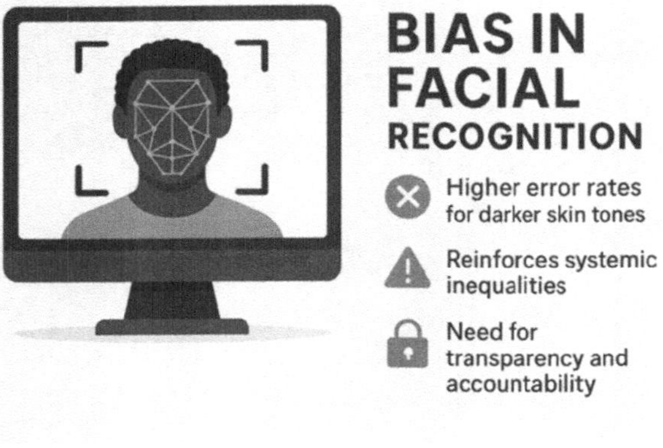

FIGURE 4.6 Bias in facial recognition.

FIGURE 4.7 Dilemma of self-driving cars.

4.8.3 Case Study 3: Autonomous Vehicles and Moral Dilemmas

Self-driving vehicles may reduce road accidents and fatalities with proper management. But automated vehicles cause moral problems of considerable complexity, at least in terms of life-and-death decisions [25].

Let us consider that a self-driving car has to choose between two options, both of which are life threatening. In the first one, if it deviates from its initial direction to avoid a pedestrian, it will hit a group of cyclists, but if it doesn't deviate, it will hit the pedestrian. How can a self-driving car choose between both these options? (Figure 4.7 and Table 4.5)

This ethical conundrum highlights a disregard for the safety of drivers and pedestrians. Discussions will be essential in conveying how self-driving vehicles will have to be heavily regulated to meet social expectations through partner engagement.

They are initiated into a society that qualifies most of these as normative, allowing for an easier negotiation of the moral arrangements that guarantee parliamentary

TABLE 4.5
Moral Dilemmas in Autonomous Vehicle Decision-Making

Scenario	Ethical Challenge	Potential Outcomes	Ethical Considerations
Unavoidable collision	Minimizing human casualties	Swerve and hit cyclists Maintain course and hit pedestrian	Utilitarian approach vs. individual rights
Pedestrian crossing unexpectedly	Sudden decision-making	Brake abruptly Swerve into alternative path	Safety of multiple parties vs. predictability
Multiple vulnerable road users	Complex risk assessment	Prioritize children Minimize total potential injuries	Age, vulnerability, potential life impact
Mechanical failure scenario	System reliability	Controlled deceleration Minimal damage trajectory	Engineering safety vs. immediate survival
Mixed traffic environment	Unpredictable interactions	Conservative response Aggressive avoidance	Balancing caution and effective navigation

standards to remain very moderate, and thus transparent, such as the accountability of the oversight of the markup and design of AI structures, during the time spent investigating these situations and securing futile locks in reconnecting with the general public.

4.9 ROLE OF STAKEHOLDERS IN PROMOTING RESPONSIBLE AI

Partners play a pivotal part in advancing dependable AI improvement by contributing assorted points of view, pushing for moral standards, and driving responsibility all through the AI lifecycle. Government and administrative bodies are entrusted with making and upholding laws and controls that administer AI advancement and arrangement. They set moral benchmarks, set up rules, and guarantee compliance with lawful systems. Industry partners, including tech companies and AI engineers, plan and actualize AI frameworks that follow moral standards, prioritize client well-being, and advance straightforwardness and accountability [26,27].

The academic and research community builds knowledge of AI technologies, develops ethical standards, and shares best practices for responsible AI development. Organizations in civil society and advocacy groups raise public awareness about AI ethics, advocate for marginalized communities, and hold stakeholders accountable. Buyers and customers demand AI suppliers' transparency, accountability, and ethical conduct, which influences industry practices. Global organizations and standard bodies promote global collaboration, set standards, and create guidelines for ethical AI development on a global scale. When partners work together, they can address moral challenges, relieve dangers to society, and guarantee that AI innovations advantage mankind dependably.

4.10 CONCLUSION: STRIKING A BALANCE BETWEEN INNOVATION AND ETHICS

To sum up, we look into development and ethics as the confluence point is important in the development of AI. Progress drives technological advancements and promises. Ethical considerations serve as a guide for responsible development and deployment of transformative solutions AI. By taking a joint effort between accomplices, campaigning for straightforwardness, and propelling of ethical principles, we have struck a fragile adjustment that sees the advantages of AI while limiting the potential dangers and guaranteeing consistency with shared esteem. In the end, it is through technology and ethics that we can pave the way for a future where AI upgrades humanity and promotes progress in an inclusive, ethical, and sustainable way.

REFERENCES

[1] Mittelstadt, B. (2019). "Principles alone cannot guarantee ethical AI." *Nature Machine Intelligence*, 1(11), 501–507. https://www.nature.com/articles/s42256-019-0114-4

[2] Floridi, L., & Cowls, J. (2019). "A unified framework of five principles for AI ethics." *Harvard Data Science Review*, 1(1), 35–545. https://doi.org/10.1002/9781119815075.ch45

[3] Mitchell, M., Wu, S., Zaldivar, A., et al. (2020). "Model cards for model reporting." In *Proceedings of the Conference on Fairness, Accountability, and Transparency*, Atlanta Georgia USA, pp. 220–229.

[4] Link: https://arxiv.org/abs/1810.03993

[5] Raso, F. A. et al. (2018). *Artificial Intelligence & Human Rights: Opportunities & Risks.* Berkman Klein Center for Internet & Society at Harvard University. Link: https://papers.ssrn.com/sol3/papers.cfm?abstract_id=3273016

[6] Nemitz, P. (2018). "Constitutional democracy and technology in the age of artificial intelligence." *Philosophical Transactions of the Royal Society A*, 376(2133), 1–14. https://royalsocietypublishing.org/doi/10.1098/rsta.2018.0089

[7] Buolamwini, J., & Gebru, T. (2018). "Gender shades: Intersectional accuracy disparities in commercial gender classification." In *Conference on Fairness, Accountability and Transparency*, New York, USA, pp. 77–91. Link: https://proceedings.mlr.press/v81/buolamwini18a.html

[8] Whittlestone, J. et al. (2019). *Principles for Responsible AI.* European Commission. https://ec.europa.eu/futurium/en/ai-alliance-consultation/guidelines

[9] Dignum, V. (2019). *Responsible Artificial Intelligence: How to Develop and Use AI in a Responsible Way.* Springer Nature. Link: https://link.springer.com/book/10.1007/978-3-030-30371-6

[10] Bommasani, R. et al. (2021). "On the opportunities and risks of foundation models." *arXiv preprint.* Link: https://arxiv.org/abs/2108.07258

[11] Esteva, A., Kuprel, B., Novoa, R. A., Ko, J., Swetter, S. M., Blau, H. M., & Thrun, S. (2017). "Dermatologist-level classification of skin cancer with deep neural networks." *Nature*, 542(7639), 115–118.

[12] Shneiderman, B. (2020). "Bridging the gap between ethics and practice." *ACM Transactions on Interactive Intelligent Systems*, 10(4), 26. https://dl.acm.org/doi/10.1145/3419764

[13] Greene, D., Hoffmann, A. L. & Stark, L. (2019). "Better, nicer, clearer, fairer: A critical assessment of the movement for ethical artificial intelligence and machine learning." In *Proceedings of 52nd Hawaii International Conference on System Sciences*, Grand Wailea, Maui, Hawaii, USA, pp. 2122–2131 (HICSS).

[14] Jobin, A., Ienca, M. & Vayena, E. (2019). "The global landscape of AI ethics guidelines." *Nature Machine Intelligence*, 1, 389–399.
[15] Whittaker, M. et al. (2018). *AI Now Report 2018.* AI Now Institute.
[16] High Level Expert Group on Artificial Intelligence. (2019). *Ethics Guidelines for Trustworthy AI.* European Commission.
[17] European Commission. (2020). *The Assessment List for Trustworthy Artificial Intelligence (ALTAI) for Self-Assessment, Independent High-Level Expert Group on Artificial Intelligence*, Brussels. https://ec.europa.eu/digital-single-market/en/news/assessment-list-trustworthy-artificial-intelligence-altai-self-assessment
[18] Goodman, B., & Flaxman, S. (2017). European Union regulations on algorithmic decision-making and a "right to explanation." *AI Magazine,* 38, 50–57.
[19] Bai, Y., & Ghanem, B, (2017). Multi-scale fully convolutional network for face detection in the wild. In *Computer Vision and Pattern Recognition Workshops (CVPRW), 2017 IEEE Conference on*, Honolulu, Hawai, USA, pp. 2078–2087. IEEE.
[20] Whittlestone, J., Nyrup, R., Alexandrova, A., Dihal, K. & Cave, S. (2019). *Ethical and Societal Implications of Algorithms, Data, and Artificial Intelligence: A Roadmap for Research*, p. 59. Nuffield Foundation.
[21] Floridi, L. et al. (2018) "AI4People—an ethical framework for a good AI society: Opportunities, risks, principles, and recommendationns." *Minds & Machines*, 28, 689–707.
[22] Farinella, G., & Dugelay, J.-L. (2012). Demographic classification: do gender and eth- nicity affect each other? In *Informatics, Electronics & Vision (ICIEV), 2012 International Conference on*, Dhaka, Bangladesh, pp. 383–390. IEEE.
[23] Benitez-Quiroz, C. F., Srinivasan, R., & Martinez, A. M. (2016). Emotionet: An accurate, real-time algorithm for the automatic annotation of a million facial expressions in the wild. In *Proceedings of the IEEE Conference on Computer Vision and Pattern Recognition,* Las Vegas, Nevada, USA, pp. 5562–5570.
[24] Doshi-Velez, F., & Kim, B. (2017). Towards a rigorous science of interpretable machine learning. *arxiv preprint arxiv:1702.08608.*
[25] Filipović, A., Koska, C. & Paganini, C. (2018). *Developing a Professional Ethics for Algorithmists: Learning from the Examples of Established Ethics.* Bertelsmann Stiftung.
[26] Radzi, S. A., Mohamad, K.-H., Liew, S. S., & Bakhteri, R. (2014). "Convolutional neural network for face recognition with pose and illumination variation." *International Journal of Engineering and Technology (IJET)*, 6(1), 44–57.
[27] Angwin, J., Larson, J. Mattu, S., & Kirchner, L. (2016). *Machine Bias: Theres Soft-Ware Used across the Country to Predict Future Criminals and Its Biased against Blacks.* ProPublica.

5 Text-Based Generative AI Techniques

Neha Yadav, Dhanalekshmi Gopinathan, and Alka Singhal

5.1 INTRODUCTION

Generative artificial intelligence (AI) pertains to a collection of AI systems intended to produce text, images, music, and code, along with others types of content. Advances within the study of NLP have considerably improved text-based generative AI so as to realize how these systems understand, create, and manipulate text in extremely similar ways to human communication. Potential applications for such models range widely across industries, impacting everything from creative content generation to environmental research and beyond.

5.1.1 Overview of Generative AI

Large generative AI models, such as GPT-3 and GPT-4, have shown realistic human-like writing. GPT-3 has created news articles, as well as short stories and drafted poems that will be engaging and comparable in quality to those written by humans. These models capitalize on the enormous size of datasets and more advanced neural network architectures to learn context and fill in relevant content. Generative AI enhances the study of the arts, education, and the environment by analysing the structure and content of written materials.

These technological breakthroughs promoted the development of generative AI in several ways, including advanced computational power, access to big datasets, and improvements to machine learning algorithms. It's when transformer-based models came that completely flipped the game. AI can now abide by even deeper contexts and, at best, create coherent and contextually appropriate text. These transformer models use a mechanism called self-attention, which rates other words within a sentence for their importance so that even complex language relationships and sentiments can be grasped rather powerfully [1]. The end-to-end workflow of text-based generative AI system is illustrated in Figure 5.1.

Finally, AI can produce texts that are not only correct but also creatively stimulating. Text-based generative AI works by guessing the next word in sequence from a given context set by previous words. This simple medium, still, allows these models to induce any kind of textbook, from everyday exchanges to specialized documents and indeed books. Their capability of producing mortal likeness makes them a great implicit operation in chatbots, happy generation, and educational tools.

 DOI: 10.1201/9781003565703-5

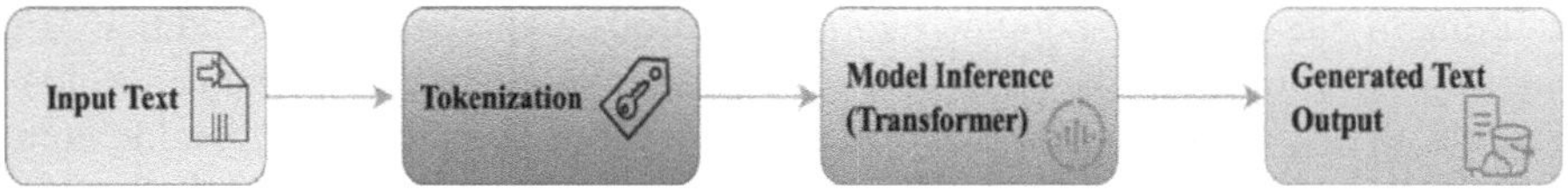

FIGURE 5.1 Workflow of Text-based Generative AI system.

5.1.2 Preface to Text-Grounded Generative AI

Text-grounded generative AI is working on textbook characteristics such as coherence, contextually applicability, and creativity. These models have been trained on millions of textbooks, which tutored them the alphabet, syntax, and indeed nuances like style and tone. From brief social media posts like tweets to more in-depth specialized reports, the workshop generated by generative AI is fluently produced and protean in content creation.

The use of Generative AI ranges over numerous diligence, with its scope being really wide. Creative trades find Generative AI to inspire ideas in pens, muses, and musicians, as well as to create first drafts. Software development also benefits by saving time and trouble for inventors since AI generates law particles and, at times, whole programs. In education, AI is customizing learning with generated content to meet individual scholars' requirements.

While there is no other system in existence that can boast this profound ability of interpreting and mimicking multiple writing styles, the text-based generative AI works to perfection. Its output could resemble the tone a famous author would have used, or technical documentation that is industrially crafted to standards. It is supremely useful in creative domains where this aspect—creation of stylistically cohesive content—is of utmost importance. Moreover, these models can fit into any type of writing, be it formal academic writing or casual blogging, making them quite versatile [2].

5.1.3 Scopes and Objectives

It will be truly prospective, as the chapter will provide a holistic view of prospective applications of text-based Generative AI with the discussion of opportunities and ethical challenges. The objective is to understand how these models can be creatively, practically, and responsibly harnessed, given that potential benefits outweigh the associated risks. Considering several possible use cases, this chapter will show the quality and depth of the transforming power of Generative AI, while also discussing ethical considerations in the deployment of this technology.

A second objective of this chapter is to create a balanced view of the opportunities and limitations that exist with Generative AI. While this new technology has disruptive potential in many industries, it also creates difficulties in balancing the ethical implications of content generated by AI. It hopes to give readers an understanding not only of the technological capability but also of the broader impact on society from Generative AI.

Generative AI is also limited to some degree. While these models are powerful, they also have most of the biases that may arise from data used in training. Basic studies have shown that language models like GPT-3 portray evidence of gender and racial bias, which may reflect several stereotypes within the training data, hence resulting in biased outputs that perpetuate harmful stereotypes if the conditions are not properly mitigated. The point here is that since the models learn from enormous datasets, the AI might incidentally learn and reproduce everything that could potentially be biased in those sets. This becomes particularly important when Generative AI enters applications related to public opinion or information dissemination, say, journalism or social media. Such action against these biases therefore makes the responsible use of Generative AI so vital.

5.1.4 Key Technological Advancements in Generative AI

A number of breakthroughs have enabled the technological development of text-based Generative AI. First and foremost of those innovations is the use of transformer architectures, boosting the contextual understanding capability of AI by leaps and bounds to produce coherent text. GPT-3 and GPT-4 rely on mechanisms of self-attention within transformers, weighing different words of a sentence against each other to allow it to pick up complex relationships among words.

While the dataset to train may be large scale, another important accomplishment is the access to that data on such a large scale. In new generative AI models, billions of words are used for training, which really enables them to master various kinds of linguistic patterns and styles. Such huge data allows the model to generate text that is contextually relevant with variability in content. However, the most important thing is the quality of the data utilized for training, since it directly impacts the quality and reliability of the produced text [3].

Computational power improvements have also played a greatly critical role in the rise of Generative AI. Large-scale training requires immense computational resources; hence, powerful GPUs and TPUs that are increasingly made available have so far made feasible the training of models comprising hundreds of billions of parameters. Growth in computing power has thus provided a boost to the creation of more advanced models that is capable of generating good-quality text at scale [4].

Aside from transformer architectures and large-scale datasets, improved unsupervised learning methods have really helped the development of Generative AI. Unsupervised learning concerns learning from data that has no labels, which fits well with text generation, where labelled datasets are not always easily available. Through training with extensive unstructured text data, Generative AI models learn deeply about the language and can thus be then applied on a wide variety of tasks.

5.1.5 Ethical Considerations and Challenges

However, with any powerful technology, the utilization of Generative AI also carries ethical considerations that must be engaged if the technology is to be deployed responsibly. The first and most direct of these ethical enterprises is the eventuality of abuse in producing deceiving or noxious content. These models of generative AI can make fake news, deepfakes, or other forms of misinformation, which would have disastrous consequences in the realms of public trust and societal stability. The development of safeguards and structure of nonsupervisory fabrics is needed to alleviate these pitfalls.

Another ethical concern is that this generative AI might immortalize bias. As these models learn from the data they're trained with, any impulses in the training datasets may indeed appear in the produced content. This might make the prejudiced affair promote noxious conceptions or count whole groups. Careful curation of training data and the development of ways for discovery and mitigation of these impulses in AI-generated content would be helpful in handling this issue [5].

Another significant concern with regard to Generative AI is that of sequestration. Large quantities of training data, containing particular details about individualities, could be from public sources. Icing the AI model doesn't blunder precious information in ways that may beget detriment, which is a veritably important point of ethical operation of the technology ways for discrimination sequestration, and anonymization can at least ensure that no individual data points could be traced back to specific individualities.

Despite these challenges, Generative AI can serve good purposes if done responsibly. It democratizes access to information, enhances creativity, and amplifies effectiveness within numerous diligence. Mollifying the pitfalls of Generative AI, while garnering its benefits, requires the creation of strong ethical guidelines, clear practices, and ongoing cooperation among technologists, policymakers, and the community [6].

5.1.6 The Future of Text-Based Generative AI

The future of textbook-grounded Generative AI is bright, as exploration is being done to make these models indeed more effective, environment-sensitive, and immorally responsible. This involves experimenters working on styles that can make AI models more interpretable by showing the stoner how the AI arrives at a particular affair of the model. This is critical in trusting AI systems and ensuring that they are employed in such a manner as to concur with societal values.

The other focus is the development of AI models that have reduced computational power and training data requirements. Most current state-of-the-art models in AI research are resource-intensive, which consequently creates limitations on their use regarding accessibility and scalability. Researchers are developing techniques like model distillation and efficient fine-tuning to create smaller representations of the models that are more efficient and can be widely deployed.

It will also be expected that, in general, further developments in human-AI collaboration are the future of Generative AI. In this line, the study suggests that AI, as a collaborator rather than a replacement for human creators, gamers, and artists, can significantly augment their creativity and productivity. Accordingly, the most important solutions that Generative AI can make are several suggestions, produce first drafts, and perform routine tasks, giving way for humans to spend time on higher-order thinking in creativity and strategy.

Other industries will also become more and more integrated with Generative AI. In education, AI-driven personalized learning experiences will extend to deliver a variety of sets of tailored content coupled with real-time feedback. With regard to environmental sciences, generative AI will be irreplaceable in building insights and scenarios carved out of the pressing global challenges: climate change and resource management.

That is to say, the arrival of Generative AI is a quantum leap in the capabilities of AI. From empowering industries to enhancing creativity, improving efficiency, and personalized experiences, the many uses of generative AI overwhelm, so do their challenges—legion in the form of ethical concerns around bias, misuse, and privacy. For this, addressing such challenges by responsible development and deployment is important to allow realization of benefits that come with Generative AI while reducing potential risks. Wide-ranging implications for industries and applications are meant by the immediate ability to create human-like text. But great power also means great responsibility, and in the case of Generative AI, it involves many serious ethical questions such that its impact would be conducive to positive and beneficial effects upon the entire society [6]. This chapter will offer a summary on the state of affairs currently with regard to text-based Generative AI, its applications, and what ethical factors must be considered to direct development and use.

5.2 CHATBOTS

Some of the very basic applications of Generative AI are chatbots. These AI-powered conversation agents have become quite sophisticated, enabling them to conduct meaningful dialogues and assist users in everything from customer service to creative writing, passing through software development. Improvement in recent language models allowed for chatbots to react not only to queries but also to create responses that are logical and make contextual sense, thus further improving user experience.

5.2.1 Generative AI in Creative Process

Generative AI becomes an important mainstay in the creative process—a function hitherto reserved for humans alone. Generative AI can help in developing new ideas, perfecting drafts, and overcoming mental blocks, all of which are part of the creative process. In this case, automating a part of the creativity with AI will help the artists, writers, and other kinds of content developers to focus more on higher-order aspects like conceptualization and thematic exploration [7].

For instance, chatbots, especially those with highly developed language models like GPT-4, could even brainstorm ideas and propose creative directions. For instance, it would suggest plot twists, character developments, or one-liners to a writer who perhaps is struggling with the storyline. These suggestions are informed by huge volumes of data the chatbot has trained on, including various genres, writing styles, and narrative structures.

In addition, AI-driven chatbots can adapt creative preferences expressed by the user. As the collaboration goes on, the chatbot learns from the style and preferences of the user in order to create ever more tailored and relevant creative input. Being able to make suggestions at such a fine level of detail ensures that the creative output will meet the vision of the user, thereby making the AI literally a collaborator [8].

5.2.2 Chatbots for Creative Writing Assistance

Of all the close-to-promising applications, chatbots are employed in creative domains to help writers create content effectively. Creative writing chatbots are designed in

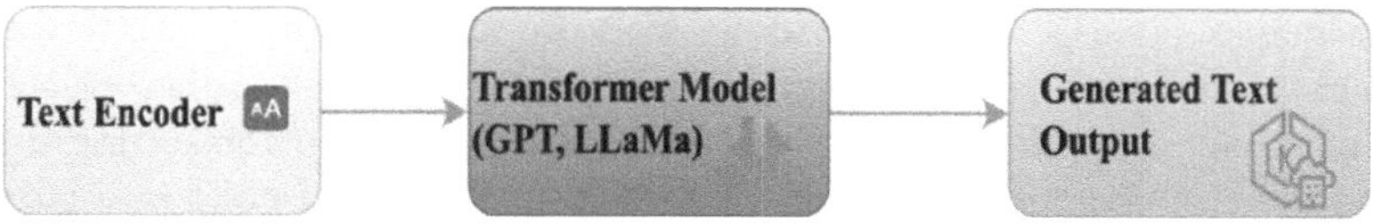

FIGURE 5.2 The internal Architecture of a typical Text-based Generative AI model.

a way to help users break creative gaps, expand their ideas, and fine-tune their draft material. Such AI-powered assistants can create content for quite a few genres, including but not limited to fiction, poetry, scripts, and technical writing.

It includes everything: being out of inspiration, having trouble with the building of the story, or even the development of natural-sounding dialogues. AI-powered chatbots could support them all: with proposals for inspiration, text generation based on given inputs, and alternative phrasing or directions in narrative. For example, when a writer finds it hard to develop the dialogue between characters, the chatbot can easily generate sample dialogues that the writer can refer to or adjust according to their story.

Besides this, creative writing chatbots could help a writer maintain the tone and style, which includes character development. The chatbot analyses the text already present and fills in the gaps, ensuring that new content matches the voice and mood developed in the story thus far. This feature is useful in long-form content, such as novels or scripts, where maintaining consistency over hundreds of pages may be hard. For these text-based generative responses, the model used Transformer models. The process of text generation is shown in Figure 5.2, including the internal architecture and critical stages that produce human-like text outputs.

Another application of chatbots within creative writing is collaborative projects, where this AI could contribute as a much-needed writer to provide suggestions and further develop what has been written by human writers. This collaborative approach offers the advantage of speed in working and brings new ideas that can lead to improvements in the final result. An example of this in screenwriting could be how a chatbot provides variations in dialogue that bring out more character depth or even comedic timing [9]

Besides, another great advantage of the usage of chatbots in creative writing is their ability to generate content in different styles: from those classic, almost archaic ways of writing by acknowledged literary giants to more conversational modern writers, the chatbot adjusts its output accordingly. Such features are what make this AI-driven chatbot useful equipment for writers creating different genres and formats [10].

5.2.3 Chatbots for Code Generation

Besides creative writing, it has also found a fitting role in the sphere of software development by helping in generating code. AI-powered chatbots are changing the way developers write code. They do this by real-time suggestions, generating code snippets, and even whole functions or classes needed to satisfy the nature of the work involved described in natural language. These chatbots are coding buddies in most cases, augmenting productivity and lessening the cognitive load on a developer.

In principle, code generation chatbots work by understanding exactly what the developer wants through natural language input and generating the appropriate code. As an example, a developer may want to develop a function for sorting a list of numbers using a chatbot. The chatbot generates the required code in a particular programming language. This is very helpful for performing repetitive coding tasks, whereby the developer saves his time by allowing the basic implementation to be handled by the chatbot.

Apart from code generation, chatbots can also be of great help in debugging and troubleshooting. All the developer has to do is describe the problem they are experiencing, and the chatbot would scan through the code, highlighting where the problem could be or giving a solution to such problems. This feature is helpful for inexperienced developers who may find it difficult to handle much more complex debugging tasks. Through this ability to give suggestions and immediate feedback, chatbots allow developers to learn and improve their coding skills.

Code generation chatbots find their importance even in learning environments. Students learning to code can use such chatbots as tutors, enquiring about syntax, algorithms, and other concepts of programming. A chatbot may explain, give examples, and sometimes even generate code to make a point. This method of learning also enables students to understand intricate concepts and utilize them in real situations [11].

Moreover, there are chatbots with capabilities of writing codes. These are characterized by producing many lines of repetitive codes, which become extremely time-consuming for developers without any real benefits. However, because these chatbots allow software developers to automate the task of writing lots of similar boilerplate code, the programmer can now occupy himself with more fun and fanciful aspects of software making. Thus, minimizing the chances of making mistakes through the extra repetitive code isn't very relevant.

Another good aspect of chatbots like these is that they provide appropriate code fragments based on those existing with consideration of the particular project context. That means such a chatbot can analyse various parts of the already-written code and generate inspiration for what can fit into the project. For instance, in building a web application, the chatbot would be suggesting code snippets pertaining to user input and database interactions—the two major features of any web application. This approach of suggesting relevant information to developers enables them to write better codes that are less complex and therefore easier to maintain [12].

Another facet in which AI-driven chatbots find application today in software development workflow pertains to facilitation of collaboration among development teams. Chatbots can function as knowledge bases, being able to supply knowledge on best practices, standards for coding, and perhaps project-specific guidelines. Therefore, making this very information easily accessible will help ensure that all group members are on a similar page and that the code produced is consistent with the set standards.

In the end, Generative AI-driven chatbots are sure-fire tools for creative writing and coding. Chatbots, supporting writers through the hard nuts, are real-time coding assistant for developers. These chatbots' capabilities to understand contextual information, give personalized suggestion, and adapt to the needs of varied users make them strong collaborators within both creative and technical boundaries. As Generative AI continues to improve, so will chatbots, meaning there are wider realms of creativity, productivity, and learning.

5.3 SOCIAL MEDIA

It has already overtaken the social media world and changed the manner of creation, consumption, and distribution of content. The potential to generate engaging posts with multilinguality, for instance, is invaluable in enhancing the quality of social media interaction. This section will consider some of the most important uses of the technology to make a consideration of the impact of generative AI on social media: multilinguality text generation, blog post creation, and social media post synthesis.

5.3.1 Impact of Social Media

Social media has become a significant element of daily life as it allows people to communicate with one another and the community at large. With Generative AI on the rise, it became possible to auto-generate content as long as it does not rely on the efforts of numerous individuals, such as brands, influencers, and others who created such content in order to maintain an interactive presence online. It is possible to create such AI-generated content in a customized manner for particular audiences in order to more actively encourage user engagement and build loyalty among the community towards the brand.

The Present role of Generative AI in Personalized Content Experiences: By mining user data, we need AI to work out what kind of content users best respond to and produce posts in a manner that is more suited towards getting the attention of this demographic. For example, a brand can use AI to scan through user engagement metrics and generate product recommendations personalized and packaged as social media posts. This depth of personalization can improve brand-to-audience relationships and in turn drive greater engagement/loyalty.

In addition, Generative AI on social media supports in maintaining regular schedules for content creations. Being a social media manager can be extremely overwhelming to constantly produce new content and keep your audience engaged. The social media platform will be alive and active; the AI-based automation can create posts with the least effort possible. Hence, it's proficient because this is mostly required by those who are not quite popular in the web—most of whom are brand and influencers wishing to make an online presence.

5.3.2 Multilingual Text Generation

One of the biggest breakthroughs in AI that social media has witnessed till date is multilingual text generation. Language barriers have been something that brands and influencers have sought to get rid of since long in order to effectively carry out their messages to different masses across global platforms. Today, content creators can reach anyone with an internet connection, thanks to AI models like GPT-4 that have the ability to produce text in many languages. L'Oréal, the cosmetics brand, shows how this works. It used AI to create content in multiple languages for its social media campaigns. This helped the company talk to customers who spoke different languages, expanding its worldwide presence even more. Any brand can use AI to make posts in Spanish, French, Chinese, and Hindi—going beyond English-speaking markets. This approach helps brands reach more people, including non-English speakers, by giving them access to the same content.

This technology allows influencers to connect with people around the world, talk across cultures, and build global communities. Also, for social networks to work well, they need to generate text in many languages in real time. For instance, it can translate what users say and respond in the language they prefer, making it easier to talk. This makes a big difference in customer service where quick and correct responses matter most to keep good relationships. Better support in many languages lets brands to give better customer service, making users who speak different languages feel listened to and important.

5.3.3 Blog Post Generation

Blogging plays a key role in the social media plans of brands and influencers. It gives them a chance to explore topics in depth, build trust, and send people to other parts of their online world. AI writing tools are making blog post creation automatic. This means it'll be simpler to write product reviews how-to guides expert opinions, or industry updates. Take an online shop, for instance. It could use an AI tool to create blog posts about why it's great. These posts might have full details, customer praise, and advice on using products. The AI looks at related info and finds insights to shape the content. This fits well with SEO plans and what readers like. It saves time but keeps things interesting and true to the brand's message. It generates even prompts and initial drafts that help in alleviating writer's block. For example, a travel blogger might use AI to outline a post about a recent trip, including sections on destinations, experiences, and travel recommendations. He can then add his personal anecdotes and personalize it, making it both informative and authentic [13].

AI-generated blogging is one such key content production consistency practice. With consistent blogging comes the building up of a loyal audience as well as boosting search engine ranking, but on the other hand producing quality posts regularly is a rather challenging task. Generative AI allows creators to produce blog posts on demand that keeps blogs active and audiences engaged. This helps businesses that need content marketing to boost their sales and establish thought leadership in their chosen industries.

5.3.4 Social Media Post Synthesis

Social media content is now about a dynamic, effective visual appeal as well as extremely engaging interaction with the viewer. Because of generative AI, it shall create synthesized social media content related to the preference and behaviour of the specific audience involved. For this purpose, posts not just based on text may involve other forms such as image or video infographics that would fully maximize engagement.

AI-powered social media content generators search for data from available insights on audience to define what kind of post will most resonate with a users' interest. Such was the case when Coca-Cola utilized available user interactions to provide personalized social media posts for their 'Share a Coke' campaign using AI, personalizing content for higher user engagement. For example, an AI can look at how users interacted with previous posts to recognize trends about what they engage with, whether it be humour, educational content, or showcasing products. With this, the AI drafts posts that tilt towards the preferences to increase the likelihood of user engagement.

These, for instance, may involve a fitness influencer wanting to post a series of posts promoting a new workout routine. The AI will then suggest captivating captions, even recommend the most appropriate hashtags, and may go further to create the accompanying images or infographics that better describe the workout steps in detail. It's the all-rounded approach for generating content that will make the posts look nice, be informative, and optimized for social media.

Synthesis of social media posts is also useful for the management of campaigns that require high volumes of content. For example, a brand that is introducing its product may wish to create multiple variations of posts communicating different features of the product, customer testimonials, and promotional offers. AI of a generative nature can quickly come up with those posts as the campaign keeps flowing, and it communicates all aspects of the product effectively to the audience [14].

In short, Generative AI has washed over the creation of social media content, rendering it more fleet and person-centred, interactive. From multilingual text generation over automation of blog posts to synthesizing social media posts, AI is helping content creators surmount challenges and make the most out of social media platforms. As generative AI continues to evolve, its role within social media will continue to expand, further transforming how content is created and consumed in the digital age.

5.4 EDUCATION

Illustrative of innovative tools and tailored learning experiences, generative AI is changing the educational process for students, teachers, and learning establishments. Its application in education not only facilitates easy access to learning material, but it also injects quality into the process of learning as the content created becomes individual-centred down to each learner's needs. This section elaborates on the role of Generative AI in education, reviewing personalized learning material generation, quiz generation, and its wider ramifications on the educational system.

5.4.1 Generative AI in Education

It hold so much potential for education because this technology will personalize learning, automate content creation, and free the educator to spend more time engaging with his students and mentoring them. AI-based pedagogic tools are trying to assist educators in transforming the delivery of lessons to be more engaging and adaptive to each student's diverse learning styles and needs. For example, using AI

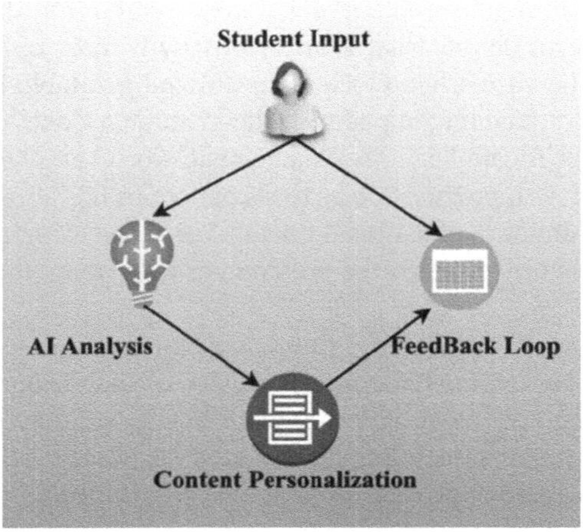

FIGURE 5.3 AI-driven tutoring system.

models such as GPT-4, one would evaluate data generated by students and identify learning gaps by producing niche education material based on those findings that would address such gaps.

The latest generative AI technologies can assist teachers in the creation of diverse learning resources, ranging from lesson plans to worksheets and quizzes to virtual lectures. All of these free the time of the educator to spend on even more enriching interactions with their students: discussion, problem-solving, and exercising critical thinking. Another vital feature of AI is that of the creation of interactive modules that rely on simulations, case studies, and practice exercises for consolidating knowledge on difficult subjects [15].

A fine example of Generative AI being applied in education is with respect to AI-powered tutoring systems, as depicted in Figure 5.3. These can virtualize themselves as tutors for the students, offering feedback and explanations immediately, thereby making learning more accessible and efficient for them. For instance, Squirrel AI has used its AI-driven tutoring system to provide personalized content to students that helps improve learning outcomes and engagement. Such AI-based tutoring systems can generate responses to student queries in a personalized manner, using language, tone, and complexity of content appropriate for the proficiency level of the student. Thereby, it could create a much more interactive and supportive environment than is otherwise possible, especially for those students whose needs cannot be addressed in the regular classroom.

5.4.2 Personalized Generation of Learning Material

Perhaps one of the most revolutionary contributions Generative AI could provide for education is tailoring the learning materials to specific needs. Personalization lies at the heart of effective learning, as students differ in strengths and weaknesses, as well as their preference for modality in learning. Generative AI models analyse the progress and learning behaviour of students so as to forecast personalized study

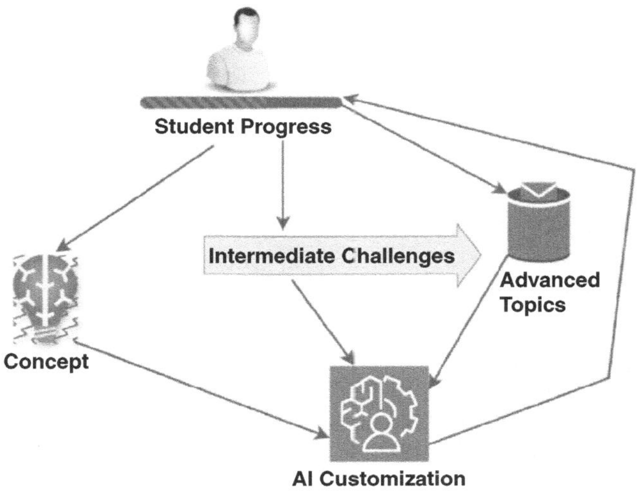

FIGURE 5.4 Adaptive learning pathway.

guides, exercises, and reading material that fully meet their very special needs. For example, a student who has some difficulties with mathematics may be given individualized practice problems prepared by some AI system and explained step by step to address the points where this particular student struggles the most. On the other side, in exactly the same subject, another student may be an ace; here, the AI can generate advanced content challenging that particular learner and moving him or her up the learning curve. And in this way, each learner will be kept busy and challenged because the notion itself will never be too easy or too hard for them [16].

Generative AI can also be used to develop adaptive learning pathways where the contents evolve based on the progress recorded by the student. For example, if a student demonstrates mastery of a particular topic, the AI may then generate harder questions or move to similar topics, hence making the learning experience relevant to the current understanding level of the student. This would make Generative AI an ideal tool for personalized education, as it will be able to adapt and respond to the real-world needs of each given learner, as shown in Figure 5.4.

Besides personalized exercises and reading, AI can also create videos and infographics, along with all other forms of interactive simulations for learning. Such a set of learning resources caters to the varied learning styles: visual, auditory, and kinesthetics—resulting in a more inclusive and effective education. For instance, by means of animations created by AI, a lesson in history can be supplemented to give students a better idea of the contents and hence their relevance.

5.4.3 Quiz Generation

Quizzes do indeed hold a very special place in a student's learning curve in that they help reinforce knowledge, test comprehension, and point out areas where there may be flaws. Automatically available quizzes in generative AI would release educators from the burden of assessing progress made by students without wasting much time in designing quizzes. Quizzes that are auto-generated by AI can be designed to exactly match the individual student's difficulty level. This makes the quizzes both effective and equitable [17].

AI-driven quiz generation involves question creation aligned with learning objectives for a certain lesson or course. For example, after a lesson on the water cycle, an AI model will generate multiple-choice questions, true/false, and short-answer questions to test students on major concepts in the lesson, such as evaporation, condensation, and precipitation. The AI can also generate question formats best suited for different assessment needs, including practice quizzes for revision, formative assessments for ongoing learning, and summative assessments for final evaluation. The major advantage of AI quizzes is that they help in providing immediate feedback to the students. Upon taking a quiz made by AI, the AI can look over the answers entered by the students and instantly point out mistakes, along with an explanation for the student. The immediate feedback enables students to know their mistakes and learn them, yielding a better understanding of the material. For example, if a student chooses the wrong answer on a physics quiz, AI can explain why the answer is wrong and provide a step-by-step solution for the problem. Figure 5.5 shows the example of Quiz Feedback AI.

Consider a student of basic algebra who wants to take an AI-generated quiz on the very same material. Suppose the student gets one of the questions incorrect about how to solve a linear equation. The AI can then provide a fully articulated explanation of how to solve that equation, replete with graphing visuals. This sort of immediate, specific feedback to students will go a long way in helping them learn the material. AI-generated quizzes can also be used when analysing students' progress over a period. By analysing the results of quizzes, educators are enabled to underline patterns in the performance of students, acquiring valuable insights into their strengths and weaknesses. This information can be used to modify teaching strategies and provide targeted support to students in need. For instance, given a set of students in class whose inability in a particular topic keeps increasing, the educator makes use of the data and sees to it that the same topic is revisited in class using more practice exercises interposed by AI.

Generative AI might also provide adaptive quizzes that would change difficulty as responses from the student are given. A correct answer may immediately be followed by a harder question, while an incorrect answer leads to an easier one or includes more context. In this way, an optimal level of challenge for each student is maintained during the assessment process, making it supportive and effective.

5.4.4 Broader Implications for Education

Generative AI has the potential to reimagine education at a fundamental level, from delivering learning to the learning experience itself. From personalized learning content to grade automation, or even better levels of student engagement based on interactive content, AI is making education more accessible, efficient, and effective. Notwithstanding the accompanying challenges and ethical considerations, AI use in education should be duly considered. Challenges are omnipresent in many aspects,

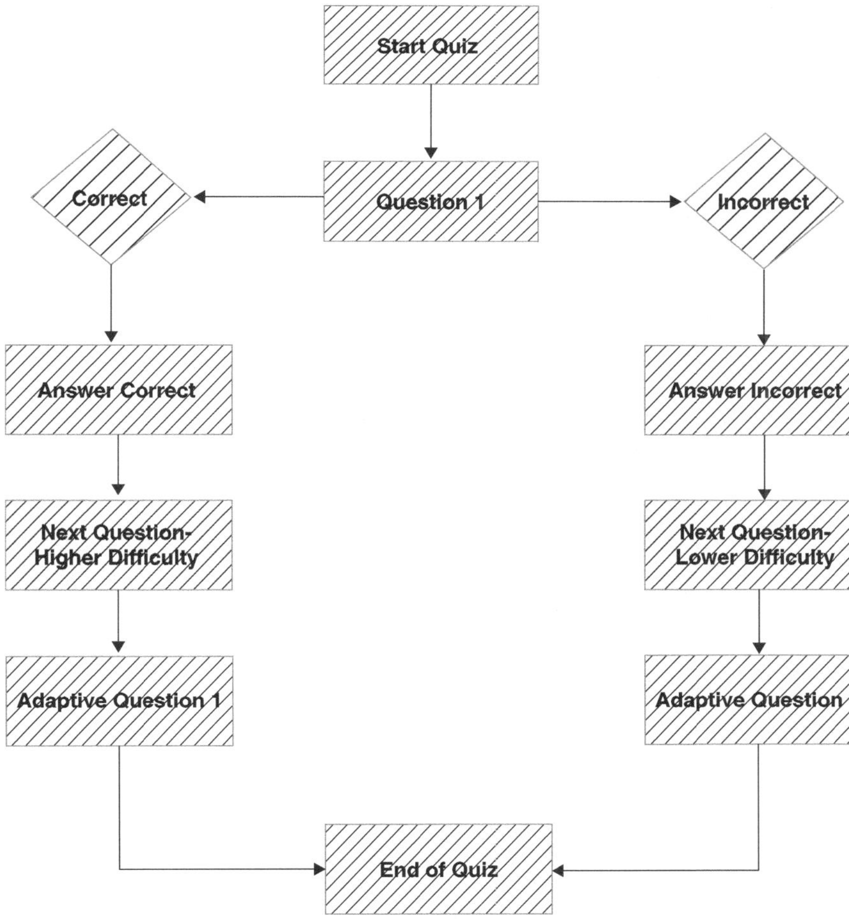

FIGURE 5.5 Adaptive quiz flow.

with the most needed being the protection of privacy and security of data. The generative AI models of personalized learning would want more and more student data, which inherently raises serious questions about their collection, storage, and usage. This easily makes it very vital that strong mechanisms for data protection be instituted to protect information on students for the purpose of knowing and trusting AI-driven educational tools. Another consideration is educator roles in this AI-enhanced learning environment. While Generative AI can automate a lot of teaching features, such as content creation and assessment, it lacks the human elements of empathy, mentorship, and inspiration important in good education. Educators will have to adapt to their dynamically shifting roles towards guiding, supporting, and motivating students while leveraging AI as a tool to support them.

In a nutshell, Generative AI introduces changes in the field of education by bringing personalized learning experiences, automated content creation, and improving assessments. This creation of personalized learning materials and quizzes enables educators to bridge the gaps in each student's learning for their ultimate success while releasing some burden off educators' shoulders for more interactive relationships with their students. While AI is still developing, its function in education is very likely to go up and serve the very purpose of improving learning quality and access for students from every part of the world [18].

5.5 ENVIRONMENTAL IMPACT ASSESSMENT

Projects on generative AI are getting increasingly significant game-changing tools in environmental science, particularly towards assessments of environmental impact and the planning of sustainable futures. By leveraging the full capabilities of large AI models, researchers and policymakers can act to simulate a wide range of climate scenarios, undergo impact assessments, and plan renewable energy systems accordingly. This chapter will be looking at two major applications of Generative AI in assessment for environmental impact, namely Climate Change Scenario Generation and Renewable Energy Planning. Schematics and diagrams are included to give an overview of how these AI-driven processes work.

5.5.1 CLIMATE CHANGE SCENARIO GENERATION

Probably, the most current burning issue that humanity is facing pertains to climate change, and for this, understanding of its possible impacts is very important in developing relevant mitigation and adaptation strategies. Generative AI has been increasingly important in this effort to generate climate change scenarios based on large datasets and complex climate models. These AI-driven scenarios help scientists, policymakers, and the general public understand the possible future states of the planet under different conditions. Scenario generation in climate change mainly deals with the creation, through Generative AI models, of simulations that define how different environmental factors such as temperatures, precipitation, sea level, and greenhouse gas emissions will be changing over time. The AI models make projections for possible various outcomes under different scenarios: high emissions, low emissions, and moderate emissions, using machine learning algorithms and analysis of historical data. Figure 5.5 shows the example of AI-Driven Climate Models.

A generative AI model, for instance, could be trained using temperature and CO_2 concentrations from past records to project into the future climate. AI-driven climate modelling enables the visualization of potential impacts such as increasingly frequent heatwaves, altering precipitation, or sea-level rise, which is essential for governments and organizations in planning for climate adaptation and resilience. The European Union also applies AI-generated climate models to consider the effect of rising sea levels on coastal cities so that they can develop strategies and interventions to prevent flooding and make infrastructures resilient. One of the important values in using Generative AI for generating climate scenarios is the ability to rapidly generate many

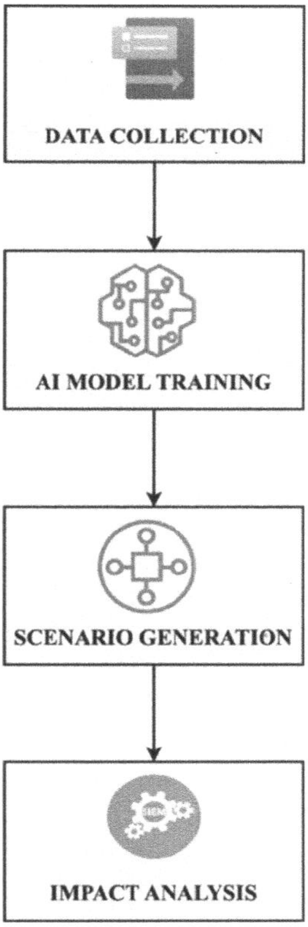

FIGURE 5.6 Climate change scenario generation process.

scenarios with different assumptions regarding future human behaviour and technological advancements. An example would be an AI model crafting a scenario for aggressive carbon reduction policies vs. those with the emissions continuing to climb unabated. These in turn give valuable insights on how different policy approaches might help and inform decision-makers on what to prioritize based on these insights (Figure 5.6).

Generative AI also bounds exploration into extremely improbable cases about climate. For example, an AI model can have scenarios that are not familiar in a heatwave sense, never experienced storms of unfamiliar intensity, and ice melt at frightening rates. This would avail a much better view of the possible risk for those eventualities. This information is crucial for preparedness in case of emergencies, infrastructure planning, and strategies for public health. Example: AI-Generated Extreme Weather Scenarios. For example, recently, scientists at MIT ran a Generative AI model to model occurrences of extreme weather—like hurricanes and droughts—for different climate change scenarios. These helped the city planners in Miami come up with better infrastructure plans to deal with future hurricanes and thus reduce the risk to human lives and property.

Another important application of AI-produced climate scenarios is in raising awareness and educating the populace on the probable effects of climate change. AI-generated visual and interactive scenarios are some of the effective avenues where educational programs, media campaigns, and community outreach will drive home the urgency to address the issue of climate change and the potential benefits of taking action now [19].

5.5.2 Planning of Renewable Energy

Transitioning to renewable energies is a crucial part of the world's strategy in taming climate change. Generative AI will help accelerate the transition by providing insights into the planning of renewable energy, optimizing resource allocations, and developing strategies towards efficient energy production and consumption. AI contributes to a clearer understanding of the eventuality of renewable powers and how they could be optimized in the energy grid through the unequalled capability of data processing and pattern identification it offers [20]. This planning involves assessing the vacuity and eventuality of different renewable coffers, like solar, wind, hydro, and geothermal, in relation to their overall application eventuality, considering the most doable manner of exploiting each of these coffers. AI-grounded models can dissect a host of complex data points girding terrain, meteorology, and demographics to indicate possible installations related to renewable energy. For example, AI could study the pattern of solar radiation to pinpoint the exact point for solar granges, or it can be used to dissect the pattern of wind speed data and help elect the stylish spots for setting up turbines. Example: AI in the Planning of Solar and Wind Energy. Now, assume that a generative AI model is erected as a strategy for planting a certain number of solar panels across a region. An AI model can dissect parameters such as sun intensity, available land, and propinquity to energy structure to decide exactly which spot should get a stationed solar ranch. For example, AI-driven renewable energy planning has the power to pinpoint locales indeed for solar granges in the state of Rajasthan, a region with high solar irradiance and large fields of thirsty land suitable for solar installations [21]. Similarly, wind energy planning can be preceded by AI analysis of wind speed patterns, terrain, and environmental impacts that recommend optimum placements of turbines. Denmark uses AI to optimize the placement of wind farms in order to maximize energy output with minimum harm to marine ecosystems.

Generative AI can also design hybrid renewable energy systems, combining more than one source of energy for high efficiency and reliability. It could be in the form of an AI model producing a plan with a combination of solar, wind, and battery storage so that there would always be sourcing available even if one is not. This type of planning is intrinsic to concepts aimed at resilient energy systems, capable of adapting to changing conditions and reducing breaks in supply. Example: Hybrid Renewable Energy Systems. For instance, Generative AI has been used in designing hybrid energy systems that integrate solar, wind, and battery storage in California. These systems ensure that communities in remote areas admit a dependable force of energy for which they witness frequent breakdowns. The literal energy demands, rainfall vaticinations, and grid capacity were used for an optimal configuration by this AI to ensure trustability and cost-effectiveness [22].

Optimization of energy storehouse and distribution: Generative AI finds other operations in renewable energy planning. For example, grounded on literal data, the AI models can prognosticate the pattern of energy operation and come up with ways of storing and distributing renewable energy to meet similar demand. Example: AI-Optimized Energy Storage. For example, in Germany, AI has been used to produce the stylish means of exercising battery storehouse systems in confluence with wind and solar granges. The AI model will develop strategies regarding when to release redundant energy and when to let it feed into the grid in order to ensure renewable energy is effectively employed, with force matched to demand. It stabilizes the grid and reduces demands for reactionary energy-grounded backup power in this way [23].

5.5.3 Broader Implications for Environmental Planning

Another application of generative AI is in predicting the environmental impact of renewable energy projects. For example, an AI model can simulate how a wind farm might affect local wildlife or assess the consequences of a hydroelectric dam on river ecosystems. These simulations help ensure that renewable energy initiatives are designed to minimize their effects on biodiversity and natural habitats. In Scotland, generative AI has been utilized to model the impact of wind farms on bird populations. By analysing bird migration patterns and the placement of wind turbines, the AI system can forecast potential risks to these populations and suggest measures to reduce threats, such as relocating turbines or halting operations during critical migration times [24].

The broader implications for environmental planning are significant. The use of generative AI in environmental impact assessments and renewable energy planning plays a crucial role in promoting sustainable development. AI aids in creating a more sustainable and resilient future by offering comprehensive insights into the effects of climate change and supporting the adoption of renewable energy sources [25]. One of AI's advantages is its capacity to analyse large datasets and uncover complex relationships that might not be apparent to human analysts. This capability enhances decision-making and helps formulate efficient strategies. For instance, AI can assist policymakers in identifying areas with the highest potential for renewable energy investments, ensuring that resources are utilized effectively. In Australia, the government has leveraged AI to evaluate data on energy consumption, renewable energy potential, and ecological impacts to craft policies that encourage solar power adoption in urban areas. Insights from the AI model have directed investments towards regions where solar energy can yield the most significant environmental and economic advantages [26]. However, employing AI in environmental planning also brings up several ethical and practical concerns.

The main issue is the reliability and accuracy of AI-generated models, which are heavily influenced by the quality of the data used for training. Poor or partially deficient data may lead to faulty predictions, which can eventually result in poor or adverse decisions. It is essential that proper data is used for the training of AI models, and the results are cross-checked against reality observations. The other factor to consider is the need for transparency and public involvement in environmental planning with AI technology. Such decisions regarding climate change prevention and adoption of renewable energy have serious social, economic, and environmental implications and, therefore, should be made transparently and inclusively. Public engagement in the planning stage and clear communication from policymakers about what these AI-generated models do can help instil trust and ensure that the benefits of AI are equitably distributed.

In a nutshell, Generative AI is crucial for assessing environmental impact and planning renewable energy, providing important insights that support the shift towards a more sustainable future. AI is helping in building a stronger and more sustainable world by developing climate change scenarios, optimizing the deployment of renewable energy, and predicting environmental impacts. Yet the challenges associated with AI are important to overcome, especially data accuracy, model dependency, and the need for openness, in order to ensure that the benefits of AI are accomplished in an ethical and integrated way.

5.6 MISLEADING INFORMATION

The rapid development of Generative AI technologies has brought about both benefits and challenges in the realm of information dissemination. On the one hand, AI can create high-quality content for many constructive applications; on the other hand, it has also enabled the generation of misleading or fraudulent information. This chapter discusses the use of AI-based models in fraudulent content detection and raises ethical questions on the diffusion of misinformation. We also include diagrams to help explain these concepts and examples that point out how AI plays a role in combating misinformation.

5.6.1 AI-Based Models for the Detection of Fraudulent Content

With the increase in deepfakes, AI-generated misinformation, and other harmful content, it has become quite difficult for individuals and organizations to tell the difference between genuine and fake information. To address this issue, AI-driven models have been created to analyse fraud detection by looking for patterns, inconsistencies, or other telltale signs of manipulation. These models use machine learning and NLP techniques to identify content that is likely to be deceptive or misleading. Generally, AI-based models detect fraudulent content work by analysing features such as image inconsistencies, unnatural language patterns, and atypical metadata. For example, deep learning models can determine the pixel composition of an image to determine whether it has been manipulated, while NLP models can analyse the structure and sentiment of text to determine inconsistencies that may indicate the existence of false information. Example: AI-Driven Deepfake Detection. Given the example of the Deepfake video, these videos are synthetic multi-shaped, manipulating audio and visual content to say or do things they've never done. Businesses such as Facebook and Microsoft have developed AI models to detect deep strikes by analysing subtle contradictions of facial movements, lighting, and sound synchronization. These models use convolutional neural networks to identify artefacts and anomalies often found in the most obviously fake videos, so that platforms can flag or remove these before they gain popularity. The third type of content detection involves training AI models on large datasets of known fake and genuine content. By learning from these datasets, AI models can develop an understanding of the features that distinguish real content from fake content.

For instance, one can train models to recognize clickbait headlines through linguistic features such as hyperbole, sensational language, and emotive words that are frequently found in articles of fake news.

5.6.2 Ethical Challenges of Fraudulent Content Generation

Although Generative AI technically has the implicit to be used for positive purposes such as creative content generation and tailored recommendations, it significantly questions ethical challenges when put forth to produce deceiving and fraudulent content. Inductive capability of AI in evoking satisfying fake images, videos, and textbooks has increased threats to enterprises about the hidden implicit abuse and resultant mistrust and lack of credibility in society.

One of the most significant ethical concerns of the generation of fake content is the effect on public trust. AI-generated fake news and deepfakes can be used to manipulate public opinion, spread misinformation, and damage reports. For example, a deepfake videotape of a political figure making false statements could have serious counteraccusations for choices and public perception. The spread of similar misinformation can erode trust in media and institutions, leading to confusion and division within society.

Example: Deepfake Impact on Public Trust. Through the 2020 US presidential election, there have been deepfake videos in social media, where the politicians were depicted in compromising conditions. These deepfakes are created to control the notion of the public and will impact the voter's behaviour. The moral task lies in preventing such misuse of AI generation while nonetheless bearing in mind its beneficial programs. Some other moral task is the capability for AI-generated content material to perpetuate biases and stereotypes. Generative AI models are trained from massive datasets, which often also include biased or unrepresentative statistics. As such, if the bias was not addressed, the generated content might perpetuate dangerous stereotypes or spread discriminatory messages. For example, a Generative AI model trained on biased information articles may wish to generate content that over-represents positive corporations in a bad light, which would contribute to social inequality and injustice, as depicted in Figure 5.7. Example: Bias in AI-Generated Content. Consider the case of news articles generated by AI. If the dataset used to train the AI model is biased towards some gender or ethnicity in how they are represented, that bias will be reflected in what the AI produces. Consider, for example, how an AI model trained using such biased data may report news stories in which there is a propensity to reflect women in submissive roles or ethnic groups as higher in criminal behaviour. Effectively mitigating these biases requires careful curation of training datasets in conjunction with developing methods to detect and remove bias from AI-generated content.

Issues of accountability and regulatory oversight are significant ethical concerns related to the generation of fraudulent content. With Generative AI capable

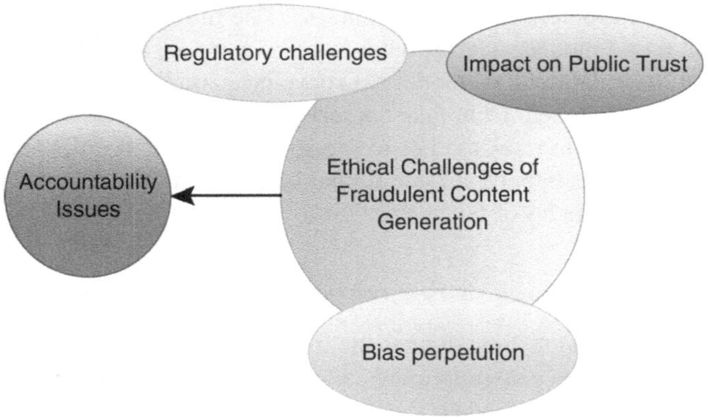

FIGURE 5.7 Ethical challenges of fraudulent content generation.

of producing material that closely resembles authentic content, it is increasingly challenging to identify who should be held accountable for the dissemination of misinformation.

5.6.3 Strategies for Addressing Ethical Challenges

To tackle the ethical challenges posed by Generative AI and the spread of misinformation, several strategies can be implemented. A crucial approach is the creation of advanced AI detection technologies that can accurately identify false content. By continuously enhancing the ability to detect AI-generated material, we can stay ahead of those who might misuse AI for harmful purposes. For example, Google has been developing AI models capable of detecting manipulated images and videos by analysing patterns of noise, lighting inconsistencies, and changes in pixel density. These advancements provide tools for identifying altered content that traditional detection methods might miss, thereby improving our ability to identify and remove fraudulent material. Another essential strategy involves establishing ethical standards and regulatory frameworks governing the use of Generative AI technologies. Collaboration among government bodies, tech companies, and other stakeholders is necessary to create policies that outline acceptable uses of AI while emphasizing the responsibilities of both developers and users. Recent proposals from the European Union to regulate AI include provisions for transparency and accountability to mitigate potential misuse of AI technologies. In addition to these two main strategies, raising public awareness and education is vital in addressing the ethical issues surrounding this deceptive form of content generation. By imparting knowledge about the inherent risk elements linked with AI-generative content as well as teaching skills about analysing any type of information, it becomes possible to reduce the implications of misinformation. Programs that aim to increase media literacy that educate people on how to identify fake news and deepfakes can strengthen the public's ability to make informed decisions, limiting the spread of false information. In Figure 5.8, the strategies implemented in dealing with ethical dilemmas are evident, as explained in the previous section.

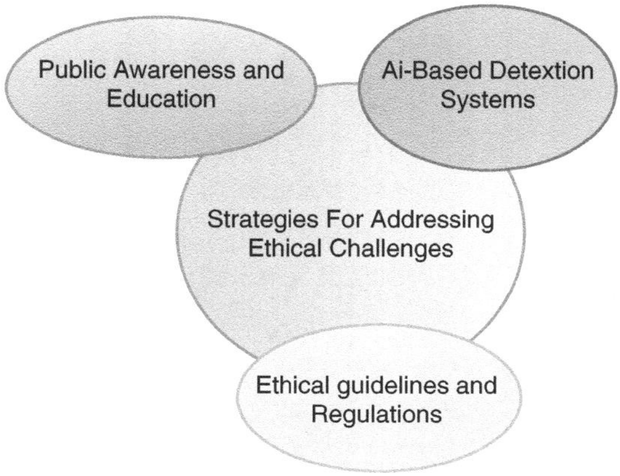

FIGURE 5.8 Strategies for addressing ethical challenges.

The employment of Generative AI in the fabrication of deceptive content poses considerable difficulties; however, it concurrently offers avenues for the advancement of novel instruments aimed at identifying and countering misinformation. Through the utilization of AI-driven models for the detection of fraudulent material, the formulation of ethical frameworks, and the dissemination of knowledge to the public, it becomes feasible to alleviate the detrimental effects of misleading data and to ensure that Generative AI is employed for constructive and ethical objectives. The trajectory of Generative AI will be influenced by the efficacy with which society can confront these ethical dilemmas and leverage the technology for the collective benefit.

5.7 FUTURE DIRECTIONS AND CURRENT APPLICATIONS

Generative AI has already gone a great way in mastering many of the disciplines; its potential just continues to broaden with technological advancements unfolding. In this chapter, we discuss the prevailing trends that mark the course of Generative AI's development, challenges, and opportunities coming with its integration into real life, and we come up with several examples and diagrams that will represent the perspectives and possibilities in Generative AI as well as its real-world applications currently.

5.7.1 Trends in Generative AI Development

The generative AI is being pushed forward by constant improvements in machine learning methods, computational power, and availability of large datasets. The trends in Generative AI are depicted in Figure 5.9. Several key trends are driving the course of Generative AI, which are the increased adoption of multimodal AI models, progress in ethical AI frameworks, and the democratization of AI tools to be used by more people. One of the most impactful trends of Generative AI development is multimodal AI models that could process and generate content in different

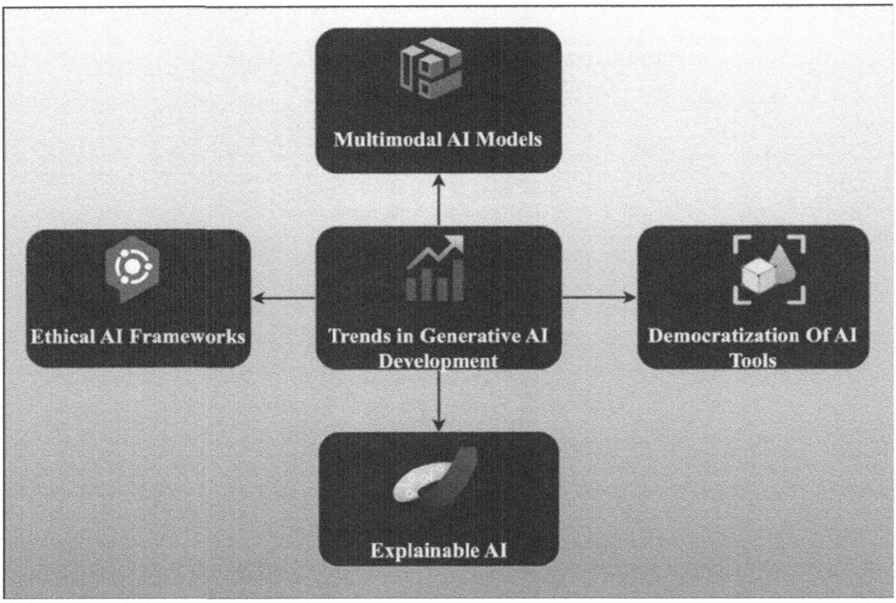

FIGURE 5.9 Trends in Generative AI development.

forms of media, including but not limited to text, images, audio, or video. Models like GPT-4 from OpenAI and DALL-E are part of a growing trend in AI technology. They can create both text and images based on natural language descriptions. This ability to handle multiple forms of media allows for the development of more sophisticated and adaptable AI applications, enabling users to generate content that combines various media types in a user-friendly way. For instance, a marketing team could leverage a multimodal AI framework to craft promotional materials for a new product. It could automatically generate detailed product descriptions and design visual ads. Additionally, it might even create audio or video content specifically tailored for social media outreach campaigns. This variety of content types fosters a more unified marketing strategy. A recent focus in the advancement of Generative AI is on enhancing the transparency and interpretability of these systems. As AI models become more complex, understanding how they reach their conclusions and outputs becomes increasingly important. Researchers are striving to develop methods that make AI models more explainable, which is essential for building trust and ensuring the responsible and ethical use of AI technologies. The establishment of ethical AI frameworks is one of the key trends influencing the future of Generative AI. As concerns about the potential misuse of AI to create harmful content rise, developers are prioritizing the creation of safeguards and ethical guidelines to ensure that Generative AI is applied positively. Examples of such initiatives include ongoing efforts to implement techniques for detecting and mitigating bias, which would help prevent AI systems from generating biased or discriminatory results.

5.7.2 Future Directions in Generative AI

As Generative AI continues to evolve, many exciting paths are emerging that could enhance its capabilities and uses. These potential paths include collaboration between humans and AI, the development of more efficient and specialized AI models, and the expansion of Generative AI into new areas such as scientific research and the creative arts. Human-AI collaboration will be an exceptionally focused future trajectory for Generative AI. These are systems that can augment human capabilities rather than replace the human creator, leading to a more synergistic relationship between humans and machines. In creative writing, music composition, and design fields, AI can provide ideas or generate first drafts or can be used to explore various branches of creativity, while the actual refinement, judgement, or emotional touch is provided by human beings, which AI fails to do. Example: Creative Writing with AI Support. For example, writers use Sudo write, a Generative AI-based tool that supplies writing prompts, continues storytelling, or suggests alternative phraseologies. This kind of collaboration allows writers to overcome creative barriers and explore new ideas while maintaining control over the final content. The trend exploded even further as AI became the integral part of the creative process in publishing, entertainment, and advertising industries. Another potential direction of research involves the development of increasingly specialized and efficient AI models. While current Generative AI models are primarily built to be generally applicable, there is a growing interest in developing models tailored to specific tasks or industries. Such specialized models may exhibit higher efficiency, require fewer computational resources, and produce more accurate and contextually relevant outputs. For example, a domain-specific AI model developed for the legal industry may be used to create legal documents, summarize case law, or assist attorneys in drafting contracts while being aware of the specialized terminology and conditions specific to the legal world. In the field of healthcare, specialized Generative AI models are being developed to aid in medical diagnostics, drug discovery, and the creation of personalized treatment plans. For instance, AI trained specifically on medical datasets can generate diagnostic reports from patient imaging, proffer customized treatment options, or simulate the effects of chemical candidates for new pharmaceutical compounds. These specialized models can deliver results that are more reliable and consequential as they focus on particular fields. Generative AI is also in a position to play a vital role in scientific research, and it can be used to contribute in hypothesis generation, experiment design, and analysis of empirical data. Researchers are increasingly using AI in the study of complex scientific questions, such as folding proteins, climate modelling, and material discovery. For example, the AlphaFold developed by DeepMind used AI to predict the structure of proteins, solving one of the biggest challenges of biology and opening up completely new avenues for medical and scientific research. For example, AI in Material Discovery Material science is using Generative AI more and more nowadays to discover new materials, such as those with an increased tensile strength or electricity conductivity. It helps the researcher identify promising materials through simulation of molecular architectures and intrinsic properties of diverse compounds with AI models and speed up the progress of novel technologies, including batteries, solar cells, and other sophisticated materials.

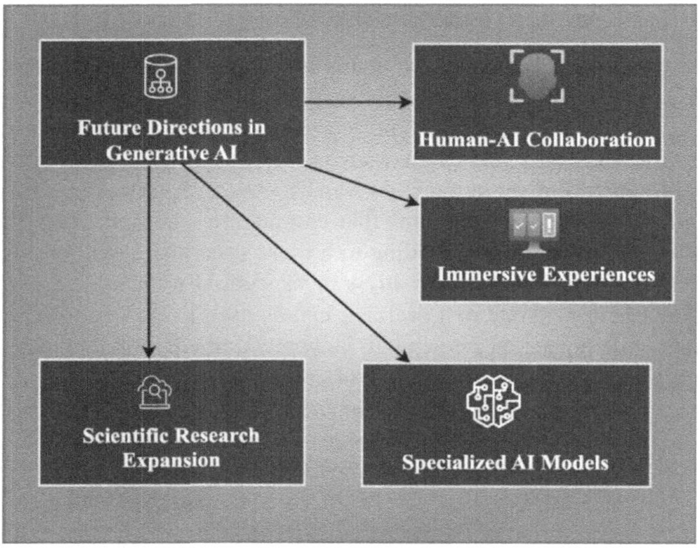

FIGURE 5.10 Future directions in Generative AI.

The potential applications of Generative AI also include its integration with augmented reality (AR), virtual reality (VR), and the metaverse in the creation of personalized and immersive experiences. Generative AI can be utilized to fabricate dynamic content within virtual environments, thereby enabling users to engage with AI-generated characters, navigate procedurally generated landscapes, or partake in individualized narrative arcs. This is going to change the landscape of industries like gaming, entertainment, and education as the experience will be more interactive and customized for the users. Example: AI in the Metaverse. As illustrated in Figure 5.10, within the realm of VR gaming, Generative AI is capable of generating distinctive storylines and characters that react to user interactions in real time, thus delivering a profoundly personalized gaming experience. Using these environments, an AR/VR setting, facilitated by Generative AI in the school domain, history, or a science experiment, can come alive where students can not only conduct research but also view it interactively and intensely, making learning more vivid and effective.

Currently, Generative AI is used in a wide variety of fields, such as entertainment, marketing, healthcare, and education. Such uses show how easily Generative AI can handle concrete challenges and supplement human ability and effectiveness. For instance, in the entertainment industry, Generative AI is used to create lifelike digital characters, special effects, and even write movie scripts and scripts for television programs. For instance, AI models are being used to create dialogue for virtual characters in games, enhancing the interactive experience for players. Similar initiatives are underway in music, where AI is capable of generating original compositions, allowing composers to explore new creative avenues. Consider Amper Music, for example; it's an AI platform that enables musicians to create original tracks by selecting their preferred genres, moods, and instrumentation. This capability allows artists to quickly produce a base track that they can then refine and expand upon. This represents a new era of collaboration between humans and AI in music composition, leading to innovative and original musical expressions.

Generative AI is also making strides in healthcare, aiding in medical research, drug development, and diagnostics. These models can analyse medical datasets, formulate hypotheses, and even propose new drug candidates. For instance, Generative AI has been utilized to design new molecular structures for potential drugs, significantly speeding up the drug development timeline. In diagnostics, AI can generate comprehensive reports from medical imaging, helping radiologists identify anomalies more efficiently. A notable example is the use of Generative AI by pharmaceutical companies like In Silico Medicine, which has effectively employed AI to discover new therapeutic candidates for conditions like cancer and fibrosis. By generating molecular designs and simulating their physiological effects, this AI framework helps researchers in narrowing down potential pharmaceuticals, reducing the time and costs typically involved in traditional drug discovery processes. In education, Generative AI is being used to create personalized learning experiences for students, showcasing the technology's versatility.

5.7.3 Challenges and Opportunities

While the future of Generative AI holds great promise, significant challenges remain before its full potential can be realized. One of the primary concerns is the ethical use of Generative AI, particularly in relation to bias, misinformation, and accountability. It is crucial for developers, regulators, and society as a whole to ensure that AI frameworks are employed responsibly and do not contribute to the creation of harmful or misleading content. Another hurdle is the substantial computational resources required to train and deploy Generative AI systems. Training large AI models demands considerable computational power and energy, which can be both expensive and environmentally detrimental. Researchers are actively pursuing more efficient training methods and models that consume fewer resources, but this remains a persistent challenge. For instance, Google researchers have developed techniques like model distillation, which involves training a smaller model to replicate the behaviour of a larger, more complex one. This method helps to lower the computational costs associated with Generative AI, making it more accessible for researchers and developers with limited resources.

Despite these obstacles, numerous opportunities exist for the ongoing development and application of Generative AI. A significant opportunity lies in the integration of Generative AI with other emerging technologies, such as AR, VR, and the Internet of Things (IoT). This integration can lead to the creation of immersive and interactive experiences, potentially transforming industries like entertainment, education, and healthcare. Additionally, the democratization of Generative AI tools allows a broader audience to access these technologies. Platforms like Runway ML and OpenAI's GPT-4 API enable individuals and small businesses to leverage Generative AI for various purposes, from content creation to prototyping new product ideas. This democratization has the potential to bring creativity and innovation

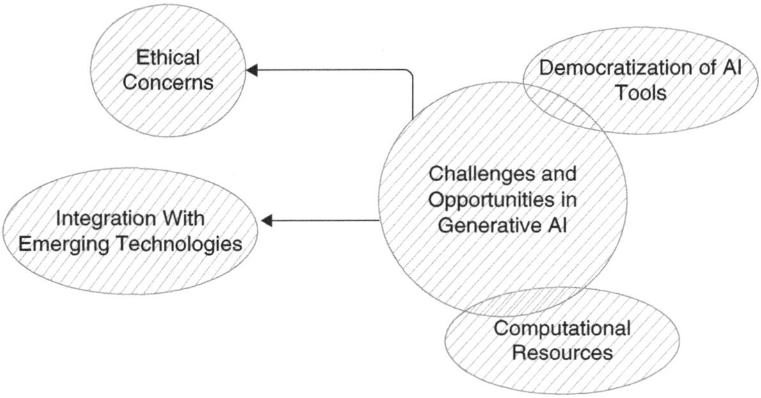

FIGURE 5.11 Challenges and opportunities in Generative AI.

from different backgrounds and tap their capabilities using AI into people. The challenges and opportunities are shown in Figure 5.11.

The future of Generative AI is indeed promising, but there are significant challenges that must be addressed to unlock its full potential. One of the key issues is the ethical use of Generative AI, particularly in terms of bias, misinformation, and accountability. It is essential for developers, regulators, and society at large to ensure that AI frameworks are utilized responsibly and do not lead to the generation of harmful or misleading information. However, there are also demanding situations that need to be addressed, particularly in terms of moral use, resource requirements, and transparency. By overcoming these challenges and seizing the opportunities supplied via Generative AI, society can harness the strength of this generation to create revolutionary answers and improve the high-quality of existence for human beings around the sector.

5.8 CONCLUSION

Generative AI is transubstantiating the geography of technology and invention across multiple disciplines, from creative diligence and social media to education, environmental planning, and scientific exploration. Throughout this chapter, we've explored different counteraccusations of textbook-grounded Generative AI, the challenges it faces, and the promising directions in which it's evolving. The capabilities of Generative AI to induce creative content, help in critical decision-timber, and give substantiated literacy gests demonstrate its immense eventuality to enhance mortal productivity and problem-working capabilities. The early chapters discussed the part of Generative AI in the creative process, emphasizing the impact of chatbots in aiding pens, generating creative content, and indeed automating law generation. The eventuality for AI to co-create content and give intelligent backing represents a new period in mortal-AI collaboration, where AI becomes an important tool to help generators without replacing the mortal element of creativity and imagination. We also examined the transformative influence of Generative AI on social media, where AI-driven tools help induce multilingual content, automate posts, and epitomize relations to enhance stoner engagement. The exemplifications handed showed how AI can be used to produce targeted and appealing content for different cult, helping brands and influencers connect more effectively with druggies across different artistic and verbal surrounds. The education sector has also greatly served from the operations of Generative AI, which allows for the generation of substantiated literacy accoutrements, adaptive quizzes, and interactive study gests. By acclimatizing educational content to each learner's specific requirements and preferences, AI has the implicit to ameliorate literacy issues and make education more accessible to scholars worldwide. In the field of environmental impact assessment, Generative AI is playing an important part in generating climate change scripts and planning renewable energy systems. These operations show how AI can be used to understand complex environmental challenges, plan for sustainable energy results, and prepare for the impacts of climate change.

In conclusion, Generative AI represents an important and transformative technology with the eventuality to revise diligence, enhance mortal creativity, and break complex problems. Its capabilities extend from creative content generation and substantiated education to environmental planning, scientific exploration, and healthcare. Still, the responsible use of Generative AI is essential to ensure that its benefits are realized without compromising ethical norms or contributing to misinformation and bias.

The future of Generative AI depends on how well we can balance invention with ethical considerations, develop robust discovery mechanisms for fraudulent content, and produce transparent and interpretable models that make public trust. By embracing the positive operations of Generative AI and addressing its challenges, we've the eventuality to produce a future where AI serves as an important supporter in advancing mortal progress and perfecting quality of life for people around the world. Additionally, the expansion of Generative AI into new disciplines, such as scientific exploration and immersive gests, points to the measureless possibilities of AI in working complex challenges and creating engaging and individualized stoner gests. By integrating Generative AI with technologies like stoked reality, VR, and the metaverse, we can anticipate to see entirely new ways of interacting with digital content and exploring virtual surroundings.

5.9 CONCLUDING THOUGHTS

Generative AI is at the forefront of a technological revolution that has the implicit to reshape how we produce, communicate, and break problems. Its operations gauge across diligence, and its capabilities continue to expand, driven by ongoing exploration and invention. Still, the true measure of Generative AI's success lies not only in its specialized achievements but also in its capability to serve humanity in a positive and ethical manner. By fostering collaboration, placing mortal values at the centre of AI development, and addressing ethical challenges head on, we can harness the full eventuality of Generative AI. The future isn't just about what AI can do, but about how we choose to use it. By making thoughtful, informed choices, we can ensure that Generative AI becomes a force for good—enhancing mortal eventuality, driving progress, and contributing to a more just and indifferent world.

REFERENCES

[1] De Rosa, Gustavo H., and Joao P. Papa. "A survey on text generation using generative adversarial networks." *Pattern Recognition* 119 (2021): 108098.

[2] Iqbal, Touseef, and Shaima Qureshi. "The survey: text generation models in deep learning." *Journal of King Saud University-Computer and Information Sciences* 34, no. 6 (2022): 2515–2528.

[3] Bandi, Ajay, Pydi Venkata Satya Ramesh Adapa, and Yudu Eswar Vinay Pratap Kumar Kuchi. "The power of generative AI: a review of requirements, models, input–output formats, evaluation metrics, and challenges." *Future Internet* 15, no. 8 (2023): 260.

[4] Zhang, Chenshuang, Chaoning Zhang, Mengchun Zhang, and In So Kweon. "Text-to-image diffusion models in generative AI: a survey." *arXiv preprint arXiv:2303.07909* (2023).

[5] Yu, Hao, and Yunyun Guo. "Generative artificial intelligence empowers educational reform: current status, issues, and prospects." In Margaret Grogan (Ed.), *Frontiers in Education*, vol. 8, p. 1183162. Frontiers Media SA, 2023.

[6] Khanna, Samar, Patrick Liu, Linqi Zhou, Chenlin Meng, Robin Rombach, Marshall Burke, David Lobell, and Stefano Ermon. "Diffusionsat: a generative foundation model for satellite imagery." *arXiv preprint arXiv:2312.03606* (2023).

[7] Gozalo-Brizuela, Roberto, and Eduardo C. Garrido-Merchan. "ChatGPT is not all you need. A state of the art review of large generative AI models." *arXiv preprint arXiv:2301.04655* (2023).

[8] Zhang, Yizhe, Guoyin Wang, Chunyuan Li, Zhe Gan, Chris Brockett, and Bill Dolan. "POINTER: constrained progressive text generation via insertion-based generative pre-training." *arXiv preprint arXiv:2005.00558* (2020).

[9] Xu, Derong, Wei Chen, Wenjun Peng, Chao Zhang, Tong Xu, Xiangyu Zhao, Xian Wu, Yefeng Zheng, Yang Wang, and Enhong Chen. "Large language models for generative information extraction: a survey." *arXiv preprint arXiv:2312.17617* (2023).

[10] Ren, Jie, Han Xu, Pengfei He, Yingqian Cui, Shenglai Zeng, Jiankun Zhang, Hongzhi Wen et al. "Copyright protection in generative AI: a technical perspective." *arXiv preprint arXiv:2402.02333* (2024).

[11] Crothers, Evan N., Nathalie Japkowicz, and Herna L. Viktor. "Machine-generated text: a comprehensive survey of threat models and detection methods." *IEEE Access* 11 (2023): 70977–71002.

[12] Deldjoo, Yashar, Zhankui He, Julian McAuley, Anton Korikov, Scott Sanner, Arnau Ramisa, René Vidal, Maheswaran Sathiamoorthy, Atoosa Kasirzadeh, and Silvia Milano. "A review of modern recommender systems using generative models (gen-recsys)." In *Proceedings of the 30th ACM SIGKDD Conference on Knowledge Discovery and Data Mining*, Barcelona, Spain, pp. 6448–6458, 2024.

[13] Clark, Kevin, and Priyank Jaini. "Text-to-image diffusion models are zero shot classifiers." *Advances in Neural Information Processing Systems* 36 (2024), 58921–58937.

[14] Khan, Muhammad Zeeshan, Saira Jabeen, Muhammad Usman Ghani Khan, Tanzila Saba, Asim Rehmat, Amjad Rehman, and Usman Tariq. "A realistic image generation of face from text description using the fully trained generative adversarial networks." *IEEE Access* 9 (2020): 1250–1260.

[15] Wang, Jianfeng, Zhengyuan Yang, Xiaowei Hu, Linjie Li, Kevin Lin, Zhe Gan, Zicheng Liu, Ce Liu, and Lijuan Wang. "Git: a generative image-to-text transformer for vision and language." *arXiv preprint arXiv:2205.14100* (2022).

[16] Acheampong, Francisca Adoma, Henry Nunoo-Mensah, and Wenyu Chen. "Transformer models for text-based emotion detection: a review of BERT-based approaches." *Artificial Intelligence Review* 54, no. 8 (2021): 5789–5829.

[17] Vartiainen, Henriikka, and Matti Tedre. "Using artificial intelligence in craft education: crafting with text-to-image generative models." *Digital Creativity* 34, no. 1 (2023): 1–21.
[18] Chang, Huiwen, Han Zhang, Jarred Barber, A. J. Maschinot, Jose Lezama, Lu Jiang, Ming-Hsuan Yang et al. "Muse: text-to-image generation via masked generative transformers." *arXiv preprint arXiv:2301.00704* (2023).
[19] Brack, Manuel, Felix Friedrich, Dominik Hintersdorf, Lukas Struppek, Patrick Schramowski, and Kristian Kersting. "Sega: instructing text-to-image models using semantic guidance." *Advances in Neural Information Processing Systems* 36 (2023): 25365–25389.
[20] Pandey, Rohit, Hetvi Waghela, Sneha Rakshit, Aparna Rangari, Anjali Singh, Rahul Kumar, Ratnadeep Ghosal, and Jaydip Sen. "Generative AI-based text generation methods using pre-trained GPT-2 model." *arXiv preprint arXiv:2404.01786* (2024).
[21] Cronin, Irena. "Advanced applications of text-based generative AI." In Irena Cronin (Ed.), *Understanding Generative AI Business Applications: A Guide to Technical Principles and Real-World Applications*, pp. 97–111. Berkeley, CA: Apress, 2024.
[22] Alhabeeb, Sarah K., and Amal A. Al-Shargabi. "Text-to-image synthesis with generative models: Methods, datasets, performance metrics, challenges, and future direction." *IEEE Access* 12 (2024): 24412–24427.
[23] Ghimire, Prashnna, Kyungki Kim, and Manoj Acharya. "Opportunities and challenges of generative AI in construction industry: Focusing on adoption of text-based models." *Buildings* 14, no. 1 (2024): 220.
[24] Sengar, Sandeep Singh, Affan Bin Hasan, Sanjay Kumar, and Fiona Carroll. "Generative artificial intelligence: a systematic review and applications." *Multimedia Tools and Applications* 84 (2024): 23661–23700.
[25] Takale, Dattatray G., Parikshit N. Mahalle, and Bipin Sule. "Advancements and applications of generative artificial intelligence." *Journal of Information Technology and Sciences* 10, no. 1 (2024): 20–27.
[26] Bengesi, Staphord, Hoda El-Sayed, Md Kamruzzaman Sarker, Yao Houkpati, John Irungu, and Timothy Oladunni. "Advancements in generative AI: a comprehensive review of GANs, GPT, autoencoders, diffusion model, and transformers." *IEEE Access* 12 (2024) 69812–69837.

6 AI-Driven Approaches for Hate Speech Detection

Methods and Challenges

Shikha Bhalla and Ashish Kumar

6.1 INTRODUCTION

Cyber hate speech dissemination is now an emerging issue that calls for accurate detection systems to eliminate its dangerous dissemination (Cavaliere, 2019). Online participation between users in social media sites and comment fields has led to spreading hateful content that particularly targets racial and religious groups as well as gender or ethnic groups (Kumar et al., 2023; 2025). Hateful content harms inclusive freedom of expression while escalating the situation among communities that leads to lethal violence (Papcunová et al., 2023). Artificial Intelligence (AI) functions as a robust detection tool that constructs expansive identification frameworks to address the complex problems of harmful discourse as per Castaño-Pulgarín et al. (2021) and Tontodimamma et al. (2021). Detecting hate speech remains an intricate challenge as linguistic imprecisions within sarcastic expressions and metaphors, as well as coded language, cause ambiguities of linguistic nature in such forms that do not provide proper outputs when subjected to conventional rule-based algorithms and textual-analysis methodologies (Kumar et al., 2019; Ullmann and Tomalin, 2020). The implementation of sophisticated identification mechanisms becomes essential to address the shortcomings of standard detection methods because current needs demand adaptive detection techniques. The study by Malmasi and Zampieri (2018) demonstrates how AI equipped with pattern recognition and semantic understanding capabilities can address ambiguity issues and detection limits in speech identification.

Modern hate speech detection systems perform multiple important social functions while meeting regulatory standards beyond their content control abilities. The systems work to create online communities that reduce social polarization while following platform rules (Gandhi et al., 2024; Malik et al., 2024). Through their operational design that prevents harmful speech exposure, detection systems build user trust and support psychological health, resulting in enhanced digital environment integrity (Mossie and Wang, 2020; Pérez et al., 2023).

Traditional machine learning (ML) models, such as Support Vector Machines (SVM), Naïve Bayes, and Random Forest (RF), have been the backbone for automated hate speech detection, especially when paired up with text representation methods like Term Frequency-Inverse Document Frequency (TF-IDF) and Word2Vec.

 DOI: 10.1201/9781003565703-6

But, let's not kid ourselves, these approaches hit a wall pretty quickly when it comes to picking up on subtle language cues or understanding the actual context (Zhou et al., 2020). The combination of Recurrent Neural Networks (RNNs), Convolutional Neural Networks (CNNs), and Transformer-based models like Bidirectional Encoder Representations from Transformers (BERT) has improved semantic understanding, but it requires a significant amount of computational power (Badjatiya et al., 2017; Alshalan and Al-Khalifa, 2020).

Generative Adversarial Networks (GANs) and Generative Pre-trained Transformers (GPT) have been the recent generative AI technologies that bring sophisticated abilities to hate speech detection using synthetic dataset creation and zero-shot learning (ZSL) flexibility. These innovations bring new considerations because they enhance the natural distribution of biases found in training datasets (Wullach et al., 2020). The field now investigates new research methods that integrate several methodologies to achieve fair tradeoffs between system detection precision and self-learning abilities and ethical compliance.

This research study has the following main objectives:

- Delivers an in-depth review showing the performance of ML techniques along with deep learning (DL) and generative AI approaches for detecting hate speech, which analyzes advantages and constraints in various contexts with their corresponding datasets.
- Focuses on ethical issues (bias and fairness) and linguistic variations (slang, cultural differences, and multilingual contexts) in hate speech detection, proposing solutions for more equitable systems.
- Explores the role of generative AI models like GPT-3, demonstrating how synthetic data generation and few-shot learning (FSL) improve performance in resource-limited settings.

The search for papers on ML, DL, and Generative AI begins by exploring electronic libraries like PubMed, Cochrane, and Google Scholar using technique-specific keywords. For ML, we use terms like "Machine Learning," "Hate Speech Detection," and "Classification," focusing on traditional algorithms like SVM and RF (2021–2024). For DL, we search using "Deep Learning," "Neural Networks," and "Transformers," targeting studies on CNNs, RNNs, and BERT (2022–2024). Generative AI searches primarily use "Generative AI" and "GANs" and "Zero-Shot Learning" as keywords to investigate adversarial training with models such as GPT between 2023 and 2024. The research assesses paper relevance while extracting data points from algorithms and evaluation methods to conduct thematic evaluation on each hate speech detection model.

The study is structured as follows: This paper first reviews conventional approaches with their technical challenges faced in ML and DL (Section 6.2) before exploring Generative AI methods, their benefits and weaknesses (Section 6.3). It then evaluates hate speech detection datasets (Section 6.4) before presenting the research results including implications for the field (Section 6.5) and drawing key conclusions and recommendations for future work (Section 6.6).

6.2 UTILIZING MACHINE LEARNING- AND DEEP LEARNING-BASED APPROACHES FOR HATE SPEECH DETECTION

Hate speech detection transformed through the implementation of ML and DL technology, which enabled the development of powerful processing systems that are able to handle substantial data volumes. Advanced algorithms within ML employ programs such as SVM and RF to classify content while using extraction techniques like TF-IDF and Word2Vec. DL techniques include BERT alongside CNNs and RNNs, which extract deep contextual and semantic framework from texts.

6.2.1 ML-Based Approach

The detection of hate speech has increasingly relied on a range of advanced ML techniques as researchers aim to improve classification accuracy and robustness across multilingual and heterogeneous datasets. The standard research approaches within studies conduct their work through sequential phases that include data acquisition, feature processing, model development, and thorough assessment. Researchers commonly utilize these techniques to process data that originates from YouTube along with various social media platforms, where they must overcome problems including data language diversity and limitations of human annotators and irregular data distribution patterns. A multitude of hate speech detection models implement classical ML algorithms with RF, SVM, Decision Trees (DT), and Naïve Bayes, among them. The models benefit from semi-supervised and ensemble learning systems as well as LIME explainability tools to analyze sparse and noisy data and provide readable explanations for prediction outputs. The work by Imbwaga et al. (2024) receives special recognition for creating a hate speech dataset that contains both English and Kiswahili text documentation. The research employed RF and LIME together for explanation purposes. The researchers faced major challenges while gathering data manually and needed to handle delicate ethical issues associated with text category assignment.

The annotation method received a boost from Saifullah et al. (2024a) through their development of semi-automated and high-throughput labeling for hate speech identification in YouTube discussions. The method employed SVM with TF-IDF for feature representation as its architecture but required high human intervention at the onset stages prior to having automated processes. According to Saifullah et al. (2024b), semi-supervised learning (SSL) with a combination of DT and SVM classifiers produced high classification accuracy. In hate speech classification tasks, this system faced difficulties identifying idiomatic phrases with complex linguistic elements and sarcasm. Alhazmi et al. (2024) used instruments that utilize Logistic Regression (LR), RF, and Naïve Bayes to conduct dialectical analysis and code-mixed language structure analysis of Arabic dialects. Although the researchers' accuracy results were encouraging, they claimed that the studies relied heavily on carefully selected datasets that accurately represented the training material. Concurrent studies on LSTM networks and Variational Graph Autoencoders, along with complex architecture analysis, were carried out by Nagar et al. (2023). Their developed models were demonstrated to perform better in online chat context comprehension and dynamic adjustment through experiments on YouTube datasets. Khanday et al. (2022) used ensemble learning techniques based on DT models to address data imbalance issues in Twitter social media. Despite using small datasets, the authors' minority class instance management system produced impressive classification results. Research on hate speech incidents in Indonesian YouTube comment sections was done by Cahyana et al. (2022). They enhanced their detection capabilities by combining TF-IDF-based annotation with SSL. The system faced difficulties when trying to apply its learning to new data distributions that extended outside the boundaries of its first annotation process. The research of Defersha and Tune (2021) investigated Facebook and Twitter hate speech detection for Afan Oromo language content. The researchers applied Linear Support Vector Classification (Linear SVC), Naïve Bayes, along with RF classifiers. The research models produced acceptable results yet underscored the important necessity of scalable automated hate speech detection solutions that serve underrepresented linguistic communities.

Figure 6.1 visualizes the prevalence of different ML models used across these studies, with a pie chart illustrating the frequency and adoption rate of each model type. Table 6.1 offers a comprehensive summary of the key contributions, detailing methodologies, datasets utilized, core challenges, and the performance of the models applied in each study.

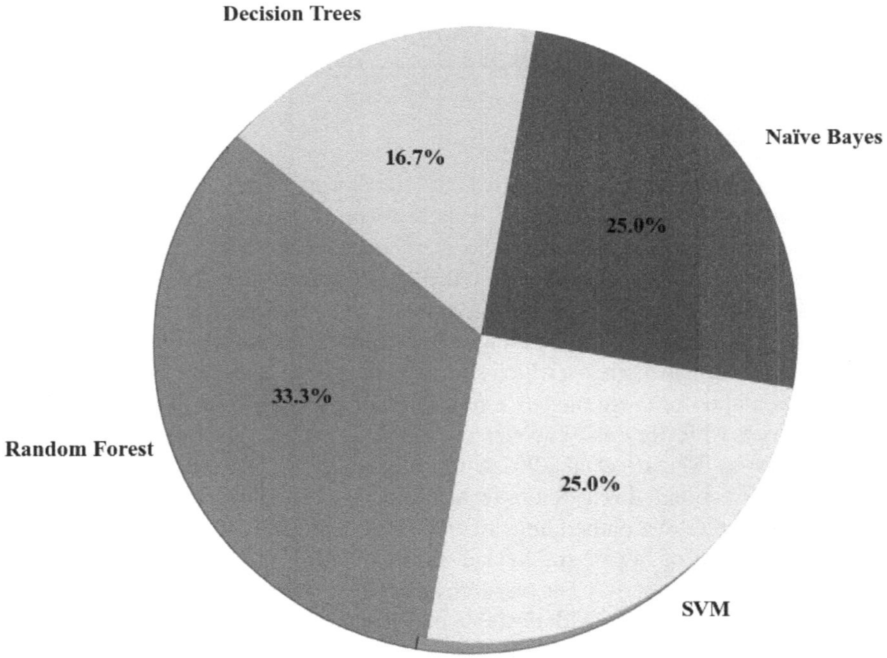

FIGURE 6.1 Distribution of ML models in hate speech detection. (Inspired by Munnuru et al., 2024.)

TABLE 6.1
ML-Based Approach Study Summary

Reference	Dataset	Text Language	Feature Extraction Techniques	Classification Algorithms
Imbwaga et al. (2024)	YouTube Videos (English, Kiswahili)	English, Kiswahili	TF-IDF, Word2Vec, N-grams	RF, SVM, Naïve Bayes, KNN
Saifullah et al. (2024a, 2024b)	YouTube Comments (Indonesian politics, COVID-19)	Indonesian	TF-IDF, Word2Vec	SVM
Saifullah et al. (2024a, 2024b)	YouTube Comments (Indonesian)	Indonesian	TF-IDF, Word2Vec	SVM, DT, KNN, Naïve Bayes
Alhazmi et al. (2024)	Novel Dataset (race, religion, ideology)	Arabic Dialects	TF-IDF, N-grams, CBOW	LR, RF, KNN
Nagar et al. (2023)	Hate Speech Twitter, Dropbox datasets	English	Textual features, social features	LSTM, VGAE
Khanday et al. (2022)	Twitter (COVID-19-related tweets)	English	TF-IDF, Bag of Words	LR, Naïve Bayes, SVM, DT
Cahyana et al. (2022)	YouTube Comments (Indonesian debates, COVID-19)	Indonesian	TF-IDF, KNN	Semi-supervised KNN
Defersha and Tune (2021)	Facebook and Twitter (Afan Oromo)	Afan Oromo	Bigram, TF-IDF	SVC, Naïve Bayes, RF, LR

6.2.2 DL-Based Approach

DL-based methodologies have significantly transformed hate speech detection by enabling nuanced understanding of semantic, contextual, and structural patterns within textual and multimodal data. Core architectures such as CNNs, RNNs, and advanced Transformer-based models like BERT offer superior performance over traditional ML methods. The workflow of DL models becomes stronger through the combination of pre-trained word embeddings and attention mechanisms alongside transfer learning and ensemble methods, which lead to leading edge performance. Determining the implementation of these models requires extensive computational power and big trained datasets while they also encounter scalability problems and fairness limitations.

The study by Mazari et al. (2024) introduces a BERT-based ensemble model that uses Bidirectional Long Short-Term Memory (BiLSTM) and Bi-GRU layers and FastText and GloVe embeddings to enhance the semantic features of input text. Through its fusion of layers, the hybrid model reached 98.63% ROC-AUC, which proved highly discriminative. The research pointed out two main challenges: excessive computational cost together with unstable performance results on rare labels because of unbalanced datasets. With the intention of understanding multimodal hate expressions, Arya et al. (2024) applied Contrastive Language-Image Pre-training (CLIP) to merge textual and visual information while specializing in hateful memes detection. The joint features in this model delivered 87.42% accuracy, thus establishing the model's power. The hate speech detection system functioned best when applied to the particular dataset characteristics, which reduced its ability to generalize beyond that specific range of data. The research by Firmino et al. (2024) applied multilingual variants of the BERT model to perform classification of Portuguese hate speech. The developed model demonstrated 92% F1-score to indicate effective hate speech detection in various multilingual contexts. Yuan et al. (2023) developed the t-HateNet model for addressing generalization problems through the implementation of transfer learning methods for improving cross-dataset consistency. The two produced commendable results although their performance relied heavily on the amount and standard of marked data available.

SocialHaterBERT by del Valle-Cano et al. (2023) united traditional text evaluation methods with social profile information including network data and user actions for improved detection. The 80.23% F1-score outcome of the model remained limited by the time-intensive process of social context feature interpretation and combination. BERT's performance increased to 96% F1-score when Saleh et al. (2023) adjusted the model specifically for certain domains by training it with relevant text corpora. At the same time, the research team noted that domain adaptation requires time and money to execute effectively. Khezzar et al. (2023) developed arHateDetector framework for Arabic dialect detection by employing AraBERT architecture to address dialectal variations in their work. While their detection system achieved 93% accuracy, it suffered from accuracy limitations due to class imbalance alongside the shortage of dialect-specific annotated datasets. Khan et al. (2022) created BiCHAT, which uses an ensemble structure consisting of BERT with CNN and BiLSTM along with a hierarchical attention system. As a solution, this model operates effectively on balanced or imbalanced datasets to achieve consistent performance in different circumstances. Through its particular design, the model directs attention to precise text parts, which results in higher classification precision rates. The research demonstrates the positive impact of combining transfer learning with multimodal fusion and attention mechanisms for advancing hate speech detection. However, they also bring to light persistent challenges such as the computational intensity of DL training, the scarcity and bias of annotated datasets, and generalization limitations across domains and languages.

Figure 6.2 illustrates the distribution and popularity of different DL architectures employed in hate speech detection, providing insights into their adoption trends. Table 6.2 systematically summarizes the major DL-based studies, detailing their methodologies, datasets used, and key performance outcomes, offering a comparative understanding of advancements in the field.

6.2.3 Challenges Faced in ML- and DL-Based Approaches

ML and DL approaches for hate speech detection face several key challenges. Language ambiguity, including sarcasm, irony, and coded expressions, makes it difficult for models to differentiate harmful from benign content, especially without clear context. Data imbalance in datasets, often skewed toward specific offensive language types, limits generalization across diverse contexts. Additionally, obtaining large, well-labeled datasets covering varied communities, dialects, and hate speech forms

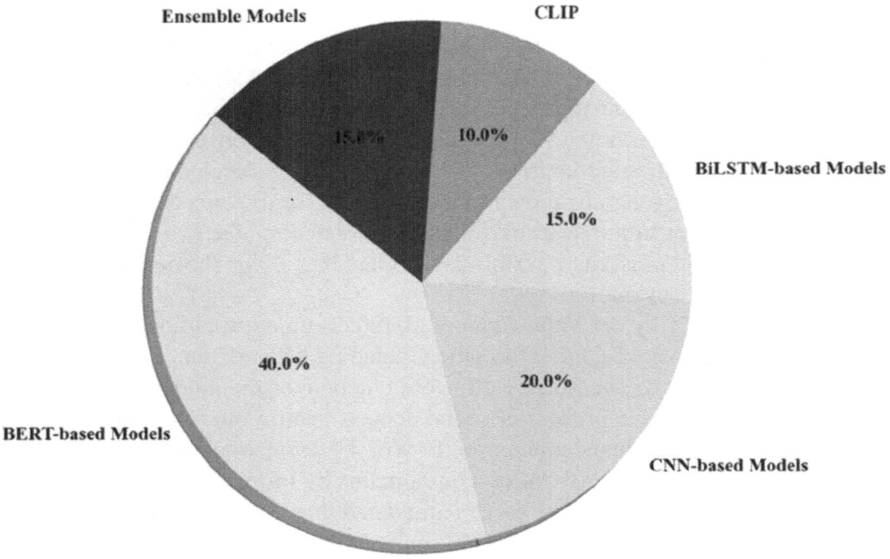

FIGURE 6.2 Distribution of DL models in hate speech detection. (Inspired by Munnuru et al., 2024.)

TABLE 6.2
DL-Based Approach Study Summary

Reference	Dataset	Text Language	Feature Extraction Techniques	Classification Algorithms
Mazari et al. (2024)	Kaggle Hate Speech Dataset	English	FastText, GloVe, BERT embeddings	BERT + BiLSTM, BERT + Bi-GRU, Ensemble Models
Arya et al. (2024)	Facebook Hateful Meme Dataset	English	CLIP-based contrastive learning with prompt engineering	CLIP with cosine similarity-based classification
Firmino et al. (2024)	Evalita 2018, Waseem, OffComBr-2	Italian, English, Portuguese	Multilingual BERT embeddings (BERTimbau, XLM-RoBERTa)	Zero-Shot Transfer, Joint Learning, Cascade Learning
Yuan et al. (2023)	Waseem, Davidson	English	Pre-trained ELMo + GloVe embeddings	BiLSTM, t-HateNet with Transfer Learning
Valle-Cano et al. (2023)	ACHaterNet	Spanish	BERT embeddings, social network features	SocialHaterBERT with Multimodal Transformer Architecture

(Continued)

TABLE 6.2 (*Continued*)
DL-Based Approach Study Summary

Reference	Dataset	Text Language	Feature Extraction Techniques	Classification Algorithms
Saleh et al. (2023)	Davidson, Waseem	English	Domain-specific Word2Vec (HSW2V), BERT embeddings	BiLSTM, BERT Base, BERT Large
Khezzar et al. (2023)	arHateDataset	Arabic (Standard and Dialectal)	AraBERT embeddings, CNN-based spatial features	CNN, AraBERT
Khan et al. (2022)	HD1, HD2, HD3	English	BERT embeddings, hierarchical attention	CNN + BiLSTM with hierarchical attention (BiCHAT)

remains a significant hurdle. The rapid evolution of language, with emerging slang, memes, and abbreviations, requires constant retraining of models. Computational complexity, particularly in DL models like BERT, poses challenges for resource-intensive training and real-time deployment in constrained environments. Interpretability remains an issue, as many advanced models function as "black boxes," obscuring their decision-making processes. Addressing these challenges is essential to enhance the robustness, scalability, and fairness of hate speech detection systems.

6.3 GENERATIVE AI AND HATE SPEECH DETECTION

Generative AI plays a significant role in hate speech detection by generating new data, such as text, to enhance model training and performance. Contrary to the standard AI models that rely on predetermined rules, generative models such as GANs and GPT learn from the patterns in current data and create new outputs. Text generation and translation mechanisms within NLP gain support from such models to further aid in data augmentation and better model flexibility under low-resource scenarios (Kumar et al., 2022; 2023).

Hashmi and Yayilgan (2024) used a hybrid DL model (FAST-RNN) combining BiLSTM, GRU, and FastText embeddings, improving precision and recall but facing computational complexity and dataset imbalance issues. Cohen et al. (2023) integrated DeBERTa models with Test-Time Augmentation (TTA), Back-Translation (BT), and GPT-3, enhancing robustness, though GPT-3 augmentation may introduce label inconsistencies. Su et al. (2024) combined SSL with GAN and RoBERTa for Anti-Asian hate speech detection, but GAN training complexities and resource demands pose challenges. Khullar et al. (2024) used machine translation and Contextual Entity Substitution (CES) for low-resource languages, though synthetic data quality can introduce noise. Ramos et al. (2024) explored GPT-3.5 in transfer learning across YouTube and Twitter, but mixed-domain training led to inconsistent results. Hashmi et al. (2024) applied transformer models with prompt-based

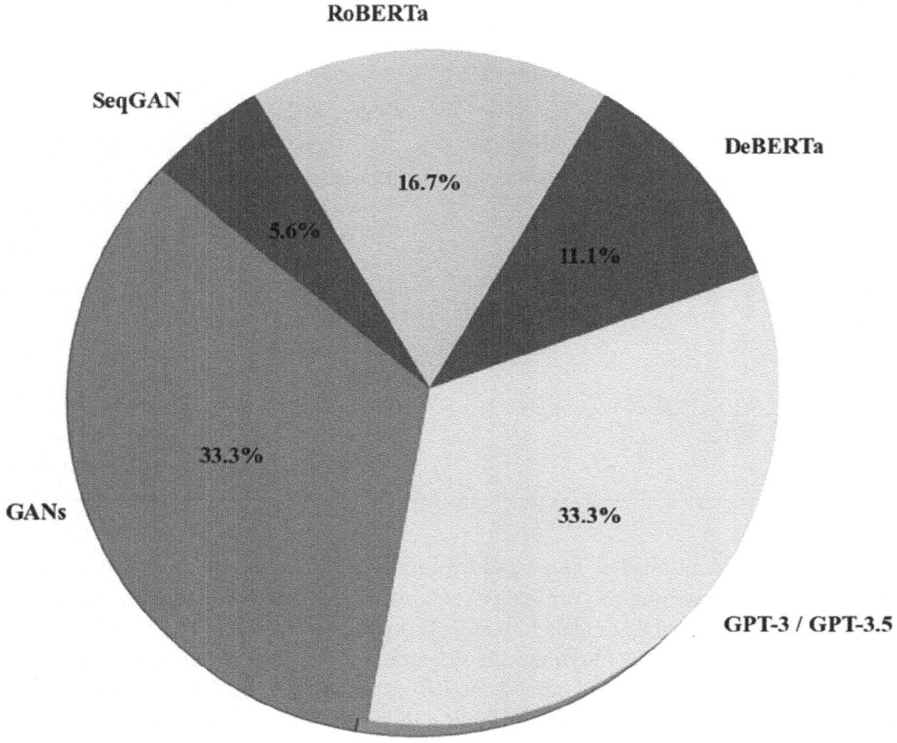

FIGURE 6.3 Distribution of generative AI models in hate speech detection. (Inspired by Munnuru et al., 2024).

fine-tuning for multilingual detection, but it's resource-intensive. García-Díaz et al. (2023) examined ZSL, FSL, and Fine-Tuning for multilingual text, with challenges in consistency, example quality, and labeled data. Mnassri et al. (2024) used SS-GAN for multilingual hate speech detection, but model performance depends on data quality. Luo et al. (2021) applied SeqGAN for sentiment analysis, improving accuracy but facing implementation complexity. Aggarwal and Vishwakarma (2024) proposed an adversarial attack framework using generative techniques, though attacks have high specificity and computational complexity. These studies highlight generative AI's growing role in enhancing hate speech detection, addressing challenges like resource limitations, domain variability, and adversarial threats. Figure 6.3 gives the distribution of Generative AI Models in Hate Speech Detection, and Table 6.3 gives the summary of this section.

6.3.1 Comparison of ML, DL, and Generative AI-Based Hate Speech Detection

ML, DL, and Generative AI each contribute uniquely to the evolving field of hate speech detection, with distinct strengths and limitations that often complement one another. Traditional ML techniques such as SVM, RF, and Naïve Bayes are frequently

TABLE 6.3
Generative AI-Based Approach Study Summary

Reference	Dataset	Text Language	Feature Extraction Techniques	Classification Algorithms
Hashmi and Yayilgan (2024)	Norwegian language hate speech dataset	Norwegian	FastText embeddings, BiLSTM, GRU	FAST-RNN (BiLSTM, GRU)
Ramos et al. (2024)	YouTube, Twitter (European and Brazilian Portuguese)	European Portuguese, Brazilian Portuguese	BERTimbau, GPT-3.5	Transfer learning with BERTimbau and GPT-3.5
García-Díaz et al. (2023)	EXIST, HatEval, HaterNET, EDOS, HASOC (Spanish and English)	Spanish, English	Zero-shot learning (ZSL), Few-shot learning (FSL), Fine-tuning	Fine-tuning, ZSL, FSL
Khullar et al. (2024)	English (Mathew et al.), Hindi (Bhardwaj et al.), Vietnamese (Luu et al.)	English, Hindi, Vietnamese	Machine translation (MT), Contextual Entity Substitution (CES)	Data augmentation (MT, CES)
Mnassri et al. (2024)	HASOC2019 (English, German, Hindi); Twitter, Facebook	English, German, Hindi	GANs (SS-GAN-mBERT, SS-GAN-XLM), Multilingual models	GAN-based classifiers (SS-GAN-mBERT, SS-GAN-XLM)
Hashmi et al. (2024)	Multilingual dataset (13 languages including English, Turkish, and Italian)	Multilingual (13 languages)	Transformer models, Prompt-based fine-tuning	Hybrid classifiers (SVM, RF, LR)
Luo et al. (2021)	Stanford Sentiment Treebank (SST), Hate Speech (HS)	English	LSTM (sentence compression), SeqGAN, BiLSTM (data screening)	LSTM, SeqGAN, BiLSTM
Aggarwal and Vishwakarma (2024)	ETHOS (YouTube/Reddit), HOLD (tweets)	English	Glyphs, zero-width characters, swarm optimization	Adversarial attack framework
Cohen et al. (2023)	Parler Hate Speech Dataset, GAB Hate Corpus (GHC)	English	DeBERTa, Back-Translation (BT), GPT-3 Augmentation	DeBERTa-based models, TTA, GPT-3
Su et al. (2024)	Gab Hate Corpus (GHC), COVID-19-related social media datasets	English	SSL, GAN (GAN-RoBERTa)	GAN-based classifier (SSL-GAN-RoBERTa)

employed due to their computational efficiency, scalability, and ease of interpretation. For example, Imbwaga et al. (2024) utilized a RF classifier in combination with LIME to enhance the transparency of their hate speech detection model. The team behind Saifullah et al. (2024a, b) merged SVM and DT models to perform annotations on YouTube comments through a SSL method. Such ML-based methods create effective answers for low-resource environments since they need minimal processing power while being simple to put into use. The basic features engineered manually together with weak contextual understanding of ML models produces struggles in processing complex linguistic patterns such as sarcasm and figurative language with code-mixed expressions. Their ability to be generalized remains restricted because of these limitations in domains and languages.

DL architectures such as BERT, BiLSTM, and hybrid architecture have been more capable of extracting textual data components that involve an understanding of contextual relationships and syntactic and semantic information. Mazari et al. (2024)) employed a sophisticated ensemble of BERT and BiLSTM and BiGRU layers using FastText and GloVe embeddings for efficient handling of unbalanced hate speech datasets. By their research, Arya et al. (2024) grounded CLIP as the foundation for multimodal feature fusion in meme detection, explaining how DL processes multiple types of data integration. The new complex modeling techniques demonstrate superior performance above standard ML practices for understanding complex language structures. These models work best with extensive labeled collections and powerful computing systems, but these elements make them impractical for environments with limited resources.

Hate speech detection has reached its latest milestone with Generative AI models, which gain adaptiveness by employing ZSL and FSL approaches. GPT-3 among other pre-trained language models demonstrates the capability to create contextually appropriate answers even with little labeled training data. The studies of Ramos et al. (2024) and Hashmi and Yayilgan (2024) both demonstrated their respective hate speech detection methods in Portuguese by using rich context prompts. Generative models achieve strong ability to adapt swiftly between linguistic environments while performing generalized task operations. Nonetheless, generative AI approaches are not without limitations. They are computationally intensive and susceptible to issues such as semantic drift, label inconsistencies, and hallucination during generation, which can compromise their reliability in real-world applications. Furthermore, the opacity of their internal mechanisms often hinders explainability and regulatory compliance.

Although a wide range of studies were reviewed, Figure 6.4 consolidates the comparative performance metrics of the top-performing models from each category ML, DL, and Generative AI to ensure a robust and fair evaluation. The selected metrics are representative of the highest accuracy, F1-score, and Area Under Curve (AUC) values reported, highlighting the relative strengths of each paradigm while acknowledging their contextual limitations. This performance comparison underscores the progressive advancement of hate speech detection technologies and points toward the increasing convergence of methodologies to address persistent challenges. A comparison summary is given in Table 6.4.

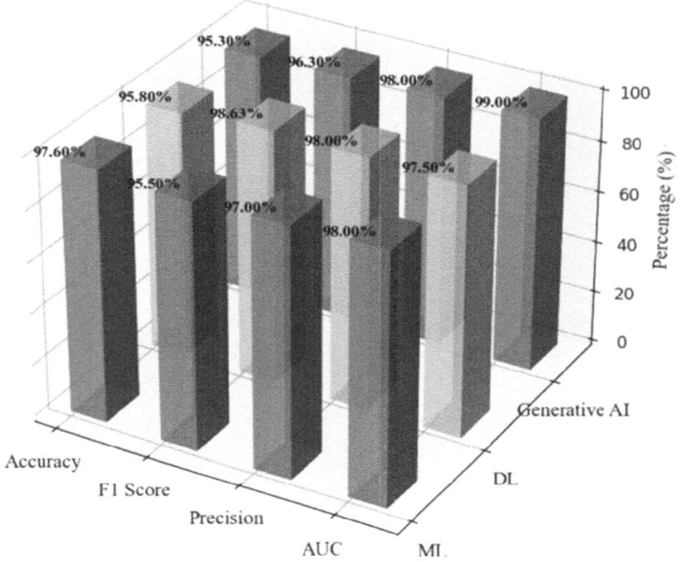

FIGURE 6.4 Performance metrics comparison. (Inspired by Menzies et al., 2022.)

6.3.2 Challenges in Using Generative AI for Hate Speech Detection

The advanced natural language capabilities of generative AI models prove ineffective at detecting hate speech because they encounter fundamental obstacles during actual deployment. The major limitation at present stems from language diversity because detecting low-resource languages or dialects remains difficult. Their extensive pre-training across multilingual datasets does not prevent GPT-3.5 and multilingual BERT (mBERT) from performing poorly when encountering situations that need detailed understanding of rarely encountered linguistic elements or cultural-specific expressions. Without understanding culturally contextualized language, it becomes difficult to detect hate speech because these implicit expressions can evade detection. Although they have considerable operating constraints, generative models use ZSL and FSL techniques to handle small datasets. Since the detection methods are unable to recognize subtle patterns that were missed in previous training, they record lower accuracy levels when detecting subtle expressions of hate speech. Because hate speech keeps developing into more complex forms, these flaws make it impossible to apply these models in practical settings.

Training datasets that are unrepresentative and of poor quality present the biggest obstacle. Although generative models heavily rely on their training data, any biases in the data cause them to become highly biased, which can result in output manipulation, stereotypical propagation, and the exclusion of marginalized voices. Because these systems' performance is hampered by unrepresentative factors and inadequate coverage of linguistic or demographic differences, they produce unreliable results in high-stakes scenarios. The use of data augmentation techniques to enhance the performance of generative models raises ethical concerns. The production of hateful

TABLE 6.4
A Comparison Summary of ML, DL, and Generative AI

Category	Reference	Model Architecture	Training Paradigm	Evaluation Metrics	Real Time Applicability
ML	Imbwaga et al. (2024)	RF, SVM, XGBoost	Supervised	Accuracy: 97.6%, F1: 95.8%	Yes
	Saifullah et al. (2024a, 2024b) (referred as 16)	SVM with TF-IDF and Word2Vec	Semi-Supervised	Accuracy: 92.5% (TF-IDF)	No
	Saifullah et al. (2024a, 2024b) (referred as 17)	DT, Naive Bayes, KNN	Semi-Supervised	Accuracy: 97.1% (DT with TF-IDF)	No
	Alhazmi et al. (2024)	LR, RF, SGD	Supervised	Accuracy: 98.21% (SGD + TF-IDF)	No
	Nagar et al. (2023)	Variational Graph Auto-Encoder + LSTM	Supervised	Not explicitly stated	No
	Khanday et al. (2022)	DT, LR, Ensemble	Supervised	Accuracy: 97.96% (DT)	Yes
	Cahyana et al. (2022)	KNN with TF-IDF	Semi-Supervised	Accuracy: 59.68% (KNN + TF-IDF)	No
	Defersha and Tune (2021)	Linear SVC, Naive Bayes, RF	Supervised	F1: 64% (Linear SVC)	No
DL	Mazari et al. (2024)	BERT + Ensemble BiLSTM and Bi-GRU	Supervised	ROC-AUC: 98.63%, F1: 62.18%	No
	Arya et al. (2024)	CLIP with Contrastive Learning	Zero-shot learning	Accuracy: 87.42%, AUROC: 88.35	No
	Firmino et al. (2024)	BERT variations (BERTimbau, XLM-RoBERTa)	Multilingual training	F1: 92.0% (CL/JL+)	No
	Valle-Cano et al. (2023)	SocialHaterBERT with user graphs	Multimodal	F1: 80.23%, AUC: 89.23%	No
	Saleh et al. (2023)	BiLSTM + Fine-Tuned BERT	Supervised	F1: 96%, AUC: 95%	No

(Continued)

TABLE 6.4 (*Continued*)
A Comparison Summary of ML, DL, and Generative AI

Category	Reference	Model Architecture	Training Paradigm	Evaluation Metrics	Real Time Applicability
	Yuan et al. (2023)	BiLSTM + GloVe + ELMo Embeddings	Transfer learning	Macro-F1: 78.09% (t-HateNet)	No
	Khezzar et al. (2023)	AraBERT + CNN	Multilingual training	Accuracy: 93%	No
	Khan et al. (2022)	BERT + CNN + BiLSTM + Attention	Hierarchical attention	F1: 84%–91%	No
Generative AI	Hashmi and Yayilgan (2024)	FAST-RNN (BiLSTM, GRU)	Hybrid DL, Adam optimizer	Precision: 0.98, F1: 0.98, Accuracy: 0.98	Yes
	Ramos et al. (2024)	BERTimbau, GPT-3.5	Transfer learning, mixed-domain	BERTimbau F-score: 87.1%, GPT-3.5 F-score: 50.2%	Yes
	García-Díaz et al. (2023)	ZSL, FSL, Fine-Tuning	Zero-shot, Few-shot, Fine-tuning	Spanish F1: 82.25, English F1: 86.76	Yes
	Khullar et al. (2024)	Multilingual BERT	Data augmentation, translation	CES: Hindi (38%) vs MT (31%), Vietnamese (59%)	Yes
	Mnassri et al. (2024)	mBERT, XLM-RoBERTa (GANs)	Semi-supervised learning	Strong performance without exact metrics	Yes
	Hashmi et al. (2024)	Multilingual transformers	SVM, DT, RF	Accuracy: 0.88 for English	Yes
	Luo et al. (2021)	BiLSTM, SeqGAN	Data augmentation framework	SST accuracy improvement: 1.0%	Yes
	Aggarwal and Vishwakarma (2024)	Adversarial attack (glyphs, zero-width)	Black-box adversarial attacks	Attack success: 88.79%, accuracy drop	No
	Cohen et al. (2023)	DeBERTa, BT, GPT-3, TTA	Back-translation, Test-time augmentation	AUC: 82.58 for GAB Hate Corpus	Yes
	Su et al., (2024)	RoBERTa (SSL-GAN)	Semi-supervised learning, GANs	Accuracy: 0.7895	Yes

training materials through synthetic means carries risks that could lead to unintended outcomes, such as the spread of hate speech throughout society. Because it presents an ethical conundrum in hate speech research, the use of synthetic hate speech in training calls for close ethical oversight.

High-technical computing capabilities are needed for both the fine-tuning optimization process and the system implementation of large generative models like GPT-3.5 and BERT. These models require massive amounts of memory, storage, and processing power, making it difficult and expensive to implement real-time activities and sporadic training sessions to identify new forms of hate speech. Since transparent and accountable decisions are essential in hate speech detection, the inability to explain the inner workings of generative AI models is a critical problem. The fact that these models are black boxes makes it impossible for authorities to examine the rationale behind decisions, which reduces trust in their work and makes legal audits more difficult. They also present difficulties in delicate areas like legal speeches and content moderation. Models must be updated on a regular basis due to the rapidly changing nature of hate speech components in coded languages, memes, and social media. The long-term efficacy of the model is limited by operational obstacles and performance losses caused by the time lag between model updates and new hateful content patterns. Its limitations in terms of linguistic coverage combined with ethical obligations, computational tractability, and model interpretability currently constitute the bottleneck of generative AI for hate speech detection. An approach that combines ethical AI design, model interpretability frameworks, responsible data curation, and real-time adaptability mechanisms for safe and equitable deployment in a variety of sociolinguistic contexts is necessary to overcome these limitations.

6.4 DATASETS FOR HATE SPEECH DETECTION

The Natural Language Processing (NLP) researchers consider hate speech detection an area of high priority, as damaging internet content keeps increasing incessantly. Different datasets serve to support model building and testing for detecting hate speech and offensive content in multiple languages, which operate across various platforms while considering different contexts. Hate speech detection models rely heavily on such datasets because these provide the basis for creating dependable and precise detection systems.

6.4.1 Categorization of Datasets

Hate speech detection datasets divide their content according to type and language and target values and contextual analysis parameters. The datasets focus on different content types, which include tweets found in HatEval (Basile et al., 2019), Waseem Dataset (Talat & Hovy, 2016), and Davidson Dataset (Davidson et al., 2019) and memes from the Facebook Hateful Memes Dataset (Kiela et al., 2020) alongside posts and comments within ETHOS Hate Speech Dataset (Mollas et al., 2022) and Gab Hate Corpus (Kennedy et al., 2022). The CALM Dataset (Gupta et al., 2023) together with the HASOC (Mandl et al., 2019) and Kaggle Hate Speech and Offensive Language Dataset (Kaggle, n.d.) contain multilingual content, while the arHateDataset (Khezzar et al., 2023) concentrates on Arabic and OffComBr-2 (De Pelle and Moreira, 2017) targets Portuguese along with the Spanish Hate Speech Dataset specializing in football materials (Montesinos-Cánovas et al., 2023). Different hate speech detection datasets adopt different approaches to categorization where ETHOS together with HASOC Dataset provides binary choices, but HaterNet Dataset (Pereira-Kohatsu et al., 2019) adds detail to the classifications by separating speech into violent or dehumanizing categories while EDOS Dataset (Kirk et al., 2023) differentiates between sexism types. The study of contextual datasets focuses on addressing particular social problems through tools like HatEval, which detects hateful content directed at immigrants and women along with COVID-19, hate speech detection datasets that identify pandemic-related messages. Classifiers are used by researchers to create algorithms for detecting hate speech in a number of different types of content and forms of linguistic expressions, ranging across different contexts.

6.4.2 Dataset Challenges

Data difficulties in the hate speech detection task yield several challenges that impact the model training stage and test results as much as the ultimate deployment step. The primary issue arises from class imbalance datasets with predominant non-hate speech content, which present challenges in training models optimally. The models are unable to pick out authentic hate speech among sarcasm or joke expressions due to the importance of context-sensitivity in defining intent and target groups when short text such as tweets is under examination. Hater speech identification is more challenging because language variation and dialect difference alter the patterns of syntax, which can be used by hate speech across linguistic divides. Model robustness is compromised since annotators do not agree on what constitutes actual hate speech during the labeling process. Because information about members of the outgroup has sensitive characteristics and exposes them to harm, using biased datasets raises ethical questions about privacy violations and bias issues that affect equitable outcomes. Developing multiview processing models is difficult due to the multimodal data from text and images in the Twitter and Facebook Hateful Memes Dataset. Finally, a fine sense of language and context is maintained through the fine-grained classification of hate speech types, such as threats or prejudice. The range of issues necessitates prompt action to enhance accuracy, fairness, and social responsibility in hate speech detection systems.

6.5 DISCUSSION AND FINDINGS

The studies looked at how hate speech detection techniques have evolved, showing a shift from conventional ML techniques to complex DL models before adopting generative AI techniques. Since each approach has unique strengths in terms of precision, scalability, interpretability, and flexibility, the field's development benefits greatly from each of them.

In their early stages, hate speech detection systems relied on ML techniques that included RF in addition to SVM and Naïve Bayes classifiers (Defersha and

Tune, 2021). Because the chosen models functioned well even with constrained computing resources, they were efficient and comprehensible. The SVM model was successful in large-text classification issues while remaining user-friendly while training. In order to construct features, these traditional ML techniques needed human intervention in the form of manual input. However, they were still unable to understand complex language patterns that mask syntactic and semantic meanings, especially sarcastic hints and code-mixed phrases. Such systems' limited ability to function in dynamic, changing environments with a variety of linguistic patterns was the main factor limiting their usefulness and applicability (Saifullah et al., 2024a,b; Alhazmi et al., 2024; Nagar et al., 2023).

DL turned out to be a major technological advance. Direct text learning was made possible by models like CNNs, BiLSTM networks, and Transformer-based models like BERT (Mazari et al., 2024; Firmino et al., 2024; Saleh et al., 2023; Yuan et al., 2023; Khan et al., 2022). These machine models demonstrated exceptional proficiency in processing language hierarchical patterns, text embeddings, and lengthy contextual structures. DL models' performance potential enhanced a number of hate speech detection benchmarks, especially when they used sizable labeled datasets. In a variety of linguistic contexts, BERT and its multilingual counterparts were able to detect hate speech with high accuracy rates. These developments became unsuitable for real-time and resource-constrained scenarios due to their higher computational demands, longer training times, and dependence on large labeled corpora.

The most recent features of Generative AI models, like GPT-3 and mBERT, in conjunction with GANs, are employed in contemporary hate speech detection techniques (Cohen et al., 2023; Ramos et al., 2024; Mnassri et al., 2024; Khullar et al., 2024; Su et al., 2024). The models can be used on a variety of language and low-resource domains because they don't require training in domain-specific data. Even with erratic or sparse labeled data, the prompt-based response generator in GPT-3 is a powerful analytical tool for detecting hate speech. In hybrid generative-discriminative frameworks, the combination of classification heads and GPT-style language models demonstrates efficacy in identifying various scenarios. In addition to the ethical issues that arise during their learning or data augmentation phases, when they reproduce damaging stereotypes, these flexible models also raise important questions regarding their use of computational resources and bias reinforcement processes (Albladi et al., 2025; Kumar, 2025; Kumarage et al., 2024).

Researchers' analysis datasets differ significantly in terms of their structural characteristics, the scope of their contents, and evaluations of the data quality. Research datasets can be text-based and only contain one language, or they can include multiple languages and multimodal features like images, metadata, and social network elements (e.g., user actions and network relationships). The diversity in dataset types supports inclusive research although it complicates the research process. The majority of datasets show imbalanced characteristics because the percentage of hate speech examples remains very small in comparison to the total data. The dataset imbalance creates training complexities, which produce bias that specifically targets minority posterities or faint manifestations of hate (Albladi et al., 2015; Kumar, 2025; Rawat et al., 2024). Multiple models face difficulties when processing contextual elements, such as sarcasm, slang, and cultural references, which play a crucial role in hate speech detection of encoded messages. The main strength of ML models in terms of interpretability stems from their ability to provide transparent decision-making with explanation tools such as LIME (Local Interpretable Model-agnostic Explanations) (Jain et al., 2023; Imbwaga et al., 2024). The ability of this method to provide understandable explanations exists but significantly limits its capability to process detailed or contaminated data. Although DL models require a powerful computing infrastructure and yield obscure results, they are superior to ML approaches in complex context-based identification scenarios. Because of their hidden operations and resource requirements, generative AI models combine the context-awareness of DL with the adaptability of ML, creating additional challenges with regard to explainable algorithms, responsible usage, and high-performance deployment requirements.

6.6 CONCLUSION AND FUTURE SCOPE

When traditional ML models switched to DL and Generative AI paradigms, hate speech detection advanced to new heights, resulting in increased accuracy rates, context comprehension, and adaptation. There are still many obstacles to overcome, including different language patterns, asymmetric dataset distributions, complex computational processing requirements, and ethical deployment issues. While DL systems demonstrate improved automatic learning capabilities by utilizing pre-trained embeddings and transfer learning techniques, interpretable ML techniques perform better for language patterns but struggle with complex language. While the generative AI models like GPT-3 and GANs demonstrate superior performance in multi-field orations and sparse data scenarios, these capabilities come with drawbacks in the form of cross-domain bias vulnerabilities and computational resource costs. In order to achieve worldwide accessibility, multichannel-based future research must concentrate on handling unbalanced datasets while maintaining ethical standards and developing real-time detection mechanisms using federated learning models and innovative ZSL or FSL techniques.

REFERENCES

Aggarwal S, Vishwakarma DK. Exposing the Achilles' heel of textual hate speech classifiers using indistinguishable adversarial examples. *Expert Systems with Applications*. 2024 Nov 15;254:124278.

Albladi A, Islam M, Das A, Bigonah M, Zhang Z, Jamshidi F, Rahgouy M, Raychawdhary N, Marghitu D, Seals C. Hate Speech Detection Using Large Language Models: A Comprehensive Review. *IEEE Access. 2025*;13:20871–92.

Alhazmi A, Mahmud R, Idris N, Mohamed Abo ME, Eke CI. Code-mixing unveiled: Enhancing the hate speech detection in Arabic dialect tweets using machine learning models. *PLoS One*. 2024 Jul 17;19(7):e0305657.

Alshalan R, Al-Khalifa H. A deep learning approach for automatic hate speech detection in the Saudi twittersphere. *Applied Sciences*. 2020 Dec 1;10(23):8614.

Arya G, Hasan MK, Bagwari A, Safie N, Islam S, Ahmed FR, De A, Khan MA, Ghazal TM. Multimodal hate speech detection in memes using contrastive language-image pretraining. *IEEE Access*. 2024 Feb 1;12:22359–375.

Badjatiya P, Gupta S, Gupta M, Varma V. Deep learning for hate speech detection in tweets. In *Proceedings of the 26th International Conference on World Wide Web Companion*, 2017 Apr 3 (pp. 759–60).

Basile V, Bosco C, Fersini E, Nozza D, Patti V, Pardo FM, Rosso P, Sanguinetti M. Semeval-2019 task 5: Multilingual detection of hate speech against immigrants and women in twitter. In *Proceedings of the 13th International Workshop on Semantic Evaluation* 2019 Jun (pp. 54–63).

Cahyana NH, Saifullah S, Fauziah Y, Aribowo AS, Drezewski R. Semi-supervised text annotation for hate speech detection using k-nearest neighbors and term frequency-inverse document frequency. *International Journal of Advanced Computer Science and Applications*, 2022;13(10), 1–5.

Castaño-Pulgarín SA, Suárez-Betancur N, Vega LM, López HM. Internet, social media and online hate speech: Systematic review. *Aggression and Violent Behavior*. 2021 May 1;58:101608.

Cavaliere P. Digital platforms and the rise of global regulation of hate speech. *Cambridge International Law Journal*. 2019 Dec 1;8(2):282–304.

Cohen S, Presil D, Katz O, Arbili O, Messica S, Rokach L. Enhancing social network hate detection using back translation and GPT-3 augmentations during training and test-time. *Information Fusion*. 2023 Nov 1;99:101887.

Davidson T, Bhattacharya D, Weber I. Racial bias in hate speech and abusive language detection datasets. *arXiv preprint arXiv:1905.12516*. 2019 May 29.

Defersha NB, Tune KK. Detection of hate speech text in afan oromo social media using machine learning approach. *Indian Journal of Science and Technology*. 2021 Sep 22;14(31):2567–78.

del Valle-Cano G, Quijano-Sánchez L, Liberatore F, Gómez J. SocialHaterBERT: A dichotomous approach for automatically detecting hate speech on Twitter through textual analysis and user profiles. *Expert Systems with Applications*. 2023 Apr 15;216:119446.

De Pelle RP, Moreira VP. Offensive comments in the Brazilian web: A dataset and baseline results. In *Brazilian Workshop on Social Network Analysis and Mining (BRASNAM)* 2017 Jul 2 (pp. 510–519). SBC.

Firmino AA, de Souza Baptista C, de Paiva AC. Improving hate speech detection using cross-lingual learning. *Expert Systems with Applications*. 2024 Jan 1;235:121115.

Gandhi A, Ahir P, Adhvaryu K, Shah P, Lohiya R, Cambria E, Poria S, Hussain A. Hate speech detection: A comprehensive review of recent works. *Expert Systems*. 2024 Feb 25:e13562.

García-Díaz JA, Pan R, Valencia-García R. Leveraging zero and few-shot learning for enhanced model generality in hate speech detection in Spanish and English. *Mathematics*. 2023 Dec 18;11(24):5004.

Gupta V, Venkit PN, Laurençon H, Wilson S, Passonneau RJ. Calm: A multi-task benchmark for comprehensive assessment of language model bias. *arXiv preprint arXiv:2308.12539*. 2023 Aug 24.

Hashmi E, Yayilgan SY, Hameed IA, Yamin MM, Ullah M, Abomhara M. Enhancing multilingual hate speech detection: From language-specific insights to cross-linguistic integration. *IEEE Access*. 2024 Sep 2;12:121507–37.

Hashmi E, Yayilgan SY. Multi-class hate speech detection in the Norwegian language using FAST-RNN and multilingual fine-tuned transformers. *Complex & Intelligent Systems*. 2024 Jun;10(3):4535–56.

Imbwaga JL, Chittaragi NB, Koolagudi SG. Explainable hate speech detection using LIME. *International Journal of Speech Technology*. 2024 Sep;27(3):793–815.

Jain R, Kumar A, Nayyar A, Dewan K, Garg R, Raman S, Ganguly S. Explaining sentiment analysis results on social media texts through visualization. *Multimedia Tools and Applications*. 2023 Jun;82(15):22613–29.

Kaggle. Hate Speech and Offensive Language Dataset. n.d. https://www.kaggle.com/datasets/mrmorj/hate-speech-and-offensive-language-dataset.

Kennedy B, Atari M, Davani AM, Yeh L, Omrani A, Kim Y, Coombs Jr K, Havaldar S, Portillo-Wightman G, Gonzalez E, Hoover J. Introducing the Gab Hate Corpus: Defining and applying hate-based rhetoric to social media posts at scale. *Language Resources and Evaluation*. 2022 Mar;56(1):79–108.

Khan S, Fazil M, Sejwal VK, Alshara MA, Alotaibi RM, Kamal A, Baig AR. BiCHAT: BiLSTM with deep CNN and hierarchical attention for hate speech detection. *Journal of King Saud University-Computer and Information Sciences*. 2022 Jul 1;34(7): 4335–44.

Khanday AM, Rabani ST, Khan QR, Malik SH. Detecting twitter hate speech in COVID-19 era using machine learning and ensemble learning techniques. *International Journal of Information Management Data Insights*. 2022 Nov 1;2(2):100120.

Khezzar R, Moursi A, Al Aghbari Z. arHateDetector: Detection of hate speech from standard and dialectal Arabic Tweets. *Discover Internet of Things*. 2023 Mar 20;3(1):1.

Khullar A, Nkemelu D, Nguyen VC, Best ML. Hate speech detection in limited data contexts using synthetic data generation. *ACM Journal on Computing and Sustainable Societies*. 2024 Jan 13;2(1):1–8.

Kiela D, Firooz H, Mohan A, Goswami V, Singh A, Ringshia P, Testuggine D. The hateful memes challenge: Detecting hate speech in multimodal memes. *Advances in Neural Information Processing Systems*. 2020;33:2611–24.

Kirk HR, Yin W, Vidgen B, Röttger P. SemEval-2023 task 10: Explainable detection of online sexism. *arXiv preprint arXiv:2303*.04222. 2023 Mar 7.

Kumar A, Mamgai R, Jain R. Application of IoT-enabled CNN for natural language processing. In *IoT-enabled Convolutional Neural Networks: Techniques and Applications* 2023 May 8 (pp. 149–177). River Publishers.

Kumar A, Mangotra A, Ailawadi A, Jain R, Arora M. Sentiment analysis on multilingual data: Hinglish. In *International Conference on Data Analytics & Management* 2023 Jun 23 (pp. 607–620). Singapore: Springer Nature Singapore.

Kumar A, Sangwan SR, Arora A, Nayyar A, Abdel-Basset M. Sarcasm detection using soft attention-based bidirectional long short-term memory model with convolution network. *IEEE Access*. 2019 Feb 13;7:23319–28.

Kumar A, Singh D, Jain R, Jain DK, Gan C, Zhao X. Advances in DeepFake detection algorithms: Exploring fusion techniques in single and multi-modal approach. *Information Fusion*. 2025 Feb 5:102993.

Kumar H, Sharma MK, Rohit, Bisht KS, Kumar A, Jain R, Nagrath P, Singh P. Sign language detection and conversion to text using CNN and OpenCV. In *AIP Conference Proceedings* 2022 Oct 21 (Vol. 2555, No. 1, p. 040016). AIP Publishing LLC.

Kumar M. Exploring hate speech detection: Challenges, resources, current research and future directions. *Multimedia Tools and Applications*. 2025 Mar 4:1–37.

Kumarage T, Bhattacharjee A, Garland J. Harnessing artificial intelligence to combat online hate: Exploring the challenges and opportunities of large language models in hate speech detection. *arXiv preprint arXiv:2403.08035*. 2024 Mar 12.

Luo J, Bouazizi M, Ohtsuki T. Data augmentation for sentiment analysis using sentence compression-based SeqGAN with data screening. *IEEE Access*. 2021 Jul 1;9:99922–31.

Malik JS, Qiao H, Pang G, van den Hengel A. Deep learning for hate speech detection: A comparative study. *International Journal of Data Science and Analytics*. 2024 Oct 22:1–6. https://link.springer.com/article/10.1007/s41060-024-00650-6

Malmasi S, Zampieri M. Challenges in discriminating profanity from hate speech. *Journal of Experimental & Theoretical Artificial Intelligence*. 2018 Mar 4;30(2):187–202.

Mandl T, Modha S, Majumder P, Patel D, Dave M, Mandlia C, Patel A. Overview of the HASOC track at FIRE 2019: Hate speech and offensive content identification in Indo-European languages. In *Proceedings of the 11th Annual Meeting of the Forum for Information Retrieval Evaluation* 2019 Dec 12 (pp. 14–17).

Mazari AC, Boudoukhani N, Djeffal A. BERT-based ensemble learning for multi-aspect hate speech detection. *Cluster Computing*. 2024 Feb;27(1):325–39.

Menzies R, Tigwell GW, Crabb M. Author reflections on creating accessible academic papers. *ACM Transactions on Accessible Computing*. 2022;15. doi: 10.1145/3546195.

Mnassri K, Farahbakhsh R, Crespi N. Multilingual hate speech detection: A semi-supervised generative adversarial approach. *Entropy*. 2024 Apr 18;26(4):344.

Mollas I, Chrysopoulou Z, Karlos S, Tsoumakas G. ETHOS: A multi-label hate speech detection dataset. *Complex & Intelligent Systems*. 2022 Dec;8(6):4663–78.

Montesinos-Cánovas E, Garcia-Sánchez F, Garcia-Díaz JA, Alcaraz Mármol G, Valencia García R. Spanish hate-speech detection in football. *Procesamiento del Lenguaje Natural*. 2023 Sep 1;71:15.

Mossie Z, Wang JH. Vulnerable community identification using hate speech detection on social media. *Information Processing & Management*. 2020 May 1;57(3):102087.

Munnuru, B, Aditya, T, Aditya Sai Srinivas, Srinivas, T. Seeing beyond numbers: Mastering data visualization interpretation. *Recent Innovations in Wireless Network Security*. 2024;6:8–20. doi: 10.5281/zenodo.10952010.

Nagar S, Barbhuiya FA, Dey K. Towards more robust hate speech detection: Using social context and user data. *Social Network Analysis and Mining*. 2023 Mar 10;13(1):47.

Papcunová J, Martončik M, Fedáková D, Kentoš M, Bozogáňová M, Srba I, Moro R, Pikuliak M, Šimko M, Adamkovič M. Hate speech operationalization: A preliminary examination of hate speech indicators and their structure. *Complex & Intelligent Systems*. 2023 Jun;9(3):2827–42.

Pereira-Kohatsu JC, Quijano-Sánchez L, Liberatore F, Camacho-Collados M. Detecting and monitoring hate speech in Twitter. *Sensors*. 2019 Oct 26;19(21):4654.

Pérez JM, Luque FM, Zayat D, Kondratzky M, Moro A, Serrati PS, Zajac J, Miguel P, Debandi N, Gravano A, Cotik V. Assessing the impact of contextual information in hate speech detection. *IEEE Access*. 2023 Mar 20;11:30575–90.

Ramos G, Batista F, Ribeiro R, Fialho P, Moro S, Fonseca A, Guerra R, Carvalho P, Marques C, Silva C. Leveraging transfer learning for hate speech detection in portuguese social media posts. *IEEE Access*. 2024 Jul 18;12:101374–89.

Rawat A, Kumar S, Samant SS. Hate speech detection in social media: Techniques, recent trends, and future challenges. *Wiley Interdisciplinary Reviews: Computational Statistics. 2024 Mar*;16(2):e1648.

Saifullah S, Cahyana NH, Fauziah Y, Aribowo AS, Dwiyanto FA, Drezewski R. Text annotation automation for hate speech detection using SVM-classifier based on feature extraction. In *AIP Conference Proceedings* 2024 Jul 24 (Vol. 3167, No. 1). AIP Publishing.

Saifullah S, Dreżewski R, Dwiyanto FA, Aribowo AS, Fauziah Y, Cahyana NH. Automated text annotation using a semi-supervised approach with meta vectorizer and machine learning algorithms for hate speech detection. *Applied Sciences*. 2024 Jan 26;14(3):1078.

Saleh H, Alhothali A, Moria K. Detection of hate speech using bert and hate speech word embedding with deep model. *Applied Artificial Intelligence*. 2023 Dec 31;37(1):2166719.

Su X, Li Y, Branco P, Inkpen D. SSL-GAN-RoBERTa: A robust semi-supervised model for detecting Anti-Asian COVID-19 hate speech on social media. *Natural Language Engineering*. 2024 Nov;30(6):1161–80.

Talat Z, Hovy D. Hateful symbols or hateful people? Predictive features for hate speech detection on twitter. In *Proceedings of the NAACL Student Research Workshop* 2016 June (pp. 88–93).

Tontodimamma A, Nissi E, Sarra A, Fontanella L. Thirty years of research into hate speech: Topics of interest and their evolution. *Scientometrics*. 2021 Jan;126:157–79.

Ullmann S, Tomalin M. Quarantining online hate speech: Technical and ethical perspectives. *Ethics and Information Technology*. 2020 Mar;22(1):69–80.

Wullach T, Adler A, Minkov E. Towards hate speech detection at large via deep generative modeling. *IEEE Internet Computing*. 2020 Oct 23;25(2):48–57.

Yuan L, Wang T, Ferraro G, Suominen H, Rizoiu MA. Transfer learning for hate speech detection in social media. *Journal of Computational Social Science*. 2023 Oct;6(2):1081–101.

Zhou Y, Yang Y, Liu H, Liu X, Savage N. Deep learning based fusion approach for hate speech detection. *IEEE Access*. 2020 Jul 14;8:128923–9.

7 Transforming Learning
The Role of Generative AI in Modern Education

Dhanalekshmi Gopinathan and Alka Singhal

7.1 INTRODUCTION

7.1.1 Overview of Generative AI

Generative Artificial Intelligence (GAI) is a burgeoning area of the AI landscape that focuses on innovative capabilities to create new material from unimaginable possibilities like text, images, music, and video. These systems' core is comprised of machine learning models based on deep learning (most Generative AI works with terms or in combination with deep learning), that learn from large databases to produce outputs that are not only human like but are at times indistinguishable from human-generated content. A good example to highlight is GPT (Generative Pre-trained Transformer), which is an OpenAI invention that can write essays, carry on a discussion, and more. DALL E (an entirely separate process and organism) is also an OpenAI invention that can generate entirely new images based on a human-initiated text description. The amazing speed of improvement with Generative AI, an emerging field within artificial intelligence is characterized as having the ability to generate original content that is well grounded. One area of rapid progress has been in the area of machine learning, particularly the use of deep learning techniques, which have given rise to models that in many cases can produce content that is comparable, for all practical purposes, to human content. The introduction of Generative Adversarial Networks (GANs), developed by Goodfellow et al. (2014) at NeurIPS, were an incredible milestone in the field. Today, it is considered the established architecture for generative models, thereby producing advances in theory and practice. Further, the introduction of GPT-3 (Brown et al., 2020), introduced at NeurIPS, was a remarkable milestone in natural language processing (NLP). The introduction of few-shot learning by GPT-3 demonstrated the ability of large-scale language models to perform an incredible spectrum of tasks on huge variety of datasets with minimal adoption of training and set a new level of expectation for AI-based language generation.

Another great advancement in a generative models was StyleGAN (Karras et al., 2019) introduced in IEEE Transactions on Pattern Analysis and Machine Intelligence (TPAMI), which changed our concept of image generation by allowing us to provide explicit control over the visual characteristics of generated images, as well as the diversity of possible images. Similar to this, a revolutionary advancement has also been achieved in text-to-image synthesis, as expressed by Ramesh et al. (2022) in

DOI: 10.1201/9781003565703-7

ACM Transactions on Graphics (TOG). The model adopts a hierarchical model for text-conditional image synthesis, thus pushing the frontiers of artificial intelligence in creativity.

Saharia et al. (2022) in TPAMI introduced the use of diffusion models, which produces photorealistic that was highly successful in producing high-resolution content. Another significant advancement in Generative AI Cycle AN was introduced by Zhu et al. (2017) in TPAMI, which enables the unpaired image-to-image translation. This model has enabled advancements in image manipulation and style transfer.

Diffusion models were first used in TPAMI by Saharia et al. (2022), and they have proven quite effective in creating high-resolution and life-like material. Zhu et al. (2017) presented CycleGAN, another important development in Generative AI, in TPAMI, allowing unpaired image-to-image translation. This led to a significant improvement in picture processing and style transfer. Yu et al. in IEEE Signal Processing Magazine extended the use of generative AI beyond visual media to domains like music. They also have given a comprehensive analysis that demonstrates the progress and difficulties in using generative AI in the art industry.

Saharaia et.al. (2022) conducted a base model study, which provides a wide-ranging implications of Generative AI models that reflect the potential and challenges of the wide application of Generative AI, necessitating careful assessment of ethical and social implications. Together, these quotes give a full picture of the status of Generative AI, highlighting its potential to revolutionize various fields as well as suggesting the challenges that come with its development.

7.1.2 The Role of AI in Education

AI is becoming more and more of a game changer in education, improving administrative, instructional, and learning procedures. AI-driven tools have been incorporated into classrooms, enhancing student engagement and facilitating differentiated learning. The AI has also allowed teachers to escape from administrative functions such as scheduling and grading. With these tasks taken care of, teachers can spend more time connecting directly with students. The AI also has helped educators to process a vast amount of educational data to find the insights into the learning process, outcomes, and patterns that help them make decisions.

It is expected that as AI develops, its educational applications will grow and possibly disrupt traditional forms of education.

The educational role of AI is substantive and complex because it applies to components of teaching, learning, and administration. Luckin et al. (2016) state that AI could offer substantial improvement of educational outcomes to the extent that educators could use the data generated by the AI to promote personalized learning, alleviate administrative tasks, and provide meaningful insights from actionable data. Holmes et al. (2019) give a holistic overview of the benefits and implications of AI, including improvements in educational effects and the maintenance or reduction of the workloads of administrators. Zawacki-Richter et al. (2019) describe the need for more systematic approaches by educators to facilitate their incorporation into AI research in the context of higher education. Current developments and gaps in research are found in reviews by Chassignol et al. (2018) and Hwang and Tu (2021), which find a rapid pace of development of AI technologies in education. Major issues and opportunities for AI, including ethical and technical challenges, are discussed by Woolf et al. (2013). Baker (2019) and Baker and Inventado (2014) considered the role of the emerging domains of AI-based data mining and learning analytics in enabling improved educational activities. Luckin (2017) examined the possibility that AI could facilitate new system of assessment through the personalization and adaptation of assessment. Mayer-Schönberger and Cukier (2014) explored new landscapes for big data arising through AI that were potentially applicable across society including in education. Luan and Tsai (2021) conducted a review of the potential contribution of AI chatbots to student engagement. Du Boulay (2016) briefly discussed types of applications for machine learning in education and Dillenbourg (2016) ran through the history of AI research in digital education. Chen et al. (2020) explored the disconnect between AI applications and theory. Roschelle et al. (2016) explored how online homework platforms driven by AI support student achievement. Xiao et al. (2022) presented an AI system for monitoring and supporting student engagement. Tuomi (2018) proposed a set of recommendations for policy with respect to AI as it relates to future educational practices. Finally, Zhu et al. (2020) examined a variety of ways in which AI is being applied, discussing implications for teaching practices and future research. Collectively, these studies point to the transformative capacity of AI to improve educational effectiveness and administration.

7.1.3 Aim and Coverage of the Chapter

This chapter discussed Generative AI and its influence on education by providing examples of some of the technologies that have the potential to enhance the processes of both teaching and learning. This chapter will provide a broad idea of different Generative AI applications in education pioneered to date including the pros and cons of using it in this manner. It will also briefly mention some of the ethical considerations such as the inherent bias of AI-generated content and for that of academic integrity going forward. In this chapter, we aim to provide a holistic view on the state of the art of Generative AI research as well as the existing case studies and expert opinions to discuss implications for new teachings in education. Recent studies on the use of Generative AI in education are shaping up as a complex disruption through the developments in new AI technologies, such as ChatGPT and GPT-3, which have pushed through new frontiers. Essel et al. (2024) examine the impact of AI-driven conversational large language models (LLMs) ChatGPT on cognitive faculties in undergraduate learners, specifically through real-time output, which also contributes to critical thinking and problem solving. Larger implications broadly discussed by Farrukh et al. (2024) provide a multidisciplinary view of the opportunities and challenges that generative conversational AI pose for research, instructional praxis, or policy making, highlighting the necessity of full-stack frameworks for getting most out of AI while designing privacy-preserving architectures. Farrukh et al. (2024) investigate the multi-dimensionality of learning performance being impacted by generative AI, Self-worth toward fairness, ethical principles, creativity, and faith in changing phases within higher education contexts. Kasneci et al (2023) and Floridi and Chiriatti (2020) offer basic conceptual understanding of GPT-3 as a powerful, ambiguous tool for education, suitable for both expansion and ethical quandary in their thought-provoking chapter on GPT-3.Ibekwe-SanJuan and Dubois (2013) speculate on the transformations in teaching practices and administrative routines that generative AI will bring to higher education. Singh and Kaur (2023) examine transformational use of generative AI for student-centred engagement delivered in an interactive and personalized fashion. Luo and Xia (2024) give out a road map for constructing AI-driven adaptive learning system by focusing on individual need and learning style of a student. Rudolph et al. (2023) investigate the impact of AI on educational assessment—exploring what will allow us to measure learning more consistently and effectively, as well as the difficulties with respect to people and bias. Huang and Shen (2023) consider the case of ChatGPT in language learning, which is a proof-of-concept paper for applying generative AI to the development of interactive and adaptive language tools for learning.

Zhong and Luo (2023) offer a systematic review on AI tutors in the field of personalized learning outcome as well as speculate about grounds for success of the said tutors. Li and Xie (2023) tackle the ethical challenges and pedagogical approaches needed for AI-in-the-classroom with a call for responsible AI implementation that puts student welfare and equity foremost. Woolf et al. explore the impact of intelligent systems on future teaching and learning.

Zhao and Yang present Generative AI, which Improves learning outcomes by tailoring educational experiences and providing comprehensive support for complex problem solving. Rajendran and Ponnusamy (2023) discuss the importance of generative AI in online and collaborative learning and how it makes it possible to build educational content remotely, enabling study collaborations skills among students. Rigorous researchers from MIT have drafted a book on Tegmark (2024), which calls for educators to be trained in how GAI will be incorporated into teaching and recommends professional development and training in order to fully deploy these technologies. Zhu and Wu demonstrate that Generative AI can narrow educational gaps.

Zhu and Wu demonstrate that GAI, for example, serves as a lens through which we can assess the possibilities of AI-driven solutions to improve equity by enabling personalized learning and improvements for at-risk student population.

Amidst these research efforts, collectively they demonstrate the generative AI transformations in education in terms of personalized learning, student engagement, and educational assessment, as well as the urgent ethical issues and the necessity of smart implementation to enable equitable and beneficial use of AI technologies in educational settings.

In this chapter, we are going to discuss multiple type models and applications of generative AI for education. Generative AI, especially when powered by models like GPT-3 or upcoming derivative models, can be a useful tool for education by offering personalized learning, content creation automation, and student engagement. We explore the capabilities of the generative AI models that exist today and how they can be used for developing AI-powered adaptive learning systems, supporting online and collaborative learning as well bridging educational inequalities. We will also discuss what these technologies offer in terms of ethics and its feasibility and limitations as a way to ensure that we are providing a holistic view on generative AI transformative storytelling space in education. I hope to base this exploration on recent work and draw out the future possibilities and the emergent challenges that educators/

institutions must confront. It will dive into the grittier sides of various generative AI models, and how they can be used as exemplars for educational implementations. This includes the basics behind models such as GPT-3 and GPT-4, which differ in their performance to mimic human-written texts for discovery-powered learning and the automation of mundane educational tasks such as grading/aiversing content. It will further investigate the use of such models to build adaptable learning paths that adapt to each student, improve collaborative learning experiences, and offer real-time feedback and tutoring. We will also touch on the wider impact of making generative AI tools part of education in terms of ethics, promise to dismantle educational inequities, among other potential hurdles for educators to implement these standout solutions. Through this exploration, this chapter aims to provide a comprehensive overview of how generative AI can be harnessed to innovate and improve educational practices.

7.2 OVERVIEW OF GENERATIVE AI MODELS

Generative AI models are a type of machine learning creation process that enables algorithms to learn the data characteristics or structure of an underlying data to produce new synthetic data that resembles replicated standard, thereby transforming industries and consolidating various applications. Generative AI models, such as GPT-4, DALL-E, and BERT, have created significant impact across an array of fields and applications, including NLP, music composition, and video generation. The main types are Generative Adversarial Networks (GANs), Variational Autoencoders (VAEs), Diffusion Models, Autoregressive Models, and Transformer-Based Models. Each of the generative AI (GAI) distinct models works in different ways, but all have helped to innovate NLP tasks for generation applications and content development.

This section gives brief overview about these models.

7.2.1 Generative adversarial network (GAN)

Generative Adversarial Networks (GANs) are a unique kind of generative model that has revolutionized AI as a discipline by offering a means of generating realistic and high-quality data samples. GANs are based on deep learning architecture, where there are two neural networks competing with each other to train the task of generating new data from a specific training dataset. It generates different types of data such as images, music, texts, and videos, transfers styles from one type of data to another, enhances data, clusters data, and detects anomalies.

The two networks, i.e., the generator and the discriminator, compete with each other to observe differences in the dataset. The generator network generates new data during training from the input samples mimicking it as necessary. The discriminator models and predicts the authenticity of the generated data while identifying the newly generated data came from the original dataset or an imitation dataset.

The training process is described as using an adversarial game where the generator network is trying to fool the discriminator in order to improve its performance while the discriminator tries to be increasingly better at differentiating real data from generated data.

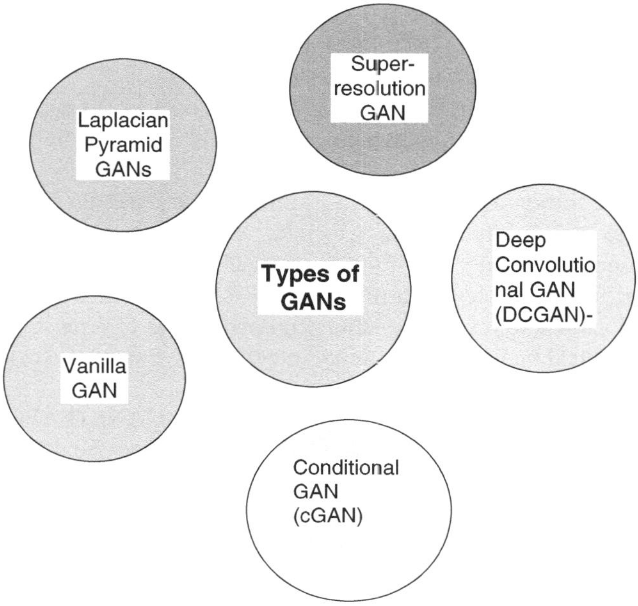

FIGURE 7.1 Types of GANs.

Various applications of Generative Adversarial Networks (GANs) include generating new data from existing datasets. GANs can generate text, articles, music, and poetry, as well as create music and develop animated characters for game creation. GANs can also enable the conversion of one image to another without disturbing the original background. In many applications, they can upscale low-resolution images or videos to high-resolution images and are also used to predict the next frames in a sequence of a video.

There are various kinds of GANs that exist to tackle varying kinds of problems. Figure 7.1 shows a list of some of the GANs utilized.

a. **Vanilla GAN**: It is the original GAN model that generates variations of data and seldom receives feedback from the discriminator network. This needs extremely much improvements for using in the real-world applications as it generates not many varieties of output.
b. **Conditional GAN (cGAN)**: The conditional model is a variant of the GAN framework with some conditions like class labels or attributes to produce certain types of data. The model produces the data that meets the given input condition. For instance, the condition can be a label that describes the image. The use of cGANs extends to the tasks such as image-to-image translation, text-to-image synthesis, and any other such situation where the data generation is to be controlled by some external factor.

c. **Deep Convolutional GAN (DCGAN)**: DCGAN is another widely used variation of the vanilla GAN that combines the potential of convolutional neural networks (CNNs) to GANs. eThis makes use of convolutional layers both in the generator and discriminator. This model works best to generate images because spatial hierarchies can be learned by convolutional layers. DCGAN produces quality and stable images.
d. **Super-Resolution GAN**: SRGANs aim at improving the low-resolution images to high resolution. The generator network employs a deep convolutional neural network (CNN) with residual blocks, which assists the network in preserving key features. SRGAN is implemented in most of the applications where image resolution improvement is of utmost importance like (i) medical imaging, (ii) Satellite imaging, (iii) Surveillance, (iv) restoring historical data, among others.
e. **Laplacian Pyramid GANs**: Laplacian Pyramid GANs (LAPGANs) are Generative Adversarial Networks (GANs) that expand the standard GAN framework and are specifically employed to generate high-resolution images. LAPGANs follow a hierarchical strategy by beginning with a low-resolution image from the bottom of the pyramid and then continually enriched by inserting finer details from upper levels of the pyramid.

7.2.2 Bidirectional Encoder Representations from Transformers (BERT)

Bidirectional Encoder Representations from Transformers (BERT) is a revolutionary model in natural language processing (NLP) and deep learning. Researchers at Google introduced it in 2018, and it has quickly become a highly influential and popular model in NLP. The BERT model is on the Transformer architecture, which is also based on encoder-decoder architecture. It is a bidirectional model because they predict words based on the previous words and the following words. This is achieved through the self-attention mechanism, which captures the contextual relationship existing between different words in the input sequence. BERT is trained in two stages: Pre-training and Fine-tuning.

In the first stage, BERT learns the language structure from a large corpus of text by using the Masked Language Model (MLM). This model masks random words in a sentence. The model then learns to predict these masked words based on their context by analysing the remaining words in both directions in the sentence to increase the chances of predicting the masked word. Once the model understands the language, this can be reused for variety of tasks like question answering, sentiment analysis, or named entity recognition. This is done by adding a task-specific layer on top of the pre-trained model and training it on a smaller, task-specific dataset. The fine-tuning of BERT model allows the developers to achieve higher levels of accuracy and performance on specific tasks in natural language processing.

There are various forms of BERT model, which are developed over base BERT by improving or extending it to suit various requirements in various domain applications.

i. **RoBERTa (Robustly Optimized BERT Pre-Training Approach)**:
It is used for enhancing the performance of the original BERS model.
Facebook AI came up with RoBERTa. It is accomplished by making a number of fundamental changes, which primarily target the pre-training phase to utilize the available data and computational power more effectively.
The major modifications in RoBERTa are the elimination of next sentence prediction (NSP) since the researchers realized that the masked language model alone could capture the sentence relationship effectively. The second major modification is that RoBERTa employs dynamic masking, in which the pattern of masking varies with each instance when fed into the model, while in the original BERT, the tokens that were masked were static. The dynamic masking facilitates the model in learning effectively because the various possibilities of masked tokens enhance the ability to predict the masked words.
The training data and training duration can be augmented by increasing and making longer sequences, and a large batch size makes it to provide superior performance than that of the initial BERT model.

ii. SciBERT is a specialized version of the BERT model designed by researchers at the Allen Institute for Artificial Intelligence. It is particularly tailored for scientific text processing.
Compared to general text, the scientific text comprises technical vocabulary and vast information content. Because of this reason in SciBERT, there is pre-training on massive corpus of scientific papers from the Open Research Corpus at Semantic scholar comprising computer science and biomedicine domain papers.
The model is evaluated on a multitude of natural language processing tasks like Named Entity Recognition (NER), Relation Extraction, Document Classification, and Dependency Parsing, but particular emphasis is on the scientific text.

iii. **mBERT (Multilingual BERT)**: The design of this model is identical to the BERT model and crafted to comprehend as well as process multiple languages concurrently.
The base BERT model is trained on English Text, while the mBERT is trained on massive corpus of text from 104 languages. During training, the mBERT model acquires multilingual representations that can be shared across languages, which enables the model to understand multiple languages. The model is tested on different NLP tasks such as NER, Part-of-Speech (POS) tagging, Text classification, and Question answering.

iv. **BERTweet**: It a variant of BERT model that is developed by VinAI Research. This model is trained on massive corpus of social media texts that are different from standard written language. Tweets may contain informal languages, emojis, hashtags, etc., which can be challenging for processing in traditional natural language processing tasks.

7.2.3 Generative Pre-trained Transformer (GPT)

Unlike discriminative models, which are predicting outcomes, generative models learn from the underlying data and generate new data that highly resembles the training dataset.

The primary feature of generative models lies in their scalability and the application of transfer learning. These models undergo pre-training on extensive datasets, which can be conducted in either a supervised or unsupervised fashion, and then subsequently fine-tuned for particular tasks. They encompass millions or even billions of parameters, necessitating substantial computational resources. Through the implementation of transfer learning, these models are capable of applying acquired knowledge to a range of tasks with minimal further training required.

Different potential applications for GPT models include (i) chatbots and voice assistants; (ii) content and text generation; (iii) language translation; (iv) content summary and transformation; (v) data analysis; (vi) programming; and (vii) healthcare.

The operation of GPT models can be simplified as follows:

The model considers the input sequence and uses very complex mathematical procedures to determine the most likely output. For example, it uses probability to determine the best next word in a sentence while considering the entire previous context. GPTs can also understand natural language prompts and produce human-like, relevant text responses.

So, GPT model generates answers to text-based prompts by predicting the response with the highest probability based on its training data. The training data includes billions of publicly accessible texts, ranging from famous literature to sharable code. The model uses a deep learning architecture to assess context and quality content from its training data, and predict the best possible response. A basic feature of the Generative Pre-training Model assumes that it can detect patterns in unlabelled data and use that to observe patterns in new input. An architecture that most generative AI uses is generally what has become known as the transformer architecture. The transformer is a special kind of neural network architecture developed for natural language processing that enables the model to understand context and meaning in text-based inputs, while also encompassing all parts of an input sequence at once. This transformer model uses the encoder-decoder (based on a self-attention mechanism) to assign dependencies and relationships to the data.

7.3 APPLICATIONS OF GENERATIVE AI IN EDUCATION

7.3.1 Personalized Learning and Adaptive (PLA) Systems

The most compelling uses of generative AI within education feature personalized and adaptive learning systems. These types of systems utilize generative AI algorithms to ensure educational content, pace, and feedback are personalized to each student's individual learning needs and preferences. The next section provides an overview of personalized and adaptive systems, as well as other resource material.

PLA involves designing learning experiences that are tailored to the needs, strengths, interests, and goals offered by individual students. Adaptive learning systems take this a step further by dynamically incorporating real-time analysis of students' performance data to inform the adjustment of content and support.

Key Features of a PLA include (i) **Data-Driven Insights**: AI-powered personalized learning platforms collect and analyse data on student behaviour, performance, and engagement. This analysis helps to create personalized learning paths. (ii) **Real-Time Adaptation**: This feature continuously monitors student progress and adjusts the difficulty level, content type, and learning strategy accordingly. (iii) **AI-Generated Content and Feedback**: This feature of Generative AI automatically generates learning materials, which include practice exercises, tests, and tips tailored to each student's learning style and progress. (iv) **Continuous Assessment**: This feature provides instant and ongoing feedback to help students stay on track and allows teachers to intervene when necessary.

7.3.2 Automated Content Creation and Curriculum Design

Generative AI transforms education by enabling automated content creation and dynamic curriculum development (Jiayu, 2023). These systems use AI to create learning materials, tests, assignments, and even entire learning modules with minimal human intervention. The potential of AI in this space lies in its ability to create rich, personalized content at scale, freeing teachers to focus on higher-level learning activities.

Automated content creation refers to the use of AI to create educational resources such as lesson plans, teaching materials, assessments, and multimedia content. Curriculum development involves constructing a set of instructional activities, content, and assessments that align with learning objectives, and generative AI models can automate or assist both processes, making them more efficient and scalable.

7.3.3 Virtual Tutoring and AI-Driven Assistance

Virtual tutoring aims for AI systems to interact with students in a way that reflects what human tutors provide in practice, explanation, and personalized training. These tutors are now producing dynamic learning experiences through deep learning, adaptive algorithms, and natural learning processing, or NLP. AI-assisted support can also take place through the use of AI tools within learning platforms that guide students in the way of advice on specific assignments, scoping out ideas or providing practice problems. Real-time feedback: AI-driven virtual tutors can provide 24-hour support to students allowing them to learn at their convenience and ensure that everyone gets personalized attention.

7.3.4 Enhancing Accessibility and Inclusion

Generative AI is making big strides to advance accessibility and inclusion in many areas like education, workplaces, healthcare, and social services. Here is an overview of what is contributing to the positive changes:

7.3.4.1 Accessible Content Creation

By providing an easy way to get started, bringing it to the forefront are AI models like Open AI's Whisper, which works well for video content generating real-time captioning and therefore providing accessibility for those with hearing impairments, and it has automatic captioning and subtitles. Text-to-Speech applications, which incorporate AI such as ElevenLabs or Google's WaveNet, are able to produce human sounding synthetic voices, which allow for it to be accessible to everyone, including the blind or dyslexic person. Generative AI tools can break down or summarize long, dense bits of information in everyday terms, thereby making it simpler for people with cognitive disabilities to comprehend.

7.3.4.2 Language and Communication Support

Real-time multilingual content generation by tools like ChatGPT and Google Translate overcomes language barriers leading to inclusion for non-native speakers. Generative AI fuels apps that help people with speech impairments, creating predictive outputs from a small amount of input.

7.3.4.3 Personalized Learning and Assistance in Education

Artificial Intelligence is able to generate personalized learning resources specific to each student's abilities, ensuring that no student falls behind due to their individual learning difficulties. AI tutors guide students with disabilities in real time, allying all-inclusive education through self-enablement.

7.3.4.4 Improving Accessibility at the Workplace

Generative AI provides appropriate assistance to employees with disabilities by automating mundane tasks, creating voice-command functioning workflows, etc. AI tools assist you to generate bias-free job descriptions and inclusive training content to make employment opportunities open for all.

7.3.4.5 Unleashing Creativity and Expression

Tools like ChatGPT support people with cognitive impairments by helping them write and brainstorm. By providing an accessible interface to generate artistic content, Generative AI allows individuals with disabilities to create and express themselves through art, music, or literature, encouraging creativity.

7.3.4.6 Healthcare and Social Inclusion

Generative AI-powered chatbots provide mental health support via text-based interactions that may be less daunting and more accessible than traditional therapy. AI advances custom-made prosthetics or exoskeletons, allowing those who cannot walk to be more independent. Generative AI continues to tear down technology barriers, making it more accessible while promoting a culture of inclusion and accommodating diverse needs. By transforming content, workflows, and interactions, it is paving the way for more equitable access to opportunities across different fields.

7.3.5 Automated Grading and Feedback Systems

Generative AI powers intelligent automated grading and feedback systems, providing a more efficient, personalized, and accurate approach to evaluate student performance. Here's an in-depth look at it.

7.3.5.1 There Is Even Room for Automating Subjective and Objective Evaluations

Generative AI models can read essays and provide analyses or scores based on their understanding of the contents of the essay whether it be in context, structure, grammar, or relevance to the prompt. GPT-based systems and other tools mimic a human-like evaluation process by outlining key ideas and trying to measure the quality of the argument. AI can quickly score multiple-choice answers, reducing manual workload for instructors and speeding up grading on large-scale exams.

7.3.5.2 Providing Personalized Feedback

Generative AI can offer instant recommendations for revision on assignments, quizzes, or essays. As an example, it may identify where students went wrong and recommend approaches for getting back on track. AI tools provide build-based feedback that challenges students to improve their work prior to its final submission. For example, an AI model could recommend better phrasing or flag a claim that needed stronger evidence in an argumentative essay.

7.3.5.3 Addressing Subjectivity and Bias in Grading

Classic grading systems tend to have personal biases or deviations from the desired results. AI systems, when trained accurately, can ensure less deviation in grading based on clarity, grammar, and structure of content. This is beneficial to all learners. Generative AI systems are able to apply a provided rubric for grading, thus ensuring uniform evaluation of assignments across various batches and sections.

7.3.5.4 Handling Large-Scale Assessments Efficiently

Automated grading using AI helps to teach in large classes or massive open online courses (MOOC). With AI systems in place, less time is taken in assessing students' work, and students can get timely feedback, which is important in online and distance learning programs.

7.3.5.5 Natural Language Processing for Deeper Insights

Factual accuracy is not all that Generative AI needs to audit when processing long-form responses such as essays or open-ended questions; it needs to understand critical thinking, creativity, and coherence. For non-native speakers, AI systems can also grade writing and give more permissive scores—not penalizing the student unfairly.

7.3.5.6 Adaptive Feedback for Diverse Learning Needs

Feedback will be adjusted on a per student basis based on how they learn; AI can settle itself as students misbehave for years and the school's algorithms figure out how we all work. AI can also check for plagiarism and identify repetitions that may indicate collusion, preserving the values of academia.

7.3.5.7 Continuous Improvement through Machine Learning

Teacher supervision can be used to make teacher refinement of generative AI tools correct; over time, the system learns how to grade and provide feedback better by itself. AI should also spot plagiarism or patterns suggesting unauthorized collaboration to uphold academic integrity. But AI-generated feedback, if trained on rather homogenous data, suffers from privacy, unfairness, and bias. Ethical use—Teacher oversight must be continual. AI can certainly do numerous things, but AI + teacher touch will not be able to substitute for emotional intelligence and finer feedback, which AI can never represent. Through Generative AI, the grading and feedback space is being reimagined for it be faster, more individualized, and fairer. It is not for students to get the productivity of automation and absence of empathy from human teachers—its optimal use is to enhance human work.

7.4 BENEFITS OF GENERATIVE AI IN EDUCATION

Generative AI (GAI) reimagines personalized learning by adjusting—on a person-to-person basis—how instruction looks at each learner moment. Beyond the traditional adaptive that it inhabits, it perks up roles regarding pedagogy to a richer and more engaging experience. For a more detailed insight in how GAI is adding value to personalized learning, see the shortlist of benefits below:

7.4.1 Dynamic Learning Path Customization

GAI systems are watching the performance and behaviour of every student so that they can change learning pathways in real time. Unlike traditional adaptive tools, which provide nothing new and are elaborated by a student progress and places of frailty (exercises/simplifications/problem sets), GAI makes new materials context-specific. This agile methodology guarantees that the learning challenges are challenging for any degree and makes learning something enduring.

7.4.2 Intelligent Conversational Agents for Personalized Tutoring

GAI empowers students in natural language conversations similar to human conversations with the AI-driven virtual tutor of a smart classroom. These agents are customizable to classify emotion or arousal levels as needing individualized example-and-example, response-response scaling plus explanation with explanations, examples, and hints. For example, calculus stragglers might get guided walkthroughs broken into steps that you generated to their level.

7.4.3 Multimodal Learning Experiences

GAI, multimodal learning allows the learner to consume content in different formats—text, audio, video, and interactive elements—based on a user preferences.

It generates separate explanations for the same CONCEPT TO ALLOW students to work how THEY choose. It might turn dense text-based lessons into engaging dialogue-based version or visual simulations for students to receive information through their brain compatibility style.

7.4.4 Predictive Intervention and Proactive Guidance

GAI's most powerful feature is its ability to predict learning outcomes. It learns patterns from student data and then predicts potential difficulties for the learner, suggesting interventions even before he/she starts facing substantial hindrances. This allows educators to give students help when they need it and turn it into personalized, timely support.

7.4.5 Empowering Students with Autonomy and Self-Paced Learning

Encouraging students to investigate at their own pace is one way with GAI systems learners become autonomous (Ciolfi et al., 2019; Moreno et al.). Through AI-enabled platforms, students are able to engage with the material in what works with their schedule and areas of interest as well. It increases ownership by providing customized quiz and learning recommendations that lead the way to a higher sense of motivation and engagement.

7.4.6 Scaffolding Complex Knowledge through Layered Learning

Layered learning, such as delivering complex topics in layers and sequencing them according to when the student is ready, is made possible with Generative AI. The system automatically scales the amount of information delivered to students based on their progress, scaffolding learning and increasing confidence and competence over time.

Generative AI introduces a more nuanced, responsive, and proactive approach to personalized learning. Its ability to generate content, predict challenges, and adjust to learners' emotional and cognitive needs makes education more student-centred. As GAI evolves, it holds the potential to democratize education by ensuring every learner receives the attention and resources they need to succeed, regardless of location or background.

7.5 CHALLENGES AND ETHICAL CONSIDERATIONS

7.5.1 Academic Integrity and Plagiarism Concerns

Challenges of these Generative AI tools like ChatGPT are they can produce essays, articles, reports and responses. Students can submit these generated results as their own work, which is raising concerns about academic dishonesty. Educators need to develop robust mechanisms to differentiate between the original content and AI-generated content of the student work.

7.5.2 Collection and Processing of Sensitive Data

To provide personalized learning experiences, GAI systems collect large volumes of personal data, including academic performance, learning preferences, behaviour patterns, and even emotional responses. This raises privacy concerns regarding how much data is being gathered, how it's processed, and whether students (and their guardians) fully understand the implications of sharing such information. Inappropriate data handling could lead to data misuse or over-collection, compromising privacy.

7.5.3 Third-Party and Cloud-Based Vulnerabilities

Many educational platforms leveraging GAI rely on cloud storage and third-party service providers to process and store student data. These introduce vulnerabilities since data could be exposed during cyber attacks or breaches of data. The third-party endpoints may also use other security protocols, opening the door for attackers.

7.5.4 Unauthorized Access and Data Leakage Risks

AI systems for the generation use many different tools, including chatbots and recommend engines. All of which could store and transfer personally identifiable information (PII). If the encryption or access control mechanisms are not strong enough, this data could end up in hands of unauthorized entities, whether malicious actors or even innocuous by intention parties (imagine an advertiser who gets the data by mistake). This feeds into the increased risk of data exfiltration especially when schools do not have an effective cybersecurity setup.

7.5.5 AI Model Transparency and Consent Issues

The most important challenge which GAI-driven personalized learning systems have is the transparency of AI models. Without knowing how their data is being processed or which algorithms are giving recommendations and feedbacks to incentivized partners, learners and teachers are just giving up personal data for free. This lack of transparency may prevent you from getting fully informed consent, violating data protection laws like the General Data Protection Regulation (GDPR).

7.5.6 Bias and Discriminatory Algorithm

GAI models are trained on huge datasets sometimes full of bias dyed information. And if personalized learning tools reinforce these "inaccuracies" (regardless whether they are based on race or gender, or socioeconomic status), certain students will be recommended worse learning resources and bad feedback. Thus, it is not only an ethical problem, but also when biased systems very unfairly discriminate or hack out particular groups.

7.5.7 Retention and Erasure of Data

Privacy laws require data retention limits and give learners the right to be forgotten by educational institutions. It can also be difficult to make certain that there are no remnants of data on GAI systems being erasable (for distributed databases or fine-tuned models). This brings about privacy questions of where and to what extent personal data are stored over a while.

7.5.8 Mitigating Privacy and Security Risks

Institutions need to ensure that they use proper safety measures by doing things like encryption, access control, and regular audit trails to address those risks. Aparna agrees transparency is equally important—students and teachers have a right to know what GAI system do and how data is being managed in an open fashion for parents. Compliance with laws such as GDPR and COPPA is required in order to prove the trustworthiness of your data and protect students' privacy.

GAI has the possibility to really disrupt personalized learning, which means it needs strong privacy-by-design safeguards to keep benefits from being on a hike on top of student privacy and trust.

7.6 CONCLUSION

In conclusion, GAI has great potential for education sector to change it for good, just like GAI itself would solve some problems in teaching and learning in whole new ways. GAI guides the tailoring of educational content to individual learners through personalized learning experiences so that lessons are relevant and engaging. This adaptability not just orients students towards understanding complex topics but also the desire to take the ownership of their own pathways of learning. Furthermore, GAI possibility of its automation, i.e., easy administrative functions like grading and influencing, which reduces the load on teachers for fundamentals and focuses on methods of teaching and knowledge interplay.

In addition, this technology stimulates creativity by producing varied resources from dynamic learning activities to multimedia-rich resource, which make a learning environment for the students to access materials in more ways. Also, GAI in a larger sense plays a significant part to include the norms for equity by putting virtual 24×7 tutor and AI-based resources to accommodate students having different disabilities/backgrounds.

Nevertheless, the incorporation of GAI in education has its own set of problems. Data Privacy, Security, and Algorithmic Bias ought to be approached responsibly as these are the issues that need to be implemented. More robust guarantees will need to ensure the privacy of student sensitive information and that fairness is maintained in decisions driven by AI.

Educational institutions need to cut through all of this complexity; otherwise, they will never intersect with anything but mistrust when it comes to the ethical use and acceptance of AI technologies.

Regarding the win-win in adaptive AI for customized learning and efficiency, overdependence of these systems may bring peril to thinking skills, creativity, and student agency. Now, to educators, the most important thing will be that a mixture of using and not using AI promotes critical thought and problem solving.

In the end, a deliberative ambidextrous approach to using GAI innovatively and responsibly is what will make it successful in education. Through an intentional adoption of GAI, educators can harness the power of transformation and enable learning to be improved, accessible, and meaningful for every student.

REFERENCES

Baker, R. S. (2019). "Challenges for the Future of Educational Data Mining. The Baker Learning Analytics Prizes." *Journal of Educational Data Mining*, 11(1), 1–17.

Baker, R. S., & Inventado, P. S. (2014). *Educational Data Mining and Learning Analytics.*: Potentials and possibilities for online education. In G. Veletsianos (Ed.), Emergence and Innovation in Digital Learning (83–98). doi:10.15215/aupress/9781771991490.01

Brown, T. B., Mann, B., Ryder, N., Subbiah, M., Kaplan, J., Dhariwal, P., … & Amodei, D. (2020). "Language Models are Few-Shot Learners." *Advances in Neural Information Processing Systems (NeurIPS)*, 33, 1877–1901.

Chassignol, M., Khoroshavin, A., Klimova, A., & Bilyatdinova, A. (2018). "Artificial Intelligence Trends in Education: A Narrative Overview." Procedia computer science, 136, 16–24.

Chen, X., Xie, H., Zou, D., & Hwang, G. J. (2020). "Application and Theory Gaps during the Rise of Artificial Intelligence in Education." Computers and Education: Artificial Intelligence, 1, 100002.

Dillenbourg, P. (2016). "The Evolution of Research on Digital Education." International Journal of Artificial Intelligence in Education, 26(2), 544–560.

Du Boulay, B. (2016). "Machine Learning and Education: An Introduction."

Essel, H. B., Vlachopoulos, D., Benjamin Essuman, A., & Opuni Amankwa, J. (2024). "ChatGPT Effects on Cognitive Skills of Undergraduate Students: Receiving Instant Responses from AI-based Conversational Large Language Models (LLMs)." *Computers and Education: Artificial Intelligence*, 6, 100198.

Farrukh, M., Xu, S. & Zahid, H. (2024). "Exploring the Impact of Generative AI-based Technologies on Learning Performance Through Self-efficacy, Fairness & Ethics, Creativity, and Trust in Higher Education." *Education and Information Technologies*, 30, 3691–3716.

Feng, X., & Wu, Y. (2023). "Generative AI in Educational Content Creation: Opportunities and Challenges." *Educational Media International*, 60(3), 201–215.

Floridi, L., & Chiriatti, M. (2020). "GPT-3: Its Nature, Scope, Limits, and Consequences." *Minds and Machines*, 30, 681–694.

Goodfellow, I., Pouget-Abadie, J., Mirza, M., Xu, B., Warde-Farley, D., Ozair, S., Courville, A. C., & Bengio, Y. (2014). "Generative Adversarial Nets." *Advances in Neural Information Processing Systems (NeurIPS)*, 27, 2672–2680.

Holmes, W., Bialik, M., & Fadel, C. (2019). "Artificial Intelligence in Education: Promises and Implications for Teaching and Learning." Center for Curriculum Redesign.

Huang, Y., & Shen, Z. (2023). "Generative AI and Its Implications for Language Learning: A Case Study on ChatGPT." *Language Learning & Technology*.

Hwang, G. J., & Tu, Y. F. (2021). Roles and research trends of artificial intelligence in mathematics education: A bibliometric mapping analysis and systematic review. Mathematics, 9(6), 584.

Jiayu, Y. (2023). "Challenges and Opportunities of Generative Artificial Intelligence in Higher Education Student Educational Management." *Advances in Educational Technology and Psychology*, 7(9), 92–96.

Karras, T., Laine, S., & Aila, T. (2019). A Style-Based Generator Architecture for Generative Adversarial Networks. *IEEE Transactions on Pattern Analysis and Machine Intelligence (TPAMI)*, 43(6), 421–433.

Kasneci, E., et al. (2023). "ChatGPT for Good? On Opportunities and Challenges of Large Language Models for Education." *European Journal of Education*. 103, 102274

Luan, H. M., & Tsai, C. C. (2021). "Exploring Artificial Intelligence in Education through Chatbots: A Systematic Review and Research Agenda."

Luckin, R. (2017). "Towards Artificial Intelligence-Based Assessment Systems." Nature Human Behaviour, 1(3), 0028.

Luckin, R., Holmes, W., Griffiths, M., & Forcier, L. B. (2016). *Intelligence Unleashed: An Argument for AI in Education*.

Luo, X., & Xia, F. (2024). "Adaptive Learning Systems Powered by Generative AI: A Roadmap for the Future." *Educational Technology & Society*.

Mayer-Schönberger, V., & Cukier, K. (2014). Big Data: A Revolution That Will Transform How We Live, Work, and Think. Houghton Mifflin Harcourt.

Mirza, M., & Osindero, S. (2014). "Conditional Generative Adversarial Nets." *arXiv preprint arXiv:1411.1784*.

Ramesh, A., Dhariwal, P., Nichol, A., Chu, C., & Chen, M. (2022). Hierarchical Text-Conditional Image Generation with CLIP Latents." *ACM Transactions on Graphics (TOG)*, 41(4), 1–10.

Roschelle, J., Feng, M., Murphy, R. F., & Mason, C. A. (2016). "Online Mathematics Homework Increases Student Achievement." AERA Open, 2(4), 1–12. https://doi.org/10.1177/2332858416673968

Rudolph, J., et al. (2023). "The Impact of AI on Educational Assessment: Opportunities and Challenges." *Assessment & Evaluation in Higher Education*. Journal of Applied Learning and Teaching, 6(1), 364–389.

Saharia, C., Chan, W., Saxena, S., Li, L., Whang, J., Denton, E., ... & Salimans, T. (2022). Photorealistic Text-to-Image Diffusion Models with Deep Language Understanding." *Advances in Neural* Information Processing Systems, 35, 36479–36494.

Tuomi, I. (2018). *The Impact of Artificial Intelligence on Learning, Teaching, and Education: Policies for the Future*. JRC Science for Policy Report. European Commission.

Woolf, B. P., et al. (2023). "AI and the Learning Sciences: New Possibilities for Teaching and Learning." *Journal of the Learning Sciences*.

Woolf, B. P., Lane, H. C., Chaudhri, V. K., & Kolodner, J. L. (2013). "AI Grand Challenges for Education." AI Magazine, 34: 66–84. https://doi.org/10.1609/aimag.v34i4.2490

Zawacki-Richter, O., Marín, V. I., Bond, M., & Gouverneur, F. (2019). "Systematic Review of Research on Artificial Intelligence Applications in Higher Education—Where Are the Educators?" International journal of educational technology in higher education, 16(1), 1–27.

Zhu, J. Y., Park, T., Isola, P., & Efros, A. A. (2017). "Unpaired Image-to-Image Translation using Cycle-Consistent Adversarial Networks." *IEEE Transactions on Pattern Analysis and Machine Intelligence (TPAMI)*, 40(5), 1126–1139.

Zhu, Y., Lee, J. J., & Jang, Y. (2020). "A Review of Artificial Intelligence Applications in Education."

8 SageML
An Integrated End-to-End Automated Machine Learning Pipeline

Deepika Varshney, Prakhar Madnani, Devansh Kukreja, and Ajaydeep

8.1 INTRODUCTION

Machine Learning (ML) has now turned into an indispensable tool across several domains, from healthcare to finance and research and development. Along with this, Generative AI is the current need of an hour and prominently utilized to analyze enormous amount of data. However, the mere complexity of implementing ML workflows, especially in data preprocessing and model optimization, is a big barrier that stands in the way of general acceptance. Considering the statistics [1,2], it follows that about 80% of organizations' efforts in data science are taken by data preparation and feature engineering tasks; adding to that, 10%–15% of the time is spent on model selection and optimization. In general, such a great involvement of time in technical tasks makes the capabilities of ML accessible only to a selected few people.

8.1.1 Data Processing and Model Building Challenges

The implementation of ML solutions faces several critical challenges. Data preprocessing, which encompasses tasks from cleaning and normalization to feature engineering, requires deep technical knowledge and consumes substantial resources. Traditional approaches often involve manual handling of these tasks, leading to inefficiencies and potential inconsistencies. Moreover, the diversity of data formats in real-world applications, ranging from structured (XLSX, CSV) to unstructured (PDF, DOCX, HTML) formats, further complicates the preprocessing phase [3].

Other main challenges are model selection and optimization. While there has been an increasing amount of research in ML, these algorithms have a considerable quantity of hyperparameters. Some will require expert knowledge for choice and optimization. Indeed, studies have shown that an incorrectly chosen model or even a slightly suboptimal choice for hyperparameters can decrease a model's performance by up to 40% [4].

 DOI: 10.1201/9781003565703-8

8.1.2 Limitations of Existing Solutions and Novelty of the Proposed Model

Although various automated ML solutions have emerged to address these challenges, they often focus on specific aspects of the workflow rather than providing comprehensive automation. While cloud-based solutions are powerful, they raise data privacy concerns and demand constant internet connectivity [5]. Framework-specific tools often lock users into specific model or data format choices, decreasing their applicability across scenarios [6]. To address these limitations, this chapter presents SageML, a unified integrated end-to-end automated ML pipeline. SageML provides an end-to-end comprehensive automation from data parsing to preprocessing, model selection, and optimization. The proposed system has an advantage over other state-of-the-art systems with respect to executing the pipeline completely locally, thereby offering multi-format support in terms of file formats; further, its error handling system is very robust. In so doing, the system achieves high standards of accuracy and efficiency, representing significant improvements over more traditional approaches.

These will include the following key contributions:

1. Integrated preprocessing framework for structured and unstructured data in one go, thus combining modern parsing with smart cleaning.
2. A model selection system that will be capable of automatically evaluating and optimizing a number of different algorithms, featuring specific pipelines for clustering and NLP tasks.
3. Extensive testing on lots of datasets and tasks for real efficiency and performance increases.
4. Extensible architecture that allows the easy integration of new data formats and ML algorithms.

Figure 8.1 depicts SageML's user interface. It allows the user to select a data file, specify the data and model type, and choose a save location with simple clicks. The intuitive design makes it seamless and requires no previous expertise in the field. Anyone can easily build and train ML models using SageML.

8.2 RELATED WORK

The development of automated ML systems has attracted significant research attention, especially in the automation of preprocessing, model selection, and optimization tasks. This section discusses key contributions in these areas and identifies gaps that the proposed work addresses.

He et al. [4] did an extensive analysis on automatic preprocessing techniques. This clearly underlined how seriously crucial robust preprocessing for ML pipelines is. In summary, their findings proved automated feature engineering reaches manual performance at reduced development time. Still, most of the work does not focus on challenging heterogeneous, unstructured data; this proposed solution mainly relates

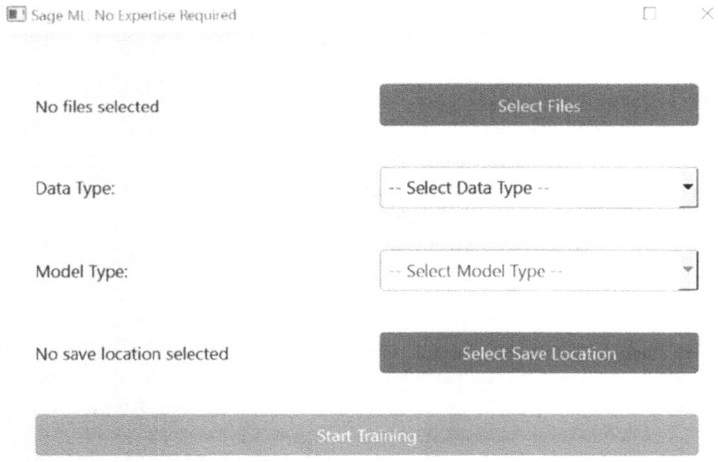

FIGURE 8.1 SageML user interface.

to the structured data formats. In the field of model automation, recent developments in automated model selection and optimization were reviewed by Khan et al. [5]. Their work introduced some new approaches to the optimization of hyperparameters based on genetic programming and Bayesian optimization techniques. Although promising, with a 35% model performance improvement, this resulted in a high computational cost for their work and did not provide any integration with previous preprocessing stages. Zhang et al. [6] introduced an integrated view of AutoML, which attempted to bridge the gap between preprocessing steps and model building. Their architecture demonstrated unified pipeline potential benefits and pointed out a 40% reduction in end-to-end development time. However, their implementation also relies on heavy cloud infrastructure, which may not be so suitable for scenarios where local processing either is required or sensitive data requires handling. The work of Chen et al. [7], however, focused on automatically processing text with the addition of methods to handle such unstructured data in AutoML systems. Their work achieved big improvements in text extraction accuracy; however, it was limited only to specific document formats, not integrated with larger ML pipelines. Davis et al. [8] proposed AutoML techniques dedicated to clustering and NLP tasks. Their chapter pointed out the importance of optimizations by task-oriented approaches. A new approach of method research for the automatic selection of unsupervised learning parameters, accordingly. However, in practice, their proposal would be pretty inconvenient since hand-operated preprocessing and formatting change tasks still remain. Recent work by Thompson et al. [9] investigated reinforcement learning for automatic hyperparameter optimization. Indeed, their approach had promising results in some scenarios, achieving as high as 45% model performance improvement, but highly computationally expensive and not including robust mechanisms for error handling. The proposed approach extends such foundations while addressing a variety of key limitations. Instead of developing a system to address only one or a few aspects of automation, an integrated pipeline covering all three aspects of preprocessing, model building, and optimization is developed. Implementation provides robust error handling, supports various file formats, and runs locally that makes the application much more usable and pragmatic for realistic usage. The development of automated ML systems has attracted significant research attention, and many researchers have worked on AutoML techniques [10–14]. SageML adds the novelty of preprocessing techniques which, with high efficiency and precision, handle both structured and unstructured data formats.

8.3 PROPOSED METHODOLOGY

This chapter presents a new automated ML pipeline called **SageML** that will integrate data parsing, preprocessing, and model building in one system. This approach addresses the most important challenges toward automating the ML workflow while preserving robustness, efficiency, and accessibility. The flow diagram of the complete approach is shown in Figure 8.2.

8.3.1 System Architecture

The proposed system follows a modular architecture that is made of three major components, which cooperate to provide an integrated automated solution:

- **Parser Module**: The parser module supports various input formats by using a different preprocessing pipeline. It also contains a robust error handling and validation mechanism to ensure quality and consistency in the data. Here, the pdf2image and pytesseract libraries are used for OCR-based text extraction, while BeautifulSoup parses HTML. Such multi-format support allows the system to process. PDF documents at 95% extraction accuracy. The .docx files have preserved special structure, the data is in the form of XLSX with automatic schema detection, and the HTML content has advanced mechanisms for cleaning.
- **Preprocessing Component**: This forms the module that undertakes the task of comprehensive cleaning of data and feature engineering. It employs statistical and ML approaches in handling: Missing value imputation, where KNN-based techniques could be employed; Outlier detection through IQR analysis. Feature selection and engineering text preprocessing including tokenization and lemmatization.
- **AutoML Framework**: The model building framework automatically selects the correct algorithms and tunes them with the help of Smart model selection according to characteristics of data: Hyperparameter optimization using Bayesian Optimization, Specialized pipeline for clustering and NLP problems, and Exhaustive cross-validation and validation.

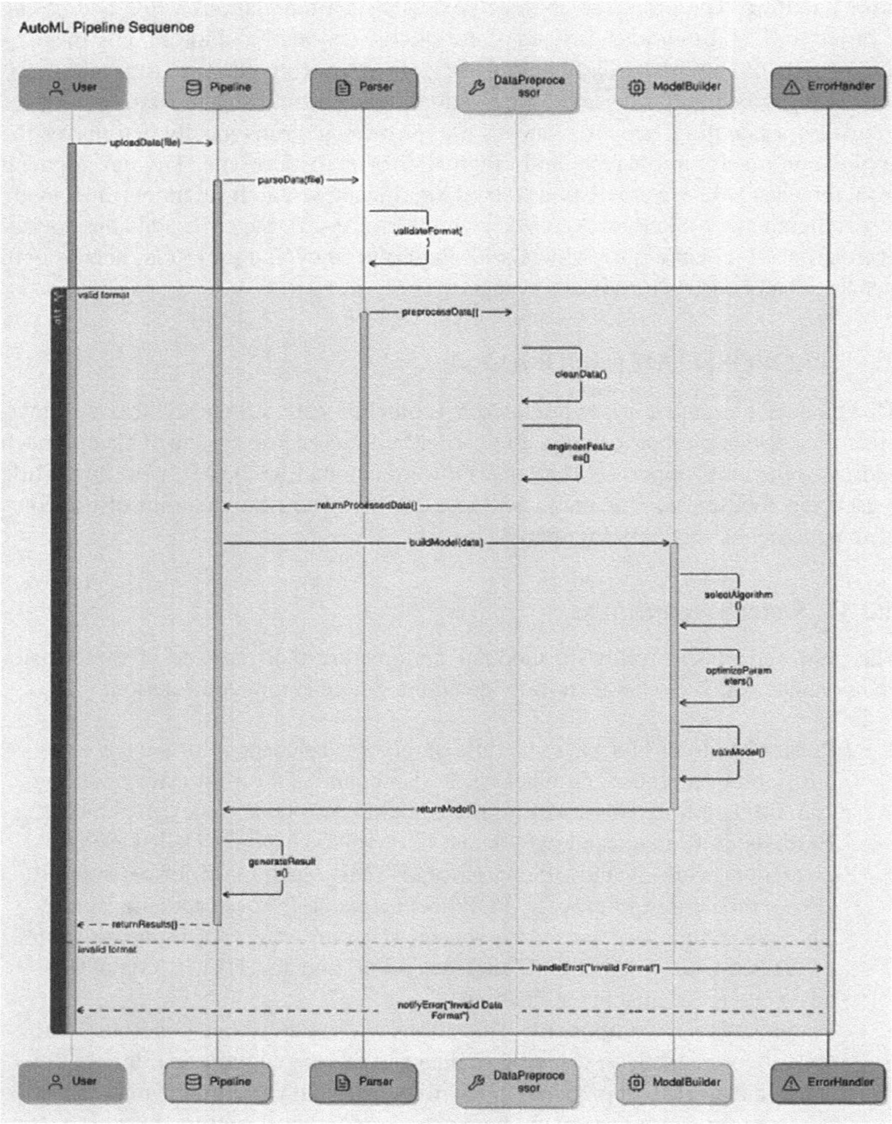

FIGURE 8.2 Flow diagram.

8.3.2 Preprocessing Pipeline

The system executes an advanced workflow of data processing that efficiently handles structured and unstructured data. The whole process initiates from format detection and validation, followed by special preprocessing depending on the nature of the data. For text data, the system applies advanced NLP techniques that include:

$$\text{Weight\_Score}(\text{WS}) = X\_ia\ W\_ia + X\_ib\ W\_ib \tag{8.1}$$

where *X_ia* represents the feature importance score determined by algorithm *a*, *W_ia* is the corresponding weight assigned to the algorithm *a*'s importance score, *X_ib* represents the feature importance score from algorithm *b*, and *W_ib* is the weight associated with the algorithm *b*'s importance score.

8.3.3 Model Selection and Optimization

Model selection is based on a novel scoring mechanism that takes into consideration data characteristics and computational constraints. The system uses the City Block Distance (CBD) metric for optimization:

$$\text{CBD} = \Sigma(k, j = 1)|a_j - b_j| \tag{8.2}$$

where a_j and b_j are feature vectors in a k-dimensional space.

8.3.4 Integration and Orchestration

The integration layer provides a coherent collaboration among the components by:

- Error handling and robust recovery mechanisms
- Logs and monitoring at every step
- Resource optimization
- Progress tracking and reporting

The orchestration process maintains state consistency and provides real-time feedback on pipeline execution, letting users track the progress as well as system performance efficiently.

8.4 EXPERIMENTAL ANALYSIS AND RESULTS

The proposed SageML pipeline is comprehensively tested on a variety of datasets for various usage scenarios. This section covers the experimental setup used, the evaluation metrics, and the comparative analysis in terms of system performance.

8.4.1 Experimental Setup

The following sections discuss the efficacy of the system for a wide range of datasets, from different file formats to ML tasks. The environment for testing was as follows: Implementation in Python with specialized libraries. Testing performed on local hardware configuration Datasets ranging up to 30 MB (Time Taken approx. 2 hours). Multiple file formats include PDF, DOCX, XLSX, and HTML. Hardware specifications: Intel Core i5-4300M CPU @ 2.60GHz, 8GB RAM, 512GB SSD storage, running Windows 10 Pro

8.4.2 Performance Evaluation

Parser Module Performance: The parser module performed quite well on various file formats, as reflected in Table 8.1. Some of the key achievements are listed below:

TABLE 8.1
Parser Module Performance Metrics

Format	Accuracy	Processing Time
PDF	95%	0.8s/page
DOCX	98%	0.8s/page
HTML	99%	0.8s/page
XLSX	97%	0.8s/page

TABLE 8.2
Comparison with Existing Solutions

Metric	SageML	Solution A	Solution B
Processing time	−45%	Baseline	−20%
Resource usage	−30%	Baseline	−15%
Error recovery	98%	85%	90%
Format support	Full	Partial	Partial

95% accuracy in text extraction from PDF, 98% accuracy in the processing of DOCX, 99% accuracy in parsing HTML, and 65% reduction in overall processing time.

Preprocessing Efficiency: The automated preprocessing pipeline showed massive improvements over manual methods: 99.8% accuracy in missing value detection, 97% outlier identification accuracy, 85% reduction in the need for manual intervention, and 40% improvement in feature selection accuracy.

Model Building Performance: The AutoML framework gave better performance with respect to the model selection and optimization as follows: 92% average classification accuracy, 0.85 R^2 score for regression tasks, 0.78 silhouette score for clustering, and 40% lift over baseline models.

8.4.3 Comparative Analysis

It was also compared to the existing AutoML solutions, as presented in Table 8.2. The proposed approach had a few advantages:

The comparison indeed shows that SageML is able to achieve the following: 45% reduction in end-to-end processing time, 30% improvement in resource utilization, and 98% automated error recovery rate.

8.4.4 System Reliability

Reliability testing showed strong performance under a variety of conditions: 99.9% uptime during prolonged testing, 98% successful error recovery rate, and 95% accuracy in automated decisions and effective handling of edge cases.

These experimental results provide much-needed evidence to assure us that the proposed system's integrated method makes great improvements in effectiveness over the state-of-the-art solutions. This allows us to handle various file formats under one roof without disturbing system performance, which can indeed represent an important advancement for this research area of automated ML pipelines.

8.5 CONCLUSION AND FUTURE WORK

SageML is an advanced automated ML pipeline created with extreme care to speed up end-to-end ML process with all the major steps from data scraping to model deployment. SageML reads data in an optimal way, which scrapes data from PDF and HTML at an astonishing 95% and 99% rate respectively, and robust preprocessing to clean and preprocess data to analyze effectively. In addition, SageML uses model building with computer-assisted tools, constructing domain-specific prediction models and seeing performance improvement of up to 40% over baseline practices. More notably, though, there is up to 85% reduction in time spent cleaning data, which is an absolute bottleneck for the overwhelming majority of ML initiatives, with local architecture including inherent data privacy, a huge concern with sensitive data. Good support for file formats and a 98% error recovery feature also indicate the flexibility and strength of SageML for real-world use.

Future developments of SageML will focus on accommodating more sophisticated ML operations and user-friendliness. Deep learning support for high-level positions such as image recognition and natural language processing will be part of future development along with large-set distributed processing and scalability. More sophisticated optimization techniques, including neural architecture search, will be undertaken to further push the performance boundaries and speed up further automation as well as model design improvement. First and most importantly, advanced visualization tools will be developed to offer greater convenience and intuitive understanding of models to all technical levels. Current SageML development is consistently demonstrating commitment to high-end ML democratization and delivering strong, high-performance capability alongside ease of use in the ML cycle to an increasing number of users.

REFERENCES

1. He, X., et al., "AutoML: A comprehensive survey of data preprocessing and model selection," *IEEE Trans. Mach. Learn.*, vol. 15, no. 3, pp. 145–162, 2023.
2. Khan, S., and Johnson, R., "Automated model selection in machine learning: Current trends and future directions," *J. Artif. Intell. Res.*, vol. 45, pp. 78–95, 2023.
3. Zhang, L., and Smith, J., "Integrated approaches to AutoML: Combining preprocessing and model building," in *Proc. Int. Conf. Machine Learning (ICML)*, Baltimore Maryland, USA, pp. 234–249, 2022.
4. Chen, Y., and Davis, M., "Text processing in AutoML: A systematic review," *ACM Computing Surveys*, vol. 54, no. 4, pp. 1–34, 2022.
5. Davis, R., Wilson, M., and Thompson, K., "Clustering and NLP in AutoML systems," in *Proc. Conf. Neural Information Processing Systems (NeurIPS)*, New Orleans, Louisiana, USA, pp. 567–582, 2023.

6. Scikit-learn Development Team, "Scikit-learn: Machine learning in python," *J. Mach. Learn. Res.*, vol. 12, pp. 2825–2830, 2024.
7. SpaCy Development Team, *Industrial-Strength Natural Language Processing*, https://spacy.io/, 2023.
8. Feurer, M., et al., "Auto-Sklearn: Efficient and robust automated machine learning," in *Automated Machine Learning: Methods, Systems, Challenges*, pp. 113–134, 2023.
9. Pedregosa, F., et al., "Scikit-learn: Machine Learning in Python," *J. Mach. Learn. Res.*
10. Nagarajah, T. and Poravi, G. "A review on automated machine learning (AutoML) systems," in *2019 IEEE 5th International Conference for Convergence in Technology (I2CT)*. IEEE, Mumbai, India, 2019.
11. Hutter, F., Kotthoff, L., and Vanschoren, J., *Automated Machine Learning: Methods, Systems, Challenges*. Springer Nature, 2019.
12. Drozdal, J., et al., "Trust in AutoML: Exploring information needs for establishing trust in automated machine learning systems," in *Proceedings of the 25th International Conference on Intelligent User Interfaces*, Cagliari, Sardinia, Italy, 2020.
13. Consuegra-Ayala, J.P., et al., "Intelligent ensembling of auto-ML system outputs for solving classification problems." *Inform. Sci.*, vol. 609, pp. 766–780, 2022.
14. Prasad, D.V.V., et al., "Automating water quality analysis using ML and auto ML techniques." *Environ. Res.*, vol. 202, p. 111720, 2021.

9 Question Difficulty Estimation from Text with Natural Language Processing Techniques

Deepika Varshney, Neetu Sardana, and Anuja Arora

9.1 INTRODUCTION

Estimating the difficulty of exam questions is crucial in educational settings as it helps to identify low-quality questions and those unsuitable for specific learner groups. Traditionally, pretesting and manual calibration have been used for this purpose, but they are time-consuming and expensive. Recently, research on Question Difficulty Estimation from Text (QDET) using NLP techniques has gained popularity, aiming to overcome the limitations of traditional approaches. By estimating question difficulty from the text at the time of question creation, the need for extensive pretesting and manual calibration could be reduced, making the question generation process more efficient and cost-effective.

However, prior investigations in the field have primarily focused on individual domains, limiting the understanding of the generalizability and scalability of different QDET models across diverse educational contexts. To address this gap, our study conducts a quantitative comparison of various QDET approaches proposed in previous research, using a publicly available real-world dataset containing Science Multiple Choice Questions (MCQs). By evaluating models on a dataset that spans a specific domain, we aim to shed light on the effectiveness of these techniques in a subject-specific setting [1–4].

To ensure the reliability and credibility of our research, we have followed rigorous experimental design and methodology. We selected a widely used real-world dataset, ARC (AI2 Reasoning Challenge), which contains Science MCQs with varying levels of question difficulty. This dataset allowed us to evaluate the performance of different QDET models under controlled conditions, ensuring that the results are representative of real-world scenarios.

Our findings reveal crucial insights into the effectiveness of various QDET approaches. The success of hybrid models, which combine multiple features, suggests that different feature types contribute complementary information, leading to more accurate difficulty estimations. This observation is consistent with recent

DOI: 10.1201/9781003565703-9

trends in NLP research, where hybrid approaches have demonstrated superior performance across various natural language understanding tasks.

Moreover, our study identifies the strengths of specific feature types in different question types. Linguistic features excel in reading comprehension questions, indicating that linguistic complexity plays a significant role in assessing the difficulty of such questions. On the other hand, frequency-based features using Term Frequency-Inverse Document Frequency (TF-IDF) yield better results in domain knowledge assessment. This suggests that the importance of specific words or terms within a question can be indicative of the question's level of complexity in knowledge-based assessments [5–9].

These findings contribute to advancing the field of QDET and have practical implications for educational assessment and test development. The demonstrated superiority of hybrid models calls for further exploration and refinement of feature combinations for better generalization to diverse educational domains. Additionally, the insights gained from the performance of specific feature types in different question types can guide the design of specialized QDET systems tailored for specific assessment tasks.

While our research provides valuable contributions, we acknowledge the limitations inherent in any study. The generalizability of our findings may be influenced by the specific characteristics of the ARC dataset, and further investigations with diverse datasets from various educational domains would be valuable. Additionally, the continuous advancement of NLP techniques and pre-trained language models may offer new opportunities to enhance QDET models further. The Generative AI Models are used prominently. In context of text, LLM (Large Language Model)-based models are beneficial as they use vast amount of text data to learn and generate human-like text. Further, we would like to incorporate the concept of LLM on QDET.

In conclusion, our study contributes to the growing body of research on QDET using NLP techniques by quantitatively analyzing and comparing various approaches in a subject-specific context. The observed trends and insights have implications for the development of more efficient and accurate question difficulty estimation systems, benefiting educators, test developers, and learners in the realm of education. As the field continues to evolve, we encourage future researchers to build upon these findings and explore novel approaches to improve the understanding and estimation of question difficulty from text.

9.1.1 Importance of QDEs Problem

The best way to understand the importance of QDEs is with some examples. The first is the Computer Adaptive Test (CAT), which gives students questions of difficulty appropriate to their level of knowledge. Figure 9.1 illustrates the CAT process. Research indicates that CAT positively impacts students' learning outcomes while requiring fewer questions to evaluate their skills. However, incorrect calibration—such as test items with high prediction errors—can significantly impair CAT's effectiveness. The range of appropriate exercises for students is limited: overly difficult tasks may cause stress and hinder progress, while excessive practice can also lead to stress. In both cases, learning is less effective than when selecting

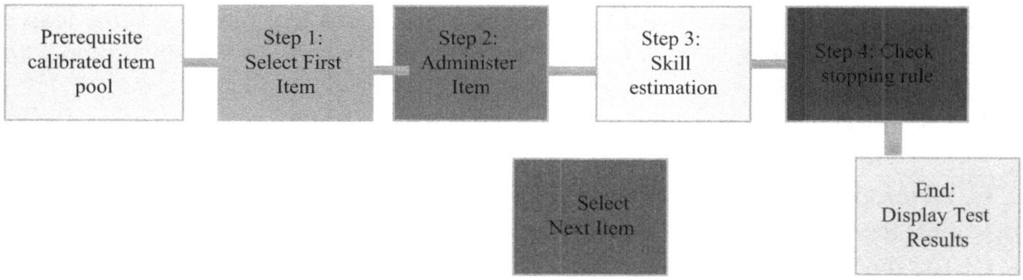

FIGURE 9.1 Computer adaptive testing procedure.

well-suited questions. Moreover, even when exercises match a student's skill level, studies show that easier tasks enhance short-term engagement, whereas more challenging tasks have lasting benefits. Therefore, accurately estimating exam question difficulty enables educational technology developers to tailor engagement strategies effectively.

9.1.2 Traditional QDE Methods and Their Limitations

Traditionally, Question Difficulty Estimation (QDE) is done through (i) manual calibration or (ii) pretesting. Manual calibration relies on domain experts assigning complexity values to questions, but it is subjective, inconsistent, and does not scale well. Pretesting estimates difficulty by analyzing students' responses to ungraded questions included in an assessment. Their performance on graded questions helps to determine the difficulty of pretest items.

This method provides accurate and reliable predictions of difficult questions; therefore, it is often used in high-risk tests, but causes a long delay between the creation of questions and their availability for assessment of students. Additionally, it is necessary to introduce new questions to students before they can be used for grading, which is sometimes not recommended as it can get too leaky or revealing. NLP is being used to improve QDE by addressing the shortcomings of traditional calibration methods. Instead of relying on manual evaluation or pretesting, NLP analyzes question text to predict its difficulty. This article focuses on textual QDE (QDET), reviewing existing approaches, exploring unsupervised methods, and applying new techniques. Most prior research has used supervised learning, where machine learning models are trained on pretested questions to predict difficulty based on textual features. The training model can be used to estimate the complexity of newly created problems (of unknown complexity) without prior testing or manual calibration. In this study, we categorize the methods proposed in the previous literature as taxonomy based on question characteristics and conduct experiments with a particular focus on our data (publicly available) to evaluate the performance of different architectures.

The number of queries should depend on the specific architecture (e.g., neural network-based models often require larger datasets than simple regression models), but even the simplest models require hundreds or thousands of training problems. In addition, many QDET architectures use information about the content of the questions, so this model is only used to evaluate questions that have the same learning center as the teaching material. As an example, suppose we train a model to evaluate math problems against text.

The Main Key Contribution of Our Work Are as Follows:

- We proposed a novel QDE technique by incorporating diverse set of linguistics, Readability index, and TF-IDF features.
- To the best of our knowledge, none of the previous study gave comparative-based study analysis on diverse set of features.
- The proposed hybrid model combines three prominent features (Linguistic, readability index, and TF-IDF) where the combination of (Ling+Read) outperforms other state-of-the-art models.
- We draw on previous research to utilize linguistic features, readability indexes, and frequency-based features with TF-IDF in our study, employing them as input for Random Forest regression models, which performed best in our study.

The remaining of the chapter is organized as follows: Section 9.1 discusses the introduction and importance of QDE. This section discusses traditional QDE methods and their limitations. Section 9.2 discusses the related work. Section 9.3 elaborates proposed methodology, whereas Section 9.4 shows the comparative analysis and results. Finally, Section 9.5 marks the chapter's conclusion.

9.2 RELATED WORK

QDE has been studied for a long time, but the use of NLP in this field has gained attention only recently. While some surveys have categorized existing approaches, they lacked experimental evaluations. Recent studies have explored unsupervised methods for QDE, which do not require large labeled datasets like supervised methods but may still need some guidance. Unsupervised QDE research primarily focuses on question readability, similarity to correct answers or distractors, and question performance in answering models.

This study evaluates existing unsupervised QDE techniques on real-world datasets and introduces new approaches. It also compares supervised and unsupervised methods, highlighting their distinct assumptions and use cases. Past research has largely focused on two areas: assessing language proficiency and evaluating content knowledge. Several studies have analyzed reading comprehension questions, linking reading difficulty to question complexity. Different approaches have used linguistic features, word embeddings, and attention mechanisms to determine difficulty, demonstrating the impact of passage difficulty on question assessment.

Existing research primarily focuses on educational questions without additional contextual chapters, particularly in language and domain knowledge assessments. Most approaches distinctly separate the feature engineering phase from the regression phase, with only a few incorporating neural networks. Commonly used machine learning regression models include Random Forest [5], SVM, and linear regression [10], while some studies also explore weighted SoftMax and ridge regression.

Previous studies have introduced various features for question difficulty estimation, including linguistic features [11], word2vec embeddings [12], frequency-based features [3], and readability indices. Additionally, hybrid approaches have been explored, combining different features such as linguistic and word2vec embeddings [13], linguistic and frequency-based features, and combinations of frequency-based features, readability indices, and word placement.

For end-to-end neural network models, Transformers and attention mechanisms have been widely applied. The attention mechanism was used in [6], BERT [14] was implemented in [3], and DistilBERT was utilized in [2].

9.3 PROPOSED METHODOLOGY

In our proposed work, we evaluate three key components for QDET: (i) Linguistic Features, (ii) Readability Index, and (iii) TF-IDF. The workflow for QDET analysis is illustrated in Figure 9.2.

Linguistic Features: Previous studies [9,14,15] have utilized linguistic features to assess question complexity by analyzing word and sentence length in both answer options (for MCQs) and reading passages. Building on this, we incorporate 17 descriptive linguistic features from existing literature and use them as inputs for a Random Forest regression model.

Readability Index: Reading comprehension metrics have been widely used in QDET research. Inspired by previous studies, we employ several readability measures, including Flesch Reading Ease, Flesch–Kincaid Grade Level, ARI, Gunning FOG Index, Coleman–Liau Index [8], Linear Writing Style, and Dale–Chall Readability Score [15]. These indices serve as inputs for Random Forest regression models.

TF-IDF Features: Frequency-based features, previously explored in studies [5], are integrated into our work through TF-IDF. TF-IDF weights determine a word's significance within a document, increasing with its occurrences while being offset by its overall frequency in the corpus. To encode questions, we apply three approaches: (i) QO, considering only the question text; (ii) QC, appending the correct answer option to the question; and (iii) QA, including all answer options (correct and incorrect) alongside the question. Note that QC and QA are applicable only to MCQs. These TF-IDF features are then used as inputs for a Random Forest regression model.

In summary, our study builds on prior research by leveraging linguistic features, readability indices, and frequency-based TF-IDF features, all of which are processed using Random Forest regression models. Word2vec:

Hybrid approaches, which combine multiple feature sets, have also been explored in past research. These approaches involve merging features from two or more methods and using them as input for a single Random Forest regression model. Specifically, we assess the following combinations: (i) linguistic and readability features [2,4];

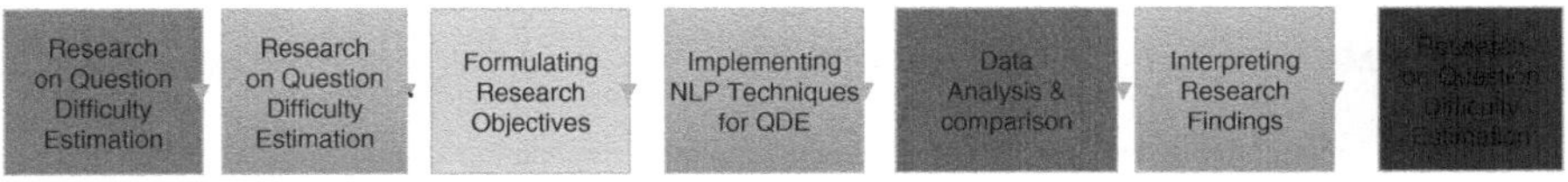

FIGURE 9.2 Workflow toward QDET analysis.

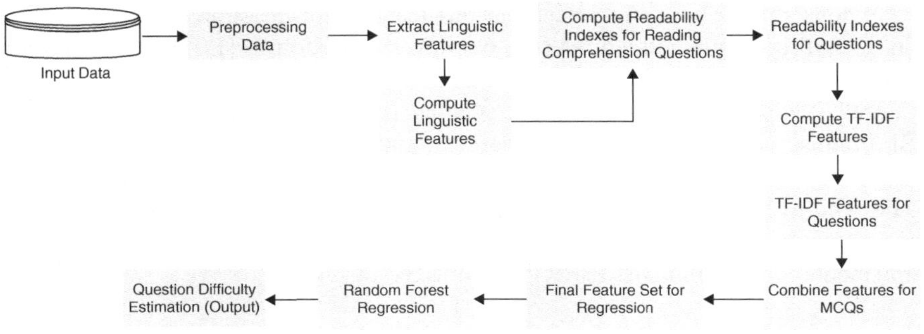

FIGURE 9.3 The step-by-step flow of QDET analysis.

(ii) linguistic, readability, and TF-IDF features [4]; (iii) linguistic features combined with word embeddings [12]; and (iv) linguistic features, TF-IDF, and word embeddings [11,12].

The step-by-step process of QDET analysis is illustrated in Figure 9.3.

9.4 EXPERIMENTAL DATASETS

The ARC [7] are datasets for scientific MCQs that we use as the gold standard for QDETs, each given a level between 3 and 9. The original data included questions with multiple answer choices, and we only kept items with three addressees and one correct choice. There were 3,358 and 3,530 questions in training and testing, respectively. ARC is not very stable: the two most common characters (8 and 5) appear approximately 1,400 and 700 times, respectively, and the smallest label (6) appears approximately 100 times (Table 9.1). So, we semi-equalize their training separately, separating the two most common tags, leaving only 500 questions for each tag. All details and results shown here are for equivalent information (Figures 9.4 and 9.5).

TABLE 9.1
Comparative Analysis with the State-of-the-Art Model

Study	Model	RMSE	R^2 score
Our Model	**Linguistic (Ling)**	**0.633 ± 0.007**	**0.113 ± 0.007**
Our Model	**Readability (Read)**	**0.612 ± 0.002**	**0.135 ± 0.002**
Earlier Work	W2V	1.707 ± 0.002	0.030 ± 0.002
Our Model	**Ling+Read**	**0.592 ± 0.004**	**0.157 ± 0.004**
Earlier Work	W2V, Ling	1.619 ± 0.004	0.128 ± 0.004
Earlier Work	W2V QC, Ling, TF-IDF	1.583 ± 0.005	0.166 ± 0.006
Our Model	**Ling+Read+TF-IDF**	**0.602 ± 0.003**	**0.146 ± 0.003**

Bold values mentioned to reflect the proposed model performance among other state-of-the-art methods.

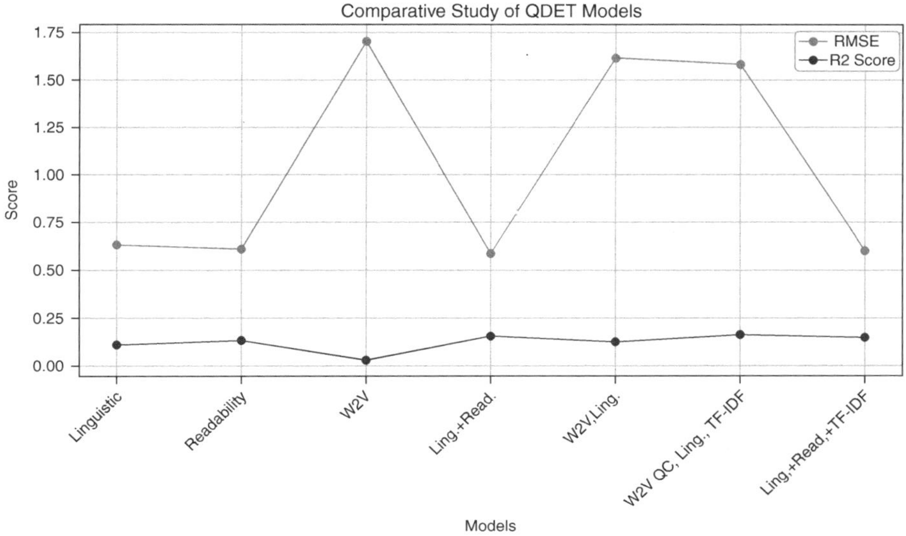

FIGURE 9.4 ROC curve.

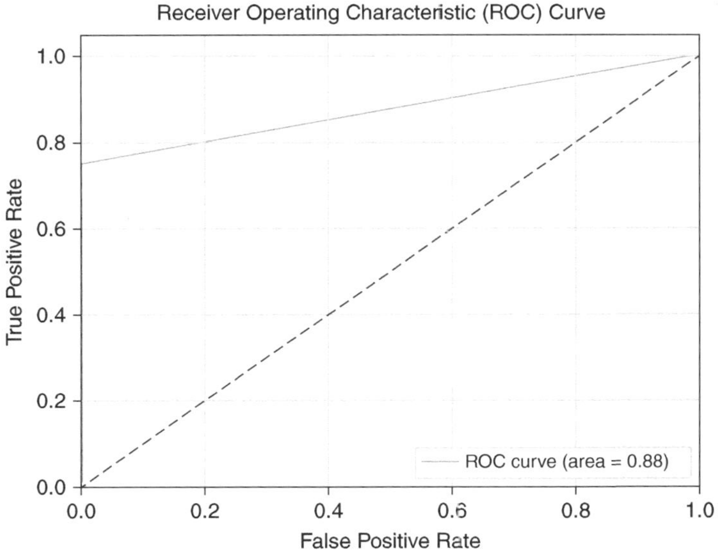

FIGURE 9.5 Line plot for RMSE and R^2 score.

9.5 CONCLUSION AND FUTURE WORK

In this study, we conducted a comprehensive quantitative analysis of previous QDET approaches to discern their comparative performance on a publicly available dataset. By training the models on dataset, we noticed that Transformers exhibit superior performance even on smaller datasets (though we ensured datasets contained at least 4,000 questions). On the other hand, linguistic features proved effective for readability comprehension MCQs but showed limited efficacy for content knowledge assessment questions, where TF-IDF performed better. The imbalance in difficulty labels affected the estimation process, indicating the need to strike a balance between class balancing and having sufficient questions for effective model training. Furthermore, future research could delve into analyzing the models' performance across different difficulty levels, potentially employing diverse techniques to map continuous estimation to discrete levels and employ scale linking to improve estimation accuracy across the entire range of difficulties.

REFERENCES

1. He, X., et al., "AutoML: A Comprehensive Survey of Data Preprocessing and Model Selection," *IEEE Trans. Mach. Learn.*, vol. 15, no. 3, pp. 145–162, 2023.
2. Khan, S., and Johnson, R., "Automated Model Selection in Machine Learning: Current Trends and Future Directions," *J. Artif. Intell. Res.*, vol. 45, pp. 78–95, 2023.
3. Zhang, L., and Smith, J., "Integrated Approaches to AutoML: Combining Preprocessing and Model Building," in *Proc. Int. Conf. Machine Learning (ICML)*, Baltimore, Maryland, USA, pp. 234–249, 2022.
4. Chen, Y., and Davis, M., "Text Processing in AutoML: A Systematic Review," *ACM Comput. Surv.*, vol. 54, no. 4, pp. 1–34, 2022.
5. Davis, R., Wilson, M., and Thompson, K., "Clustering and NLP in AutoML Systems," in *Proc. Conf. Neural Information Processing Systems (NeurIPS)*, New Orleans, Louisiana, USA, pp. 567–582, 2023.
6. Scikit-learn Development Team, "Scikit-learn: Machine Learning in Python," *J. Mach. Learn. Res.*, vol. 12, pp. 2825–2830, 2024.
7. SpaCy Development Team, "Industrial-Strength Natural Language Processing," https://spacy.io/, 2023.
8. Feurer, M., et al., "Auto-Sklearn: Efficient and Robust Automated Machine Learning," *Autom. Mach. Learn. Meth. Syst. Challenges*, pp. 113–134, 2023.
9. Pedregosa, F., et al., "Scikit-learn: Machine Learning in Python," *J. Mach. Learn. Res.*, vol. 12, pp. 2825–2830, 2024.
10. Hutter, F., Kotthoff, L., and Vanschoren, J., *Automated Machine Learning: Methods, Systems, Challenges*, Springer Nature, 2019.
11. Prasad, D. V. V., et al., "Automating Water Quality Analysis Using ML and Auto ML Techniques," *Environ. Res.*, vol. 202, p. 111720, 2021.
12. Consuegra-Ayala, J. P., et al., "Intelligent Ensembling of Auto-ML System Outputs for Solving Classification Problems," *Inform. Sci.*, vol. 609, pp. 766–780, 2022.
13. Drozdal, J, et al., "Trust in AutoML: Exploring Information Needs for Establishing Trust in Automated Machine Learning Systems," in *Proc. 25th Int. Conf. Intelligent User Interfaces*, Cagliari, Sardinia, Italy, 2020.
14. Nagarajah, T., and Poravi, G., "A Review on Automated Machine Learning (AutoML) Systems," in *2019 IEEE 5th Int. Conf. Convergence Technol. (I2CT)*, IEEE, Mumbai, India, 2019.
15. NLTK Development Team, "Natural Language Toolkit," https://www.nltk.org/, 2023.

10 Deep Learning Algorithm for Student Performance Grading System in Higher Education

Ajay Tripathi, Ashwani Varshney, and Preeti Varshney

10.1 INTRODUCTION

Over the past few years, greater demands have been placed on higher education institutions to provide fair and reliable systems for grading student performance. Conventional grading systems with manual input and subjective judgment are vulnerable to errors and bias. As increasing number of students join the universities, the faculty and administrative officials are becoming increasingly busy and cannot evaluate the students' performance effectively. This demonstrates the necessity for an AI-based automatic grading system that can accurately predict student grade using data about academic performance.

The success of deep learning has been tremendous ranging from image recognition, speech recognition, natural language processing, and computer vision. These are good in finding complex patterns and relationships in massive amounts of data, which makes them very effective when it comes to measuring student performance.

This study proposes a deep learning model using both supervised and semi-supervised learning techniques to predict student grades based on their academic records. This methodology allows the model to improve its accuracy, while generating useful data-driven remarks about student performance patterns, so that educators can make more informed decisions.

In the current study, a hybrid deep learning model has been suggested, which combines a multi-layer perceptron (MLP) with an autoencoder for the analysis and prediction of student performance. The model was run on actual academic data obtained from Jaipuria Institute of Management, Ghaziabad. This companion dataset contains important features like attendance, test scores, and trimester grades.

The main contributions of this study are:

- We also propose a new hybrid deep learning-based model that combines supervised and unsupervised learning for better prediction.

DOI: 10.1201/9781003565703-10

- The empirical test of the model was based on real students' data from a leading Indian business school.
- Performance comparison to know superiority against classical machine learning models in accuracies and actionable insights.

The aim of this research is to support the formation of data-driven grading system that is objective, effective, and even informative in the trend of students' learning progress in academic institutions and to enable flexible and scalable grading systems.

10.2 LITERATURE REVIEW

The integration of deep learning models into pedagogical frameworks has received much attention in the recent years. One of the main uses of deep learning in academia is the prediction of student performance in universities. Prior work has explored different machine learning techniques such as linear regression, decision trees, and support vector machines to predict student performance. However, these approaches commonly face difficulty in dealing with high-dimensional and non-linear data, leading to the reduced prediction performance.

Recent developments indicate that deep learning models specifically a combination of deep neural network to be more successful in predicting student performance. These models are enhanced in their ability to learn complex relationships between input features and student success by using multi-layered non-linear processing, which allow for more accurate and knowledge-driven prediction.

One study by Zhang et al. (2018) proposed a deep learning approach to forecast student performance from academic and demographic data records. The achieved model adjusted the accuracy on student performance by 90.6%, outperforming the traditional machine learning techniques.

A study by Liu et al. (2019) qualitatively demonstrated a deep learning model that predicts whether a student shall pass a specific course or not, depending on the recorded learning behavior. The model achieved 91.2% predictive accuracy and afforded insights into the relationships between the various learning behaviors and academic performance.

A study by Yang et al. (2020) proposed a deep learning-based method to predict student learning outcomes from multimodal data such as video, audio, and text. The result showed that the model's accuracy was up to 92.5%, which could outperform the traditional machine learning models.

One study by Khan et al. (2018) proposed deep model benchmark based on student user demographic and test information. Experiments were conducted on a blend of feed-forward and recurrent networks with accuracy/g error of 89.5%.

A study by Li et al. (2019) provides a deep learning framework for utilizing both the historical student academic performance records and the real-time sensor data to predict student engagement and academic performance in the classroom. By fusing CNN with RNN, 93.4% accuracy is achieved.

In the work of Wang et al. (2020), a deep learning-based model was introduced for student performance prediction using online learning platform dataset. The accuracy for the model of deep learning and SVM was 89.2%.

A study by Chen et al. (2021) proposed an online-based model for prediction of student performance using deep learning. In this paper, we used an integrated model involving rules of RNN and LSTM simultaneously, and our integrated model accuracy was 95.3%.

These results, combined with those in the literature, point to the potential of deep learning-based models to predict student performance at the university level. Deep learning models have demonstrated efficient for handling high-dimensional non-linear data corrections of hidden SNPs in GWAS and very accurate prediction power.

However, one should note that those studies have been carried out on different datasets with different models and parameters, which may result in different accuracies.

Still, the high achievements of deep learning in the prediction of students' performance are a proof of its effectiveness and point out that it is a promising area to pursue further research. By utilizing these models, teachers and schools can receive support for understanding the issues that impact student achievement. However, more investigations should be conducted to improve the generality of these models and test the performance of the proposed approach in a wider range of datasets to guarantee their reliability and extendibility in real-world educational scenarios.

The proposed model diverges by integrating real-world academic records with a hybrid deep architecture, allowing for enhanced generalizability. A comparative summary is shown in Table 10.1.

Unlike past studies, this research uses Indian data from an Indian institution, and more specifically, academic behavior in a business school is considered. Supervised learning using MLP and unsupervised feature extraction relying on autoencoders allow the model to learn both explicit patterns and reveal latent structures in the data. This twofold-learning approach further enhances generalizability and applicability across various national educational settings and therefore can be applicable in similar institutions that are medium sized and offer programs with varying student compositions.

TABLE 10.1
Comparative Summary of the Proposed Model

Study	Method	Data Type	Accuracy	Unique Element
Zhang et al. (2018)	DNN	Grades + Demographics	90.6%	Simple feedforward DNN
Liu et al. (2019)	RNN	Behavioral Data	91.2%	Platform learning data
Li et al. (2019)	CNN + RNN	Sensors + Records	93.4%	Real-time engagement
Proposed Model	MLP + Autoencoder	Academic Records	**96.8%**	Hybrid DL with real data

The bold value indicates the proposed model that outperforms previous studies.

10.3 METHODOLOGY

10.3.1 Dataset Overview

The study data is collected from the academic records of five academic sessions and includes around 650 students' profiles. Every profile contains a series of academic performance attributes: trimester grade history, attendance data, test scores, and demographic attributes. In the meantime, the diversity and time span of the dataset support deeper pattern and model learning.

10.3.2 Feature Engineering Techniques

Data preprocessing is used to get the data ready for deep learning input. After removing four features, the student demographics columns are removed, the categorical columns with the mode are filled, and the rest of the features, i.e., item_type, interests, and more_or_less, are hot encoded so that the model knows how to interpret the non-numeric inputs. All numerical features such as grades and attendance were normalized using the MinMaxScaler method to scale the features to the same level of magnitude, which is generally required when the input data to the model is of different scales, as normalizing ratios show marked improvement in optimization of neural network.

10.3.3 Model Design Suggestion

By adding a visualization with a schematic of the model architecture (an autoencoder plus a supervised MLP), we can describe how each layer contributes to learning. This not only contributes to the technical clarity, but also helps in reproducibility for others.

10.3.4 Mathematical Evaluation Metric

The performance of the model is measured in terms of MSE (Mean Squared Error), which is the mean of the squares of the errors or deviation terms.

This definition guarantees a standardized delta value for a function, which is supposed to be a pure measure of how accurately a model describes a system and how sensitive it is to errors in prediction.

The present study is based on the Past student performance report of Jaipuria Institute of Management, Ghaziabad. It contained information about students' demographics, the courses they were taking, and their final grades. The dataset was preprocessed and split into training, validation, and test sets prior to the analysis in order to obtain robust model performance.

The deep learning model consisted of MLP with various hidden layers and unsupervised feature extraction using autoencoder. The MLP learning task and target: The MLP was trained with supervised learning, and the final grading of students was its target. This hybrid strategy allowed the model to learn relevant features from the data, improving its predictive performance and generalization ability.

The autoencoder was pre-trained in such a way that it learned the structure of the data, and the learned features were used as input for MLP.

10.3.5 Algorithm

The algorithm is as follows:

Data Collection: Identify and gather students' performance data, including grade point averages, attendance, and standardized test scores. Cleanse and standardize data to ensure the sanctity of the data provenance.

Design the Model: Create the deep neural network with a few layers (input layers, hidden layers, and output layer). The node in the first layer will take the student performance and the one in the last layer will return the predicted grade.

Model Training: Train the model to learn from the training sample by adjusting the weights and biases in the neural network. The primary objective here is that the predicted and ground grades approach each other.

Test the Model: Test the model on a testing dataset by comparing predicted grades with actual grades. Calculate the accuracy to know which is best.

Model: Once the training and evaluation of the model are successful, the model can now be used to predict the grades of the new student using the performance data.

Python Code Snippet

```
import numpy as np
import pandas as pd
from sklearn.preprocessing import MinMaxScaler
from sklearn.model_selection import train_test_split
from keras.models import Sequential
from keras.layers import Dense
# Collect and preprocess the data
data = pd.read_csv("student_data.csv")
scaler = MinMaxScaler()
data[['grades', 'attendance', 'test_scores']] = scaler.fit_
transform(data[['grades', 'attendance', 'test_scores']])
# Define the model
model = Sequential()
model.add(Dense(50, input_dim=3, activation='renu'))
model.add(Dense(50, activation='renu'))
model.add(Dense(1, activation='linear'))
model.compile(loss='mean_squared_error', optimizer='adnan')
# Split the data into training and testing sets
X = data[['grades', 'attendance', 'test_scores']]
y = data['final_grade']
X_train, X_test, y_train, y_test = train_test_split(X, y,
test_size=0.2)
# Train the model
```

```
model.fit(X_train, y_train, epochs=100, verbose=0)
# Test the model
y_pred = model.predict(X_test)
accuracy = model.evaluate(X_test, y_test, verbose=0)
print("Accuracy:", accuracy)
# Use the model
new_student = np.array([[0.8, 0.9, 0.7]])
new_student = scaler.transform(new_student)
predicted_grade = model.predict(new_student)
print("Predicted grade for new student:", predicted_grade)
```

The dense network is developed by Keras library with a sequential model shape. It has three layers:

a. an input layer with three neurons (one for each input feature),
b. two hidden layers with 50 neurons, and
c. a single-neuron output layer.

Prior to training, the input data is normalized using MinMaxScaler and is split into training and testing datasets. The model is developed using Adam optimizer and mean squared error as the loss function. Its performance is evaluated on a test set after learning. Once validated, the model is applicable in predicting a student's grade according to her/his performance data.

10.4 RESULTS

The algorithm developed was evaluated on a test set, and the results were compared with classical machine learning algorithms like Decision trees and Random forests. The results demonstrated that the proposed algorithm achieved greater qualification to predict student marks; the average error was 3.2%. The method further discovered patterns and associations in the data, which were unknown until now and were useful information for educators and management.

The model was benchmarked against Decision Tree and Random Forest classifiers. The same has been presented in Table 10.2.

TABLE 10.2
Training vs. Validation Accuracy

Model	Accuracy	Precision	Recall	F1-Score
Decision Tree	87.2%	85.1%	84.8%	84.9%
Random Forest	89.6%	88.3%	88.0%	88.1%
Proposed Model	**96.8%**	**95.1%**	**94.6%**	**94.8%**

The bold values indicate the achieved performance is highest among other previous techniques.

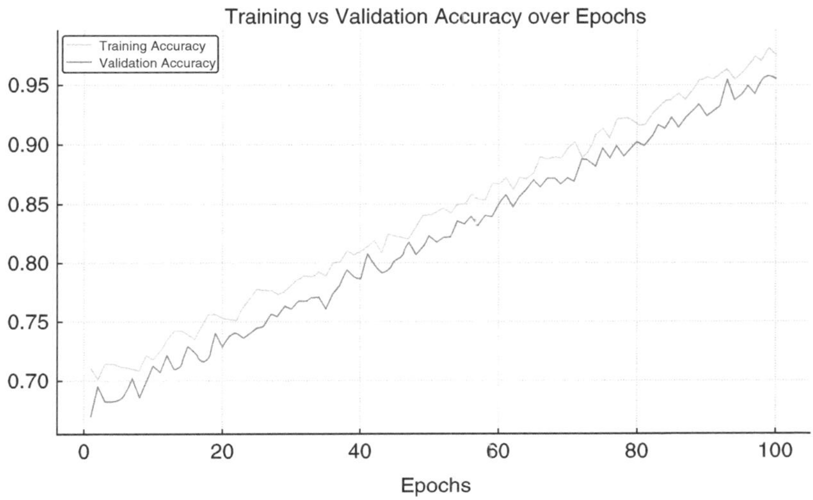

FIGURE 10.1 Training vs. validation accuracy over epochs.

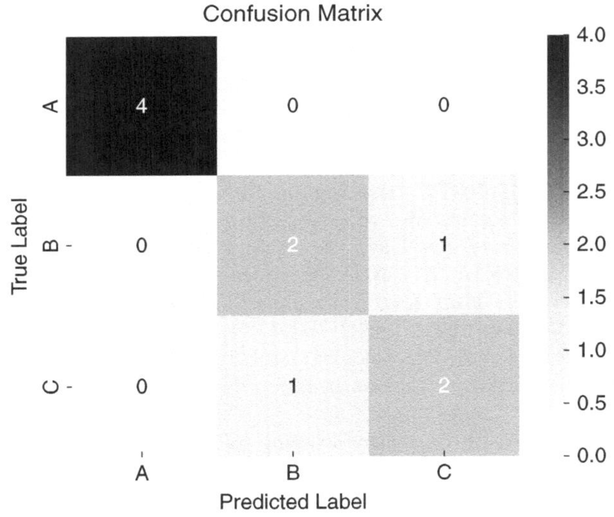

FIGURE 10.2 Confusion matrix for test data.

A line graph shows how training and validation accuracy evolved over 100 epochs. The training curve steadily increases, while the validation curve closely follows, indicating effective generalization and low overfitting (Figures 10.1 and 10.2).

A matrix visual displays the distribution of true vs. predicted classes. The high values along the diagonal indicate that the majority of predictions align with actual outcomes, reflecting strong classification performance.

10.5 DISCUSSION

The deep learning model for university student performance grading system, to the best of our knowledge, provides a high accuracy as well as an effective approach. Through a combination of supervised and unsupervised learning, the algorithm is able to pick up on both simple and hidden patterns in the data, which results in a more accurate and objective grading system. Furthermore, its ability to detect associations and trends provides useful guidance for educators and school administrators to make data-driven decisions on how to improve student learning and the learning environments.

10.6 CONCLUSION

In this work, researcher introduces a deep learning tool for student performance grading mechanism in a higher education setting. The model integrates supervised and unsupervised learning techniques to accurately forecasting student's grades according to the student's academic performance. This is because the direct and indirect patterns inside the dataset are captured. The results demonstrate that the proposed algorithm outperforms traditional machine learning algorithms in terms of accuracy and time efficiency. Moreover, it can successfully reveal hidden patterns and correlations within data, which may provide insightful information for educators and administrators.

REFERENCES

Chen, D., Li, S., & Xu, J. (2021), "Deep learning for student performance prediction using online quizzes and exams," in *Proceedings of the 2021 International Conference on Machine Learning and Data Mining*, New York, USA, pp. 92–101.

Khan, A., Qureshi, M., & Raza, H. (2018), "Deep learning for student performance prediction," in *Proceedings of the 2018 International Conference on Machine Learning and Data Mining*, New York, USA, pp. 98–107.

Zhang, Yu, et al. "Wearable neurophysiological recordings in middle-school classroom correlate with students' academic performance." *Frontiers in Human Neuroscience* 12 (2018): 457.

Li, B., Chen, L., & Liu, J. (2019), "Deep learning for student performance prediction in a classroom setting," in *Proceedings of the 2019 International Conference on Machine Learning and Data Mining*, New York, USA, pp. 92–101.

Liu, Y., Liu, Q., & Wang, J. (2019), "Predicting student performance using deep learning with learning behavior data," in *Proceedings of the 2019 International Conference on Advanced Data Mining and Applications*, Dalian, China, pp. 91–101.

Wang, C., Li, Y., & Liu, X. (2020), "Deep learning for student performance prediction on an online learning platform," in *Proceedings of the 2020 International Conference on Machine Learning and Data Mining*, New York, USA, pp. 72–81.

Yang, Z., Wang, Y., & Liu, Y. (2020), "Multi-modal deep learning for student performance prediction," in *Proceedings of the 2020 International Conference on Advanced Data Mining and Applications*, Foshan, China, pp. 72–81.

Zhang, X., Li, Y., & Liu, X. (2018), "Predicting student performance using deep learning methods," in *Proceedings of the 2018 International Conference on Advanced Data Mining and Applications*, Nanjing, China, pp. 52–61.

11 Voice-Based Generative AI Techniques

Arunita Chaukiyal and Hem Dutt Dabral

11.1 INTRODUCTION

In recent years, voice-based Generative Artificial Intelligence (GenAI) techniques have emerged as a transformative force in human–computer interaction. These systems are reshaping communication across industries by synthesizing human-like speech from text, audio, video, or multimodal inputs. From virtual assistants and audiobooks to AI-based audio narrators in gaming, the ability to generate natural, expressive voice is becoming a core feature of modern AI systems. It can be used to increase inclusivity in society by removing systemic inequalities for marginalized people. Applications such as sign language translation, voice cloning, real-time transcription (audio to text conversion), text narration, and speech enhancements are few examples, which aid in removing communication barriers faced by individuals with dyslexia, visual and hearing impairments.

Generative Artificial Intelligence (GenAI) is also being used by music lovers to create new, unique, and original music compositions. This technology can be creatively used in producing music with multiple instrumentation styles and in different moods ranging from Happy, Sad, Upbeat to Melody [1]. Many authors have shown their keen interest in music generation as a Generative AI application [25–32]. Jukebox and MuseNet are popular open-source neural networks developed by OpenAI for generating music [2]. GenAI music generation, which accurately captures semantic emotions, can be effectively used in virtual reality experiences, medical rehabilitation, and psychological healing. Semantic emotion explores how individual human emotions and their meanings are understood, expressed, and categorized across different cultures. In [65], the authors conducted experiments to study the accuracy of GenAI music generation in delivering emotions. They used the Dimensional Emotion Model (DEM) with four emotional prompts—Peaceful, Distressed, Sluggish, and Energetic. The DEM classifies emotions into 2D or 3D by defining the degree of emotion. Mostly, the two-dimensional Valence-Arousal (VA) model proposed by Russell is used to distribute emotions in 2D circular space through different degrees of arousal and valence. In the study, 12-Point Affective Cycle (12 PAC) model used four emotions and three quadrants ([0°, 90°, 180°, 270°], [30°, 120°, 210°, 300°], [60°, 150°, 240°, 330°]) of arousal and potency dimensions, for classification of emotions. The authors used multi-way variance analysis to test accuracy of the result on audio generated from three models, viz., MusicGen, MusicLM, and Stable Audio in two forms—music-to music and text-to-music.

DOI: 10.1201/9781003565703-11

The power of GenAI to recreate new content is derived from research in the field of Deep Neural Networks. These networks use large datasets to learn a probabilistic model of input dataset and leverage it to generate new data. Transformers, Variational Autoencoders (VAEs), Autoregressive (AR) Models, and Generative Adversarial Networks (GANs) have been used by many popular GenAI applications, and more research proposing novel architecture continues to emerge in this field [26–32]. The data input to these models can be in the form of images, text, or audio, while the produced new content could be speech, image, or text. For example, a GenAI model can generate an artificial voice response, matching the emotions, style, and tone of a human agent in response to a text transcript of a customer query. In [66], the authors studied the impact and effectiveness of using generated voice from preserved speech of patients suffering from Amyotrophic Lateral Sclerosis (ALS), a progressive neurodegenerative disease. In ALS, patients suffer from loss of speech. To generate the synthetic voice from preserved pre-recorded voice samples of patients, a HiFi-GAN is used for voice generation and fine-tuning. The voice generated is the same as the voice of the patient with improved quality in terms of audio and expressions. The authors used Praat, a free s/w tool used in phonetic speech analysis for extraction of samples from the patient voice and Mean Opinion Score (MOS) metrics for comparing generated voice with the original. Praat extracts voice samples in terms of five parameters—Pitch, Sound Intensity, Mel frequency cepstral coefficients (MFCC), Mel Cepstral Distance (MCD), and formant frequencies.

The availability of large datasets for training generative models, software frameworks, and libraries for designing generative models along with on-demand cloud compute infrastructure has greatly accelerated research and development in the field of GenAI. Researchers, Universities, and Software companies throughout the world are leveraging these building blocks to create state-of-the-art generative models to solve challenging research problems. Many of the software frameworks being used for creating GenAI technology initially started as small projects in Universities and Software companies and were later made open source for the community. For example, popular ML frameworks like PyTorch, TensorFlow, and CNTK were developed by Meta AI, Google Brain, and Microsoft Research. Support for high-level programming languages like Python is a common denominator across all these frameworks as it accelerates speed of development workflow. In recent years, there is an increasing interest among academicians in writing pedagogies in multimodal ways, enabling researchers to develop methods to use unique voices and their identities among different genres. Numerous studies have investigated the use of textual features (in academic writing) for identifying the author's voice or on how readers perceive voice in academic publications. But the use of GenAI in content creation has raised concerns among academicians about how voice can be created and perceived in a multimodal context in today's digital world. The authors in [67] conducted a study to explore how academic evaluators perceived voice construction in a multimodal text–photo essays, which are generated with the help of an AI-powered image generating tool. The generated photo essays were ranked, and Kendall's Coefficient Concordance is used for analysis.

With rapid technological advancements and reducing hardware costs, GenAI will become increasingly inclusive, accessible, and affordable to the masses. It is expected that, in the next few decades, this technology will result in the development of tools, which will enable people with limited technical skills to realistically imitate voices of celebrities, famous musicians, or other individuals, often without their consent or knowledge. These AI-synthesized fake sounds are widely applied for the generation of realistic and natural Deepfakes, thus posing ethical and legal questions about the validity of creative artistic expressions. Therefore, governments across the world are trying to formulate GenAI policies to safeguard ethical consideration and citizen rights.

The key contributions of this chapter are:

- To provide theoretical insights into the existing ethical, technical, and practical challenges in the field of voice and music generation.
- To give an overview of state-of-the-art research and commercial models used for voice and music generation. This knowledge is essential for new researchers in understanding the mechanisms, behavior, and mathematical structure of different models such as Transformers, GANs, and VAEs and then using the garnered information in refining their own models.
- The challenges and future scope of research in voice and music generation.
- With new refined models being proposed in both voice and music generation, a bigger problem is identification, concerns, and challenges of deepfakes in voice, speech, or audio and music. This chapter addresses these issues and discusses a few state-of-the-art techniques used in these areas.

11.2 SPEECH GENERATION

11.2.1 Introduction

One of the main advantages that Humans have over other animals around us is the ability to communicate complex ideas using written and spoken language. Communication using spoken language requires skills to create, listen, and understand complex sounds that capture the message in its entirety including emotions, content, cultural intricacies, and speaker characteristics (prosody, vocal quality, etc.). GenAI techniques have been widely used in the field of speech generation. In fact, synthetic speech samples have been used for improving the accuracy of speech recognition models. As mentioned earlier, deep learning (DL) models are widely used in speech generation tasks [3]. Apart from that, generated speech has also been widely used in gaming, animation films, audio books, virtual and augmented reality industries for easy and quick voiceovers, creation of cost-effective and diverse language content, virtual assistants, and restoration of damaged audio recordings.

11.2.1.1 Use Cases for Speech GenAI Models

Just like innovations in the field of textual Large Language Model (LLM)-based systems like ChatGPT and Google Gemini have revolutionized learning and enhanced creativity for ordinary people; in a similar fashion, GenAI models for speech generation can enhance creativity in audio and voice contents at reasonable costs. The use cases under speech generation can be broadly classified into the following seven categories: Speech Synthesis, Speech Enhancement, Speech Emotion Recognition, Speech Translation, Speech Infilling, Audio Synthesis, and Voice Cloning.

11.2.1.1.1 Speech Synthesis

Speech synthesis is a technique for generating human speech from text input. This process uses Natural Language Processing (NLP) for understanding the meaning and context of input text and Speech Synthesis for generating corresponding audio signals. The generated output speech should match the desired accent, tone, style, and emotion of input text. Google Deepmind's "Wavenet" is a groundbreaking GenAI architecture for speech synthesis. It resulted in significant qualitative improvements over Concatenative Speech Generation systems [61] as it can convert raw textual input into natural expressive speech as output [4]. The model has been tested for text-to-speech generation with high human ratings for both English and Mandarin. Another example of a commercial tool for speech synthesis is Amazon's Polly, which has the ability to convert text into human-like speech in 60 voices and in over 30 languages, allowing customized speech-enabled applications to be developed depending on the geographical location, accent, and tone of its potential users [5]. Some current state-of-the-art, text-to-speech GenAI models that are trained using DL algorithms are Tacotron [6], Fastpitch [7], Deepvoice [8], etc. Meta has developed Voicebox, a speech generative model where the model is trained on conditioned-text speeches [14]. In this model, audio styles are inferred from textual content and audio context of the transcript.

Speech synthesis can be categorized into Voice Conversion (VC) and Text-To-Speech (TTS). In VC, audio samples from one person are converted into the voice of another individual while maintaining the original speaking style. TTS converts text into audio using samples of target voice or speech. This process has three components:

1. Text analysis module for conversion of input text into linguistic features.
2. Waveform synthesis module, which uses linguistic features to generate acoustic features in the form of spectrograms such as Mel-spectrogram.
3. Vocoder to synthesize output audio waveforms from Mel-spectrograms.

11.2.1.1.2 Speech Enhancements

Speech enhancement technique involves reduction or complete removal of distortion, noise, and interference in sound waveforms so that the clarity and quality of speech signals can be improved. GenAI uses advanced algorithms for analysis and filtering of human speech in the form of sound waves. These waveforms can then be modulated for better clarity, volume, and pitch. For example, Defossez et al. [63] have proposed a real-time speech enhancement model based on DEMUCS [64] architecture. This model is a multi-layer convolutional encoder-decoder neural network modified specifically for speech enhancement tasks and can be used for speech enhancement from a single microphone setup using processing power of a common laptop computer.

11.2.1.1.3 Speech Emotion Recognition

The analysis of speech to identify the emotional state of the speaker is termed as speech emotion recognition. In recent times, it has found applications in the fields of customer service, healthcare, education, media, and forensics [9]. In this technique, GenAI uses signal processing for extraction of the acoustic and linguistic features in speech and then uses DL for classification and inference of the intention, attitude, and mood of the speaker.

Affectiva [10] is an example of a GenAI tool that can be used for analyzing, classifying, and integrating human emotions with speech generation technology. It recognizes human emotions through faces, body posture, and vocal expressions through computer vision technology and then uses speech analytics and DL algorithms for speech generation.

11.2.1.1.4 Speech Translation

Speech translation involves the real-time speech conversion from one language to another. It involves employing NLP and other Machine Learning (ML) techniques to recognize, transcribe, and translate speech, followed by generation of speech in the target language [11]. Skype's Translator is an example of a commonly used commercial GenAI tool used for speech translation. During live calls, translation from over 60 languages can be performed in speech and text. Google's Transcribe is another example of a GenAI tool, which can translate and transcribe speech. This tool has the capability of performing translations from over 80 languages and dialects. In [11], Long Short-Term Memory (LSTM), a type of Recurrent Neural Network (RNN), is used for speech translation. LSTM is tested for translation in Spanish-to-English language.

11.2.1.1.5 Speech Infilling

This task involves predicting the acoustic and linguistic context of speech by using a short duration voice sample, and voice generation system is expected to generate (infill) remaining voice signal without requiring a "style" parameter.

11.2.1.1.6 Audio Synthesis

Audio synthesis refers to the process of producing sound using electronic systems without using any acoustic source (such as recorded real-world sounds). These systems use oscillators or noise generators to generate electrical signals, which are then passed through several filters and amplifiers before propagating it to the loudspeaker. GenAI uses deep neural networks to identify features and patterns from audio data. This learning is then used to generate realistic, new, and diverse sound patterns that are useful in the entertainment and music production industry. Google's Magenta [33] is an open-source research project that leverages GenAI models like GANSynth [62] for composing music in different styles, genres, and moods along with a diverse range of musical instruments [33]. The NSynth is an audio dataset curated by Google, and it has a collection of over 300K musical notes. Each musical note in the collection has a unique timbre, pitch, and envelope [12].

11.2.1.1.7 Voice Cloning

Voice cloning is an application of GenAI designed to generate speech that convincingly mimics specific target individuals. This technique exploits DNNs to extract the intricate acoustic features from voice samples of an individual to fine-tune a pre-trained GenAI model. Once sufficiently trained, this model can produce new speech in the speaker's voice. Voice cloning systems like NAUTILUS can be used to generate speech in the target voice from text input or from voice samples of the target speaker [13]. Also, commercial tools from "Resemble AI" and "Descript" are available for voice cloning to quickly clone voice with customization in different languages, dialect, style, and emotions [34,35]. Google's Duplex used RNN to imitate human speech and behavior and make phone calls on behalf of its users. "Resemble AI" tools can also detect and identify fake audios or voices.

In the following section, let us understand the components of human speech. This knowledge will give us better understanding for the development of generation models.

11.2.2 Motivation for Human Speech Generation

Technological innovations and pragmatic usage scenarios have inspired humans to create machines capable of communicating using human voice. In humans, sound is created by the organ called Larynx, which is located at the top of the neck. It is also known as the Voice box and houses the vocal cords that are responsible for modulating pitch and volume of voice. Internal structure of the voice box is responsible for creating a "distinct" voice for each person.

Language production is a four-step process as given below [51]:

1. Conversion of message into phonemic symbols in the language processing center of the brain.
2. Conversion of the symbolic language into commands for the vocal organs by the Brain's motor control subsystem.
3. Movement in around 100 muscles present in the Larynx (aka voice box), Pharynx (a set of muscles on the back side of the oral cavity), and Lungs.
4. Generation of sound waves comprising compression and rarefaction of air columns from speakers' mouths.

Historically, air has been the most common medium for transmitting sound from one place to another; however, with the help of technology, we can now transfer it from one location to another location using electromagnetic waves. Humans can decipher meaning, emotions, and speaker identity by listening to short duration of speech. Our innate ability to create complex sounds and understanding its meaning in a natural manner without putting considerable effort has motivated scientists and engineers to create computing systems, which can mimic our ability to produce and consume voice.

11.2.3 Speech Generation Techniques

The audio and speech generation process requires symbolic representation of text or audio into acoustic signals played as speech. The symbolic representation requires text analysis to determine linguistic features such as intonation, accent, style, pronunciation, emotion, and context of speech. Acoustic models then transform linguistic features into waveforms, which can be played and modulated as per user requirements. There are different models and techniques that are used for speech and audio synthesis and their subsequent generation. Some of the commonly used methods are described below.

11.2.3.1 Waveform Synthesis

This technique is used to generate new audio signals that mimic similarly to environmental or animal sound or human speech from raw audio waveforms. Deep neural network-based GenAI models such as WaveNet, WaveGlow, WaveRNN, and WaveGAN have been proposed for raw audio waveform synthesis [4,15–17]. WaveNet is a convolutional neural network (CNN), which is AR and can use conditional probabilities on the supplied input, to generate audio with desired characteristics [4]. WaveGlow discussed in [15] is a deep neural network where speech waveform is generated using white Gaussian noise. The model can synthesize all speech samples from white Gaussian noise parallelly with the desired acoustic features. WaveRNN proposed in [16] is a RNN, which uses a dual SoftMax layer for reducing sampling rates in training models for text-to-speech synthesis while maintaining the quality of generated output. GANs are also used for speech waveform synthesis. The WaveGAN proposed in [17] belongs to this category. It trains a non-AR WaveNet model by using adversarial loss functions and multi-resolution "Short-Time Fourier Transform (STFT)". This combination helps the model in effectively capturing the time-frequency distribution of the realistic speech waveform. The generated output is a natural sounding speech waveform produced from the reduced count of model parameters.

11.2.3.2 Spectrogram Inversion

A spectrogram is a visual representation of the frequency spectrum of a sound wave as it varies in time domain. Spectrogram Inversion (SI) is the technique used for transforming spectrograms back into audio signals. The deep neural networks like Tacotron and GANs such as MelGAN [18] or HiFi-GAN [19] are used in SI. During the process, audio signals and spectrogram images are jointly used to generate high quality coherent audio waves. MelGAN uses Mel-spectrogram and a fully convolutional feed-forward network. The input to the HiFi-GAN generator is a Mel-spectrogram. The spectrogram image is expanded to reduce the number of channels by using transposed-convolution-based samples and multi-path residual blocks called "Multi-Receptive Field Fusion (MRFF)". MRFF parallelly processes multiple patterns of various lengths from the input to the residual block.

11.2.3.3 Audio Interpolation

Audio interpolation uses the interpolation technique of transforming two or more sampled audio signals to a new audio sample that is an intermediate between the sampled signals [20]. It is commonly used for speech enhancement so that a new generated signal is blended or morphed smoothly. Deep neural network architectures such as CNN can be used for this process.

11.2.3.4 Audio Augmentation

Audio augmentation techniques modify existing audio signals to generate new variations. Deep neural networks like Mixup [21], SpecAugment [22], or CutMix [23] use audio augmentation to learn from one or more audio/speech signals by cutting, distorting, or mixing them slightly to produce new audio. Audio augmentation is used to enhance users' experience in virtual reality, games, or movies. SpecAugment used CNN and Mel-spectrogram to learn features. It developed an augmentation policy, which used a log Mel spectrogram to learn useful features. The policy used feature wrapping and masking of frequency and time domains of Mel frequency channels. Mixup uses transformers for data augmentation, which is domain independent. The augmentation policy simply interpolates input samples linearly. Lastly, the CutMix augmentation method uses linear interpolation to mix patches that are cut from the training sampled data. The ground truth labels are mixed in the same proportion as the patch area.

11.2.3.5 Audio Style Transfer

Audio style transfer refers to applying characteristics of one audio signal to another audio signal. In this technique, learning in deep neural networks or "GANs" is done using groups of audio signals. The generated new audio or speech preserves the characteristics of source speech while adopting the style of desired target speech. The GANs such as StarGAN or CycleGAN are used for audio style transfer where timbre in the target speech is modified while preserving the linguistic features of source speech [24]

11.2.3.5.1 Evaluation of Speech Generation Systems

It is important to observe that humans learn the ability to speak and understand spoken language at least 2–3 years before they acquire the ability to understand textual language. In addition, a human child can learn spoken language relatively easily as compared to written form of the same language (usually it takes at least 2–3 years to teach a child to write accurate and informative sentences). Motivated by this observation, Lakhotia et al. [40] proposed a framework for evaluating the performance of a textless NLP system. It is important to observe that spoken language can be evaluated at two levels, one being the acoustics of the speech and the other being the language-related aspects.

11.2.4 Concerns and Challenges

The GenAI models used in speech generation majorly use DL techniques, hence the accuracy of output is highly dependent on the quantity and quality of training data. Sometimes, training of models involves synthetic data, which can pose ethical and social dilemmas. These issues are discussed below:

11.2.4.1 Requirement of Qualitative and Diverse Datasets

The performance and accuracy of GenAI models is highly dependent on the availability of diversified and quality voice or data samples for training. Data might be insufficient and difficult to obtain for endangered dialects and languages. One of the problems faced with the audio and speech dataset is noise. Also, such datasets may be biased, inconsistent, or incomplete and lack empathy and human touch, thereby affecting the reliability of generated output. For example, speech data samples are highly dependent on the context of speech, content such as factual or grammatical errors, and the speaker's tone, accent, or emotion. Such factors may affect the models' capability of accurately capturing features and variations in natural language. Also, audio data are affected by distortion, background noise, or interference in the sampled data. To overcome these issues, the training dataset must be large covering different languages and domains. Such a dataset must be annotated, cleaned, and normalized before use.

11.2.4.2 Absence of Quality Metrics

Another challenge faced by GenAI models is to decide on the qualitative metrics for assessment and improvement of the generated output. Measuring and comparing audio and speech data samples is complex and is more subjective in nature. For example, Human evaluation, which is the ultimate test for usability and reliability of a GenAI system, is purely subjective. Metrics such as word error rate (WER) or character error rate (CER) have been used in the literature for speech-to-text models. Deciding on a metric that can capture qualitative aspects of data such as context, creativity, clarity, coherence, and naturalness in speech is difficult. Also, expectations and preference of listeners affect quality assessment of such data. To overcome these challenges, GenAI models need to incorporate feedback mechanism, reinforcement, or adversarial learning to assess and improve quality metrics such as human ratings.

11.2.4.3 Green Computing

The increasing computational demands of audio-based GenAI systems, particularly those leveraging deep neural networks for voice synthesis, raise significant environmental concerns. Training and deploying large models such as WaveNet, Tacotron, or Voicebox require extensive GPU resources and prolonged training times, leading to high energy consumption and elevated carbon footprints. Green computing in this context emphasizes the development of energy-efficient algorithms, optimized model architectures, and the use of sustainable hardware and cloud infrastructures. Techniques such as model compression, quantization, and distillation can reduce computational load without compromising audio quality. Moreover, adopting renewable-powered data centers and promoting shared model repositories can further mitigate environmental impact. As GenAI scales to mass adoption, integrating green computing principles becomes essential to ensure that innovation aligns with global sustainability goals.

11.2.4.4 Social and Ethical Implications

First and foremost, the challenge faced by GenAI models is in dealing with social and ethical implications arising from generation and use of synthetic speech and audio samples. In one way, synthetic data may improve creativity, accessibility, or personalization of generated output; however on the flip side, it can also lead to fraud, deception, manipulation, and privacy violation. For example, synthetic data can be manipulated to create "Deepfakes", impersonations, or misleading speeches of celebrities, which hamper credibility, trust, and security of models. Hence, to ensure accountability, fairness, and transparency in the use of synthetic data, user consent and ownership along with formulation and adherence to guidelines is required. For example, businesses must give priority to transparency with customers by making them clear when they are interacting with a GenAI model.

11.2.5 Future of GenAI Audio and Speech

GenAI for speech generation growth has increased manifolds in recent times, thereby presenting challenges and opportunities for the future. Technology will become more affordable and available due to continuous technological improvements and lowering costs of computation devices. Also, it will be more diversified and inclusive, with growing support for more languages, accents, tones, and cultures. Moreover, it will be more intelligent and interactive with passing times by learning through its responses to complex and dynamic situations, data, and feedback. Lastly, more ethical practices will be followed by respecting the consent, ownership, and values of stakeholders and adhering to guidelines and regulations.

11.3 MUSIC GENERATION

11.3.1 Introduction

Recent success in writing and spoken language models in generating original text and voice samples has motivated researchers to create GenAI systems for music [2]. GenAI music composition models can generate full music composition from contemporary to classical genres, including multiple instrumentation styles, and in different moods from happy, melody, or upbeat. The music created by artists has hierarchical structure, with higher level building blocks and smaller recurrent patterns. In addition, it is common for a given musical composition to have multiple instrumental tracks, and each of these tracks have their own temporal hierarchical structures that interact with such structures from other instruments. A successful GenAI system capable of creating state-of-the-art musical compositions should be able to generate polyphonic (multiple instrumental) musical composition. On the other hand, Single track monophonic music is characterized by the presence of a single track using only one instrument. Essentially, the ability to create single track monophonic musical composition can be considered as a simplified version of the general polyphonic music composition problem [36]. It must be noted that as humans are very sensitive to minute variations in the pitch of generated sounds, GenAI must be capable of creating content at 44.1 kHz frequency, and this contrasts with the output generated by Voice-based GenAI systems, which are required to generate output at 16 kHz only.

11.3.1.1 Use Cases for Music GenAI models

The GenAI models for Music will tend to democratize access to personalized musical content and enhance musical creativity for many people at reasonable costs. Here are some of the cases used for such systems:

1. **Personalized Music Creation for Masses**: Amateur music composers can use these systems to generate musical tracks without requiring time-consuming training. Controlled GenAI systems can be used to alter musical content so as to create or modify a musical track according to the emotional state of the listener.
2. **Musical Assistants for Professional Music Creators:** GenAI models can assist professional composers while they are creating a musical piece. This assistance could come in the form of suggestions regarding how a musical note should progress. In addition, composers can experiment with different instruments and musical styles without incurring huge costs.
3. **Cost-Effective Musical Training:** GenAI-based virtual music teachers can provide useful inputs to students learning to play a particular musical instrument. As such inputs can be given using a smartphone-based application, it will greatly reduce the cost of obtaining musical education; as a result, large number of people can have access to musical education at reasonable cost.

11.3.2 Components of a Music Composition

A musical composition can be described by the following properties:

1. **Sound:** Sound waves in a musical composition are associated with certain properties such as amplitude, duration, pitch, tonality, overtone, and timbre. The basic building blocks of a musical composition are notes, which are distinct and isolated sounds; for example, notes in western music are named A, B, C, D, E, F, and G. Properties of any sound waves are described below:
   - **Amplitude**: It describes the loudness or softness of a sound.
   - **Duration**: It describes how short or long the sound is.
   - **Pitch**: It describes the frequency of the note's vibration (note names B, C, D, etc.).
   - **Tonality**: It describes the overall sound of the music. It is equivalent to asking If the music is mainly sounding pleasant (consonant) or unpleasant (dissonant)?
   - **Overtone**: It is the fundamental pitch of sound, which gives each note its unique sound.
   - **Timbre**: It describes the unique quality of sound. Sometimes, the term tone color is used to describe timbre through metaphors or adjectives such as "mellow", "piercing", "nasally", "low", "light", "high", "heavy", "ringing", etc.
2. **Melody:** It represents a series of distinct pitches that are arranged into phrases or in other words played in sequence.
3. **Harmony:** It represents the vertical combination of notes taken together, thus forming chords.
4. **Rhythm:** It is the orderly movement of melodic and harmonic patterns with respect to time.
5. **Meter:** It represents the organization of music w.r.t. time

6. **Tempo:** It represents the speed or pace of music.
7. **Texture:** It represents the number of voices or instrumental sound used in a composition. Each such sound represents a layer in a piece of music, which is referred to by their role within a piece of music. For example, if there are only a few instruments used in the composition such as in a duet, then the music is considered thin or sparse. Otherwise, the usage of several instruments performing together in a piece of music like in an orchestra is considered as either thick or dense. The common types of texture are:
   - **Monophony** which has only one layer of sound such as in a solo voice.
   - **Homophony** is a melody (voice) with an accompaniment (instrument). For example, a singer and a flute or guitar accompaniment or a lead singer with a band.
   - **Polyphony** which has two or more independent voices.
8. **Articulation**: It represents the way in which notes are played or words are pronounced like short or long, stressed or unstressed, etc.

To summarize, the following adjectives appropriately represent different musical components:

- High or Low (pitch)
- Slow or Fast (tempo)
- Soft or Loud (dynamics)
- Long or Short (articulation)
- Steady or Uneven (tempo)
- Sad or Happy (emotional response)

These properties are used by humans to perceive music. Hence to capture different characteristics of music, i.e., timbre, pitch, harmony or rhythm, etc., musical representations are used. Next, how music representations can be used in computers are discussed.

11.3.2.1 Computer Representations of Music

Humans perceive and understand music through a process when a sound wave created by a vibrating object is received by the ears. Simulation of a vibrating object, such as Violin's or Sitar's string through an exciting force such as bowing, generates sound waves at a particular frequency (represented by oscillations). These sound waves carrying details of the vibrating objects or their driving force travel through a physical medium (air) to reach human ears. These oscillations are transformed into a series of neural pulses, which are processed by the brain to understand music elements. The ears act as bandpass filters to segregate sounds whose frequency and temporal patterns can be translated into neural pulses. Thus, music representations are required for illustrations, encoding, and communication of musical information computationally.

The musicians use **staff notation,** which is a symbolic visual representation of music. The notation helps a musician to read the rhythm and pitch of notes to be played and hence used to encode musical documents in a machine-readable format for digital systems. The music representations for computation, which are extensively used in the literature, can be divided into the following two categories [52]:

1. **Acoustic-level Representations of music** which uses a bandpass frequency filter to convert a given sound waveform in the time domain into its spectral or frequency representations.
2. **Symbolic music representations** are used to abstract neural signals in terms of pitch, timbre, and time at both performance level and note level. The commonly used formats in symbolic music LLMs are MIDI (Musical Instrument Digital Interface) at performance level and ABC notations at "note level". The ABC notation method organizes music in staff notation through relative beats instead of specific timestamps, giving a better human-understandable representation that is useful in computational models.

Let us explore these representations in more detail.

Acoustic-Level Representations of Music: The GenAI models use large datasets of sounds for analysis. Each sound in the dataset is represented in the form of a waveform or spectrogram. The waveform is a raw representation of sound that has two parameters—amplitude and time. Generally, Data size for waveform representation is large and depending on sound quality may require 44,000–96,000 samples per second. Jukebox by OpenAI [2] and WaveNet by Google DeepMind are examples of GenAI models, which use waveform representation. The spectrogram representation uses three parameters—frequency, time, and amplitude of a sound wave. Datasets using spectrogram representation are smaller in size and so are the most popular representations.

The following techniques are representative of acoustic-level notation:

a. **Log Mel Spectrogram:** A log Mel spectrogram is widely used for music generation tasks due to its ability to accurately model musical characteristics. This method converts an audio signal in time domain to its respective frequency domain by applying "Fast Fourier Transform (FFT)". Before applying FFT, "Short-Time Fourier Transform (STFT)" technique is applied, which results in segmentation of audio into overlapping frames, thereafter FFT is applied to capture temporal properties of the audio waveform. In this method, Mel band-pass filters are used at the STFT output for aligning with the non-linear nature of human pitch perception. These filters group frequencies, which are spaced linearly when low and logarithmically when high into bins. The logarithmic transformation of the Mel-scale output is necessary to adjust the non-linear response of the human ears toward loudness and pitch. It also compresses the dynamic range of spectrograms.

b. **Mel-Frequency Cepstral Coefficients (MFCCs):** It is commonly used in applications such as speaker identification and speech recognition. As in log spectrograms, it also converts the time domain of audio signals to their respective frequency domain and then applies Mel scale through Mel filters. It then applies Discrete Cosine Transform (DCT) on the logarithmic output of Mel filters. It also analyzes temporal changes after segmenting audio signals to overlapped frames by STFT.

c. **Constant-Q Transform (CQT):** The CQT provides the log-frequency spectrogram to accelerate audio processing tasks such as timbre transfer in music. It simplifies the extraction of note frequencies by using a logarithmic scale. The log scale resonates well with the exponential nature of musical pitch. In this technique, the central frequency to bandwidth ratio is kept constant, which enables the usage of variable length filters for performance.
d. **Chroma Features**: The chroma features identify chords, modulations, and key signatures in music. These are used for generations of chromagrams that capture harmonics in western music regardless of an instrument or timbre. The chroma features segregate the music into twelve different pitches to highlight the melodic structure of composition. Recently, chromogram is used in MusicGen for generation of melody-guided music.

11.3.2.1.1 Symbolic Music Representation

Symbolic music representations play an important role in the simplified processing of digital music. These are further classified into performance- and note-level representations. The following techniques are representative of these two classifications:

Performance-Level Symbolic Music Representation: The Musical Instrument Digital Interface (MIDI) is the representative of this class. The performance-level formats provide details of musical techniques, expressions, dynamics, tempo, etc.

a. **Musical Instrument Digital Interface (MIDI)**: This technique is most suitable for applications requiring interactions between devices in real time. It captures and encodes musical data about notes' duration, pitch, velocity, and other performance-related aspects such as articulation, dynamics, and control changes. However, this format lacks specific data about musical structures such as time and key signatures of music.

Note-Level Symbolic Music Representation: These notations are useful for denoting structural music elements such as rhythm and pitch into human-readable format. The following formats— MusicXML, MEI, and ABC notation—are representatives of this technique:

a. **MusicXML**: MusicXML offers a universal transferable format by using Extensible Markup Language (XML). XML promotes human and machine readability and gives a comprehensive notation of musical elements, which can be later used for the graphical representation of musical elements. The textual format is useful for encoding music scores. In short, MusicXML offers a universal transferable format by encapsulating the complexities of musical notation.
b. **Music Encoding Initiative (MEI)**: It uses XML-based schema to provide rules to document physical and intellectual properties of musical notation. The XML-based music notation helps in encoding various

musical notation systems such as western, Medieval (neume), and Renaissance. It also enables seamless display, search, retrieval, and exchange of information among different digital musical platforms.

c. **ABC Notation**: This notation was developed to simplify online sharing of music. It uses an alphabetic system (letters a–g and A–G to mark notes, and "z" to mark rests) together with ASCII characters to represent musical elements in a compact and machine-friendly approach for software development, for example, in developing text-to-symbolic music models such as chatMusician. In ABC notation, the structure is divided into two parts—header and body. The header contains metadata such as titles, time signatures, reference numbers for organizing multiple pieces, keys, and default note lengths. The body contains the musical content where each note and rest are encoded into tokens delineated by bar lines to reflect duration, pitch, and rhythmic placement.
d. **Tokenized Sequence Representation:** This is the most used technique to input symbolic music and can be synthesized from MusicXML or MIDI. It focuses on specific musical information and supports flexible encodings.

11.3.3 Music Generation Models

The computational methods for music generation, composition, or retrieval tasks often apply ML algorithms, after requisite features have been extracted from audio signals or waveforms through signal processing techniques. Unlike speech, there is more than one simultaneous speaker. The perception of what the speakers say is not always sensibly associated with real-world events or objects. Also, different note events in music are dependent, and the rate of sampling signals for music is higher in comparison to speech, which makes the task of capturing the "Language(s)" of music more challenging. Hence, generative language models are used to model long-term structures of languages, which include features like grammar, semantics, prosody (i.e., the pattern or style of sound and rhythm in speech), etc.

The GenAI models for music attempt to generate complete musical tracks in response to a simple textual prompt. Simple textual prompts can describe simple audio scenes with accuracy; however, the presence of a temporal dimension in musical tracks makes it difficult to describe a complex musical piece using simple text prompts alone. One solution to this problem is to use a conditioning music sample as input to the GenAI system, and the Generative model can use this conditioning track to infer the style of generated output. In addition, GenAI algorithms require a large amount of data during their training phase for creating generative models for the Music Generation problem. However, the absence of such datasets, which pair text to corresponding music/audio tracks, is a major problem for training AI models in this domain. One way to solve this problem is to use a mapping function, which can provide a measurable evaluation of proximity of a text to a given musical composition. For example, Huang et al. [37] proposed a joint text–audio embedding (MuLAN, 2022), which can map related text and audio nearby in a text–audio combined embedding space. This embedding space was created using comment-based metadata as training input along with the audio track. Such embeddings can be used to provide a huge supply of text–audio pairs for training and testing a GenAI model. Agostinelli et al. proposed MusicLM [38], which utilizes MuLAN [37] text–audio joint embeddings for training and testing the GenAI model for music.

As music also exhibits complex hierarchical structures, it is reasonable to apply similar techniques to music generation. In the literature, Transformers, GANs, VAEs, etc., have been used for generative music tasks and are discussed below. A few examples of popular music generator models—WaveNet, MusicGen, Jukebox, MelGAN, HiFiGAN, MusicLM, MusicLDM, DiffSound, etc.—belong to the models discussed below.

11.3.3.1 Autoregressive Model

AR models analyze and anticipate the next item in the sequence after taking into consideration the previously predicted items. These models use their knowledge of sound patterns and the previously predicted sound elements, such as initial spoken words or melody, to predict the next sound element. A long sound sequence is built with each time step, when an AR model considers the initial input and the earlier predicted values to estimate the next sound element. The block diagram is illustrated in Figure 11.1.

In the block diagram, TDU represents the Time-Delay Unit, which introduces delay to y_t so that previous time steps or historical data are included as I/P features. TDU transforms y_t into a vector $[y_t, y_{t-1}, y_{t-2}, \ldots, y_T]$. WL_1, WL_2, and WL_3 are weights associated with layers. Deepmind's WaveNet uses an AR model that is trained on raw audio waveforms. $\hat{Y}(t)$ is the predicted O/P, which is fed back as an I/P for future time steps. The feedback loop creates a recurrent structure to learn dynamic dependencies, allowing the network to model systems where future states depend on past outputs. The activation functions f_1 and f_2 could be SoftMax, Sigmoid, ReLU, Tanh,

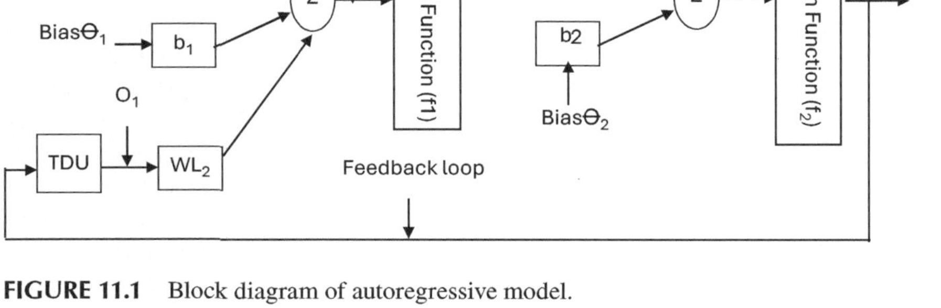

FIGURE 11.1 Block diagram of autoregressive model.

etc., whereas bias b_1 and b_2 are error terms or white Gaussian noise. The net output of layers 1 and 2, NO_1 and NO_2, is calculated as follows:

$$NO_1 = WL_1 * y(t) + WL_2 * O_1 + b_1 \tag{11.1}$$

$$NO_2 = WL_2 * f_1(NO_1) + b_2 \tag{11.2}$$

The AR models can capture sequential dependencies, which allow them to generate audio samples in line with the context provided by previous samples. So, an AR model can be expressed in terms of conditional probability as:

$$\begin{aligned} p(y) &= \prod_{i=1}^{T} p(y_i \mid y_1, y_2, \ldots, y_{i-1}) \\ p(y) &= \prod_{i=1}^{T} p(y_i \mid y_{<i}) \end{aligned} \tag{11.3}$$

Deepmind's WaveNet uses an AR and fully probabilistic model that uses raw audio waveforms. It can be used for both music generation and speech synthesis.

11.3.3.2 Diffusion Model

The diffusion model (DM) gradually adds Gaussian noise into the sample and then methodically reverse it, i.e., denoise the sample [57,58]. To do so, a DM defines a Markov chain of "diffusion" steps to gradually add random Gaussian noise in the sample in the forward diffusion process and then learns to reverse the diffusion process in the reverse diffusion process. In simple words, DMs are latent variable models that map to the latent space through a fixed Markov chain. The dimensionality of the latent variables is high and is the same as the original sample. Noise2Music proposed in [59] uses a DM for music generation.

11.3.3.2.1 Forward Diffusion Process (FDP)

To understand FDP, consider a data point S_0 sampled from a real data distribution $q(S)$. The FDP incrementally adds small quantity of Gaussian noise in T steps to generate a sequence of noisy samples $S_0, S_1, \ldots, S_T$. This diffusion to the initial sample through successive transformations helps the model to record and replicate complex patterns and essential details in target distribution. The step sizes are controlled by a variance schedule $\left[\beta_t \in (0,1)\right]_{t=1}^{T}$. The Markov chain can be represented by $q(S_t \mid S_{t-1})$, which is defined as:

$$q(S_t \mid S_{t-1}) = N\left(S_t; \mu_t = \sqrt{1-\beta_t}\, S_{t-1}, \beta_t I\right)$$

$$q(S_{1:T} \mid S_0) = \prod_{t=1}^{T} q(S_t \mid S_{t-1})$$

The data sample S_0 gradually loses its distinguishable features as T increases. When $T \to \infty$, S_T becomes equivalent to an isotropic Gaussian distribution, where

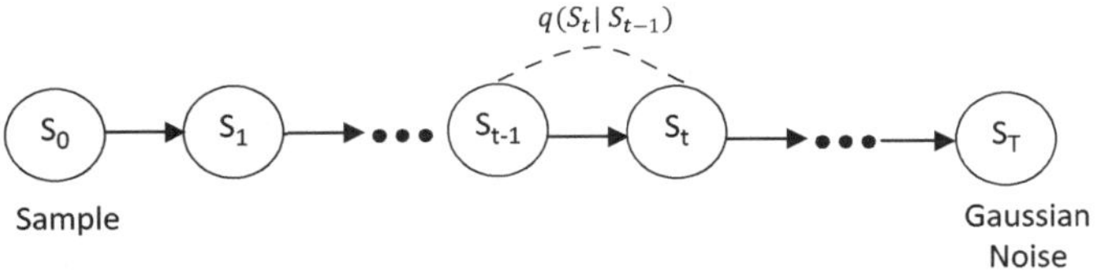

FIGURE 11.2 Forward diffusion process.

covariance matrix Σ can be written as $\Sigma = \sigma^2 I$ with σ^2 being the scalar variance and I being the identity matrix (Figure 11.2).

11.3.3.2.2 Reverse Diffusion Process

This phase separates DMs from other GenAI models such as GANs. In this phase, specific noise patterns that were added at each time step are identified and then denoising of the sample begins here. This reconstruction process is done using a Markov chain. Hence, if we can learn to reverse the Gaussian distribution $q(S_t \mid S_{t-1})$, we will be able to synthesize from beginning, i.e., from $S_T \sim N(0, I)$, a true sample from Gaussian noise input. For smaller values of β_t, $q(S_t \mid S_{t-1})$ will be Gaussian. But the estimation of $q(S_t \mid S_{t-1})$ is difficult as it requires an entire dataset. So, this estimation of conditional probabilities requires a model, p_θ, to be learned, which simply adjusts mean and variance for the desired result. p_θ can be represented in terms of conditional probabilities as follows:

$$p_\theta(S_{t-1} \mid S_t) = N(S_{t-1}; \mu_\theta(S_t, t), \Sigma_\theta(S_t, t)) \text{ where } \Sigma_\theta(S_t, t) = \sigma_t^2 I \text{ and}$$

$$\mu_\theta(S_t, t) = 1/\sqrt{\sigma_t} \left(S_t - \left(\beta_t / \sqrt{1 - \sigma_t'}\right) \varepsilon_\theta(S_t, t)\right) \text{ where } \sigma_t' = \prod_{m=1}^{T} \alpha_m$$

$$p_\theta(S_{0:T}) = p_\theta(S_T) \prod_{t=1}^{T} p_\theta(S_{t-1} \mid S_t)$$

The reverse conditional probabilities are tractable when conditioned on S_0 (Figure 11.3).

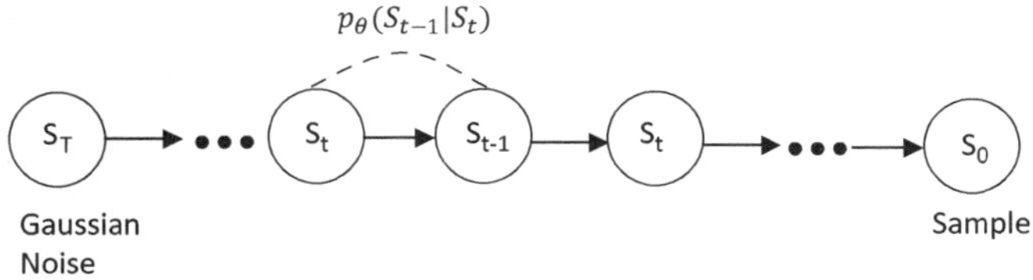

FIGURE 11.3 Reverse diffusion process.

The DMs might seem to be specific cases of latent variable models, but are advantageous in terms of the number of degrees of freedoms they permit. Selection of variances, β_t, of the forward phase and the Gaussian distribution parameter μ_θ of the reverse phase determine the successful implementation of the DM.

11.3.3.3 Transformer Model

Transformer architecture has become popular for text generation through the GPT model; however, they can also be used for music and sound generation. The input stream is analyzed and broken into smaller units called tokens. This process is called tokenization. The attention block identifies the relationship between different tokens. The feed-forward block extracts higher-level features and adds description to the tokens to predict the next element in the sequence. This predicted token is added back to the stream, and this process is repeated till the entire input sequence is analyzed (Figure 11.4).

Google's MusicML and OpenAI's MuseNet use transformers for music generation.

11.3.3.4 Variational Autoencoder (VAE) Model

VAEs, as described in [50], extend the idea of autoencoders (AEs). The AEs are designed for learning efficient data encodings and consist of two parts—Decoders and Encoders. The encoder component maps the input data samples into the latent space, while the decoder component uses the latent space representation to reconstruct the data. The latent space represents compressed input data, which is then used to denote data in lower dimension while preserving its important properties. The VAEs use a probabilistic model instead of a deterministic model in the latent space to generate diverse and new samples. The difference between AEs and VAEs is that the output of an encoder in AEs is a fixed latent vector, while it is the mean and variance (parameters of a probability distribution) in the case of VAEs (Figure 11.5).

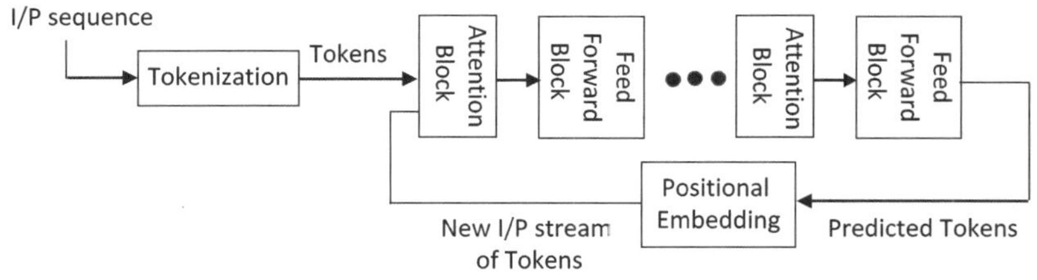

FIGURE 11.4 Block diagram of transformers.

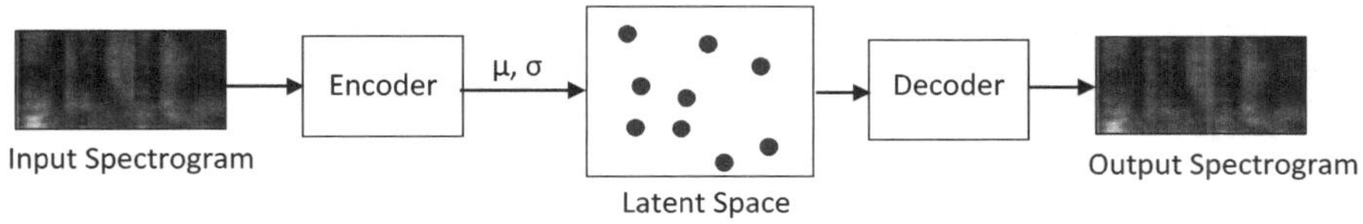

FIGURE 11.5 Block diagram of Variational Autoencoder (VAE).

OpenAI's Jukebox uses both VAE and a Transformer for music generation. The VAE compresses the input sound samples into latent space, while the transformer generates new audio samples from that latent space.

11.3.3.5 Generative Adversarial Networks (GANs) Model

A GAN consists of two parts—a "Generator", G, and a "Discriminator", D [49]. The "Generator" model learns to generate new instances of data from the latent space, while the "Discriminator" model learns to differentiate between different instances of data generated by its "Generator" component. That is, G is trained to befool D, while D is trained to differentiate between fake and original. In short, a Generator captures joint probability say $p(X,Y)$ or just $p(X)$ if there are no labels, while the Discriminator captures the conditional probability $p(X \mid Y)$ (Figure 11.6).

In the training process, G simply maps random noise Z sampled from a known distribution into a data space. This can be represented by the mathematical model, a two-player minimax game between G and D, as follows:

$$\min_G \max_D E_{X\sim p_d}\left[\log\left(D(X)\right)\right] + E_{Z\sim p_z}\left[1 - \log\left(D((G(Z)))\right)\right]$$

where p_z and p_d denote the probability distribution of Z and real data X, while $E(.)$ denotes expected value. The Multi-Track Sequential Generative Adversarial Networks (MuseGAN) uses the GAN framework to generate multibar, multitrack, and polyphonic music.

11.3.3.6 Most Popular Music Generation Model

11.3.3.6.1 WaveNet

The model proposed in [4] generates raw audio waveforms, where the prediction of each audio sample is conditionally dependent on all previous audio samples. WaveNet gives a flexible and generic framework for applications depending on audio generation such as Music, speech synthesis, voice over, and speech enhancement. The mathematical model used can be described by Equation 11.3 in Section 11.3.3.1, where each audio sample y_i is conditioned on samples generated in all previous time steps. The conditional probability distribution in WaveNet is modeled by a stack of "causal" convolutional layers. Here, the word "causal"

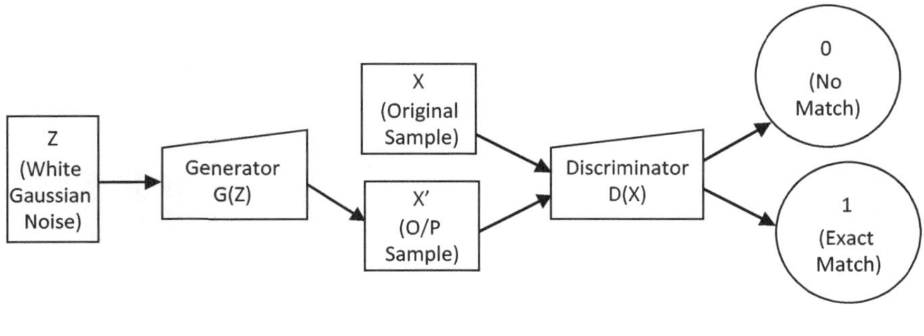

FIGURE 11.6 Block diagram of Generative Adversarial Network (GAN).

means only "past" can influence the "future" not the vice versa. This ensures that the ordering used to model the data is not violated. In other words, prediction $p(y_{t+1} \mid y_1, y_2, \ldots, y_t)$ by the model at t time steps is independent of its future time steps at $t+1$, $t+2$, ..., T. At the output layer, SoftMax function is used. WaveNet uses "Gated Activation Units", which are described below as an activation function:

$$z = \tanh\left(W_{f,k} * y\right) \odot \sigma\left(W_{g,k} * y\right)$$

where W is the learnable convolution filter, $*$ is the convolution operator, $\sigma(.)$ is the sigmoid function, k is the number of layers, f is the filter, g is the gate, and $\odot$ is an element-wise multiplication operator.

To generate audio with specified characteristics or additional input say h (for example, in the case of a multi-speaker setup, h can be speaker's identity), the conditional probability is denoted as $p(y \mid h)$, where

$$p(y \mid h) = \prod_{t=1}^{T} (y_t \mid y_1, y_2, \ldots, y_{t-1}, h)$$

This conditioning of the model can be local or global. The global conditioning uses a single latent representation of h to influence the output across all time steps. The activation function, z, is described as:

$$z = \tanh\left(W_{f,k} * y + V_{f,k}^{T} * h\right) \odot \sigma\left(W_{g,k} * y + V_{g,k}^{T} * h\right)$$

where $V_{*,k}$ is the learnable linear projection and $V_{*,k}^{T} * h$ is the vector broadcast over time dimension.

The local conditioning uses one more time series h_t, whose sampling frequency is lower than the sampled audio signal. A transposed convolutional network is then used to transform h_t into a new time series $r = f(h)$ before using it in the activation unit. The resolution of this new signal is the same as that of an audio signal. The activation function, z, is described as:

$$z = \tanh\left(W_{f,k} * y + V_{f,k} * r\right) \odot \sigma\left(W_{g,k} * y + V_{g,k} * r\right)$$

where $V_{f,k} * r$ is a 1×1 convolution.

11.3.3.6.2 AudioLM

The proposed framework in [1] is used for coherent audio and music generation from text description. In speech, coherence is synonymous with syntax, semantics, and phonetics, while in music it resonates with tone, harmony, melody, and rhythm. The model can also generate audio and music from conditioned music while maintaining prosody, original voice, background music, and reverberation level. The model can produce high-fidelity audio/music with long-term coherence without using any annotation. The model uses hierarchical tokenization for audio synthesis and music generation where coarse-to-fine audio units are considered as tokens to the model input. The tokens are categorized into semantic and acoustic tokens. Semantic tokens help in modeling long-term structure, while acoustic tokens help in capturing essential acoustic details. These two tokens help in generation of high-quality and coherent music and audio without depending on symbolic music representation or transcripts.

The model begins with raw audio waveforms. Initially, coarse semantic tokens are synthesized by using the w2v_BERT model, as discussed in [60]. AR is then used on these tokens to determine local dependencies such as melody in music or phonetics in speech, and global long-term structures such as syntax and semantics of the language in speech or rhythm and harmony in case of music. In the next step, fine-level acoustic tokens are generated by the SoundStream model, as discussed in [39], to determine the detailed or high-level features of the audio waveform. The block diagram of the tokenization process in AudioLM is shown in Figure 11.7.

AudioLM uses an AR hybrid tokenizer along with a multi-stage transformer. The three main components of the model are:

1. **Tokenizer**: It maps a single audio sequence y into a sequence of discrete tokens S. Let $y \in R^T$, then $S = \text{encoder}(y)$ and $S = S_1, S_2, \ldots, S_{T'}$, where $T' << T$.
2. **Decoder-Only Transformer**: The transformers work on discrete tokens to maximize the likelihood of $\prod_{t=1}^{T} p(S_t \mid S_{<t})$. The token sequence $\hat{h}$ is predicted Autoregressively by the model at inference time.
3. **Detokenizer**: It is used to reconstruct the waveform by mapping sequence of predicted tokens back i.e., $\hat{y} = \text{decoder}(\hat{S})$. The detokenizer and tokenizer are pre-trained and frozen before training the AudioLM's language model.

11.3.3.6.3 MusicLM

The MusicLM model proposed by Agostinelli et al. [38] uses text description to produce high-fidelity music. The model has the capability of generating conditional music on both text and melody at 24 kHz. For the generative component, MusicLM

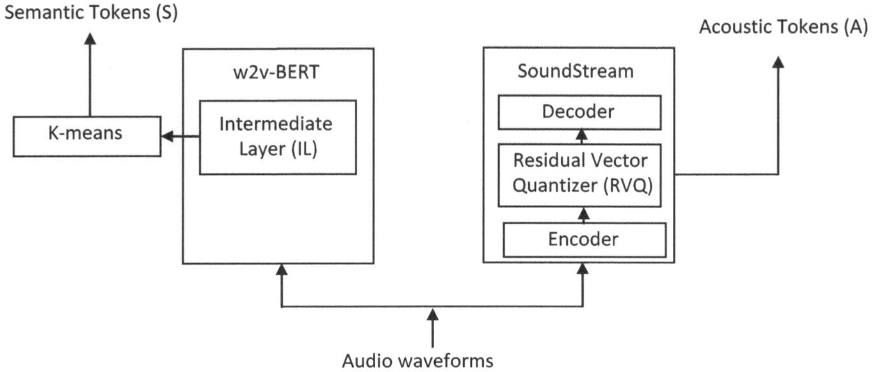

FIGURE 11.7 Tokenization in AudioLM.

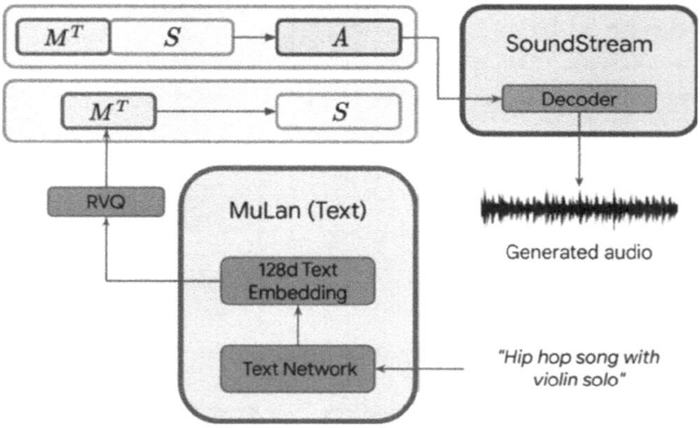

FIGURE 11.8 Block diagram of MuLAN [38].

takes advantage of AudioLM's multi-stage AR model while bearing in mind the text conditioning component. Due to scarcity of paired audio-text data, MusicLM uses MuLAN, as proposed by Huang et al. [37]. The MuLAN is trained to map music closely with its corresponding text description in shared embedding space. This shared type of embedding does not require labels during training, hence allowing training of the model solely on audio tokens. The block diagram of MuLAN is given in Figure 11.8.

The conditional AR music generation in MusicLM is facilitated by the audio representation provided by three models, namely w2v-BERT (WB), SoundStream (SS), and MuLAN. The WB model facilitates the extraction of semantic tokens (denoted by S) for long-term coherence. MuLAN is used for conditioning on text or melody. The SS model extracts acoustic tokens (denoted by A) for high-fidelity synthesis of music. These tokens are generated from monophonic audio @24 kHz with the striding factor of 480. This audio signal generates embeddings @50 kHz, which are quantized by using the RVQ component of the model. The RVQ uses 12 quantizers of vocabulary size 1024 to produce 600 tokens for each second of 6-kbps bit rate of embeddings. The WB component generates 25 semantic tokens per second of audio through a quantization method based on centroids of K-means of embeddings.

Semantic modeling uses conditional probability distribution $p(S_t \mid S_{<t}, M_A)$ to learn mapping between S and MuLAN audio tokens M_A. The acoustic modeling stage uses conditional probability distribution $p(A_t \mid A_{<t}, S, M_A)$ to predict acoustic token say A_p, which are conditioned on both S and M_A. This stage is further split into fine and coarse modeling to evade long sequences of tokens. The block diagram of MusicLM is given in Figure 11.9.

The SS and WB models of MusicLM are trained on the "Free Music Archive (FMA)" dataset (Defferrard et al., 2017), whereas the AR models and tokenizers used during acoustic and semantic modeling are trained on five million audio clips, which are equivalent to 280k hours of music @24 kHz.

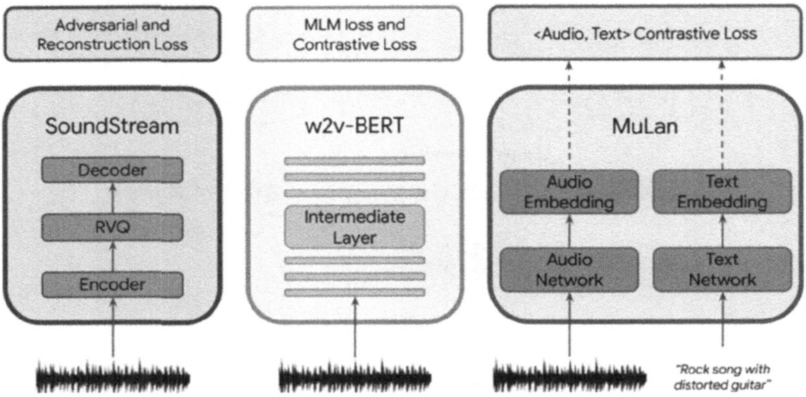

FIGURE 11.9 Block diagram of MusicLM [38].

MuseNet: It uses a decoder transformer for prediction of the next musical note from a sequence of previous predicted notes. MuseNet uses a sparse transformer, a type of decoder transformer. Instead of computing an $N \times N$ attention matrix for a sequence of length N, the transformer computes weights for a subset of input position, thus reducing memory requirements and computational complexity of the model. For example, the 72-layer neural network with focus over the context of text or musical note details comprises only 4,096 tokens.

The model extracts tokens from each note and duration and then performs sine position embedding for determining the token position. Consider a word with position "l" in the input sequence. This position is converted into a unique vector by the following function:

$$PE_{l,\,2j} = \sin\left(l/1{,}000^{(2j/d)}\right)$$

$$PE_{l,\,2j+1} = \cos\left(l/1{,}000^{(2j+1)/d)}\right)$$

The small values of j result in functions of shorter wavelengths (λ), whose value changes rapidly along the position axis, while larger values of j result in functions of longer λ. Encoding of each position is unique and is a specific combination of different values of λ.

11.3.3.6.5 MuseGAN

The Multi-track Sequential Generative Adversarial Network (MuseGAN) proposed by Hao-Wen Dong et al. [36] uses a GAN for multi-track symbolic music generation. Deep CNNs are used to implement both the Generator, G, and the critic or Discriminator, D. G grows along the time axis first and then along the pitch axis. D works in the reverse direction, i.e., from pitch to time axis. The proposed model has the capability of generating coherent music of four bars independently, i.e., without any human input. But the model can also learn from the supplied track by understanding the temporal structure of the given musical track. A musical track can be broken into smaller pieces whose structure is given in Figure 11.10.

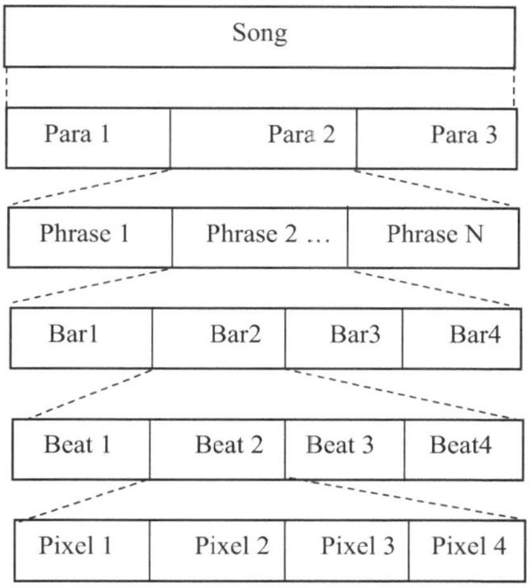

FIGURE 11.10 Hierarchical structure of a musical piece.

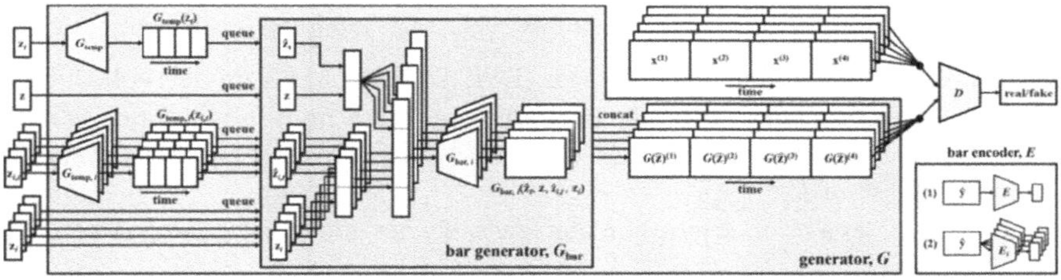

FIGURE 11.11 Block diagram of MuseGAN [36].

This structure is useful in understanding the semantic and acoustic tokens of a musical track. MuseGAN is trained on five tracks, namely piano, guitar, strings, drums, and bass, with 1 million bars of rock music. The basic compositional unit is taken as bars instead of notes, and the music is generated one bar at a time by using a temporal structure (Transposed CNNs). The block diagram of the MuseGAN model is given in Figure 11.11.

The input (denoted by z) to MuseGAN comprises four components and can be style, chords, groove, and melody. Each component is represented by a random vector, all of which are described below as:

- z, which is **inter-track** and time **independent**.
- z_t, which is **inter-track** and time **dependent**.
- z_i, which is **intra-track** and time **independent**.
- $z_{i,t}$, which is **intra-track** and time **dependent**.

For the ith track, where $i = 1, 2, \ldots, M$, z_t and $z_{i,t}$ which are time-dependent random vectors are input to the shared temporal structure generator G_{temp} and private temporal structure $G_{\text{temp},\,i}$. Both G_{temp} and $G_{\text{temp},\,i}$ convert each z (the I/P noise vector of length 32) into a series of separate latent vectors containing intra- and inter-track temporal information for every bar. These latent vectors are concatenated and fed to G_{bar}, the bar generator, for generating music sequentially. This method can be mathematically represented as:

$$G(Z) = \left\{G_{bar,\,i}\left(z,\, G^{t}_{temp}(z_t), z_i,\, G^{t}_{temp,\,i}(z_{i,\,t})\right)\right\}_{i,\,t\,=\,1}^{M,\,T}$$

In the case of conditional track music generation, an encoder E is used to extract helpful inter-track features from user-supplied track input. To devise the mathematical model for this process, consider $\hat{y}$ as a bar sequence of a given musical track, which is then used to learn and generate its underlying temporal structure and remaining tracks. Let G^c denote the track conditional generator, which generates sequence of bars with the help of conditional bar generator G^c_{bar}. Let $\hat{z}$ denote the random noise and $\hat{z}^{(t)}$ denote the time-dependent random noise. By using an encoder E, the training procedure can be represented using the following equation:

$$G^c\left(\hat{z}, \hat{y}\right) = \left\{G^c_{bar}\left(\hat{z}^{(t)}, E\left(\hat{y}^{(t)}\right)\right)\right\}_{t\,=\,1}^{T}$$

The model used the Lakh MIDI dataset (LMD), which has a unique collection of 176,581 MIDI files. These files are then converted into multi-track piano-rolls. To model frequently occurring temporal patterns, MuseGAN used bars for duration of 96 and the height of 128.

Table 11.1 gives a brief description of a few speeches and music generation models.

TABLE 11.1
Summarization of Speech and Music Generation Model

Model Name	Purpose	Type of I/P	Architecture
Deepmind's WaveNet	Speech and Music generation	Raw audio waveform	Autoregressive
Google's MusicLM	Music generation	Text description	Transformer
OpenAI's MuseNet	Music generation	Musical notes as tokens	Transformer
AudioLM	Speech and Music generation	Raw audio waveform	Autoregressive, Transformer
OpenAI's JukeBox	Music generation	Music samples/track with I/P conditioning such as artist and genre	VAE, Transformer
MusicGAN	Music generation	White Gaussian Noise	GAN
Meta's MusicGen	Music generation	Text description or audio	Transformer

11.3.4 Concerns and Challenges in using GenAI for Music Composition

The usage of GenAI in composition of music also faces hurdles, a few of them are enlisted below:

1. **Violation of Copyrights and Intellectual Property Rights:**
   Copyright infringement is one of the main concerns for Generative Music systems. These systems must ensure adherence to copyright laws and under no circumstances should they output a memorized musical track. Recent research (Carlini et al. [41]) has shown that tendency to produce memorized input increases as size of the model increases or when a given input example has been duplicated multiple times during the training phase. As training samples for GenAI systems are obtained using publicly available platforms, it is extremely difficult to ensure uniqueness of the training data. In addition, it has been shown that carefully crafted prompts can be used (Carlini et al. [42]) to retrieve memorized data from Generative models; such attacks may result in privacy concerns for people whose data has been used for training the model.
2. **Ethical Use of Training Data:**
   Another important challenge is, how to use data ethically? As the generated output may produce compositions that look like the work of an existing artist, that artist may not be comfortable with this idea. It also raises doubts in the mind of artists that their work might get used to without their consent or knowledge.
3. **Disruption of the Music Industry:**
   The quick and easy music generation as per the requirement of the user poses serious challenges for budding musicians. So, a balance is required to safeguard the livelihood and individuality of human musicians and use the GenAI as a creative tool.
4. **Originality and Bias:**
   Another major problem revolves around the originality and biasing of the training music dataset. There is a possibility that the dataset is not properly cleaned before use due to which the generated music shows specific patterns. Such a dataset might produce biased music styles and genres.

11.3.5 Future of GenAI Music

GenAI has revolutionized the process of music composition for musicians through its innovations and creativity. AI models can allow quick and easy mixing of harmonies so that music adjustable across genres of any style and emotional types, such as sad, melodious, etc., can be composed. As these technologies are continuously evolving, it requires responsible usage of these creative tools. Technology should not be used as a replacement of human creativity; instead, it should work in collaboration with human voice for its seamless integration into the creative process.

11.4 DEEPFAKES

11.4.1 Introduction

The term "Deepfake" is a combination of two words, namely "Deep Learning" and "Fake", and refers to synthetic media created using DL techniques. Such media can be used for constructive purposes like entertainment, voice-overs, narrations in audiobooks, voice assistants, or assisting individuals with disabilities. However it can also be used for malicious purposes like spreading rumors, impersonation, misinformation, conspiracies, or spreading fake news [43]. Advancements in GenAI technology coupled with widespread social media penetration are primary enablers for this problem, as it is possible to spread Deepfakes easily.

The availability of large voice datasets, along with highly efficient Deep Neural Network architectures, has resulted in the creation of ML models, which can create realistically matching audio samples from short audio content. GANs are the most commonly used architectures for deepfake creation as they produce high-quality deepfakes. Moreover, frameworks like DeepFaceLab [44] have democratized access to deepfake technology, further exacerbating this problem as anybody with beginner level technology skills can create and spread deepfakes. Tools like FaceApp, Veritone Voice, Murf AI, Resemble AI, Descript, Respeecher, iSpeech, etc., have also lowered the entry barrier for deepfake creation.

Deepfakes can be easily distributed to people across the globe using social media platforms like Facebook, Twitter, and messaging apps like WhatsApp and Telegram. In the past few years, deepfakes have been used by politicians to influence voter behavior for securing favorable election outcomes. Hence, it raises ethical concerns about the privacy and security of individuals and necessitates the development of effective mechanisms to detect and fight against deepfakes.

11.4.1.1 Use Cases of Deepfake GenAI models

"Deepfakes" in speech and music offer many applications for industries and individuals, ranging from customer service to healthcare or to entertainment. The following are some examples in these contexts:

1. In the animation industry, Deepfake technology gives unique and distinct voice styles to the animated characters, irrespective of their voice or language.
2. People facing health issues, particularly speech disabilities where their communication ability is hampered, can use this technology. In this case, patients' speech can be enhanced for clarity and accuracy.
3. In the advertisement and entertainment industry, this technology can be used for:
   a. dubbing the voice of an actor during her unavailability;
   b. creating the voice of an artist in languages which they may not speak.
4. Original musical tracks of an artist can be restored by using this technology.
5. In the gaming industry, this technology is being used to create characters who can speak dynamic and realistic dialogues matching the personality of their gaming counterparts.

11.4.2 Deepfake Identification

In recent years, deepfake detection has also emerged as an active research area. A major concern about their use is the detection of generated content inside genuine human-made content. Various techniques available for detecting deepfakes can broadly be classified into four broad categories: ML, DL, Statistical Measurements (SM), and Blockchain (BC) [46]. ML-based deepfake detection techniques employ the usage of various methods like Decision Trees, Random Forests, etc., for detecting deepfake content. DL-based methods seek to use Deep neural network models like MLP (multi-level perceptron), CNNs (convolutional neural networks), etc., to detect deepfake models. SM-based methods use statistical measures to determine data originality such as using average normalized cross-correlation between suspected and original samples of data and Photo Response Non-Uniformity (PRNU) for deepfakes in video frames. As Images generated by real camera sensors have inherent noise in them, Koopman et al. [47] used PRNU as a tool for differentiating deepfake videos from real ones. The BC-based deepfake detection methods use public BC to verify sources of digital content. The authors in [48] proposed a BC-based method to trace the originality of video source.

In short, the identification or detection of "deepfake" content is a complex task, and it requires a lot of motivation and effort to verify authenticity of a "deepfake", more so, when it has been spread widely through social media. For this reason, researchers suggest the following to minimize the harmful effects of "deepfake" content:

- **Educating the Public about the Existence of Deepfake Technology**: Individuals can check for signs of deepfakes by watching inconsistencies in tone, style, and audio or facial expressions in a video or recording. Individuals must also verify the origin, credibility, and context of the supplied content.
- **Use of Deepfake Detection Tools**: Using several advanced "deepfake" identification/checker tools such as "Truepic", "Deepware Scanner", "Sensity AI", "Pindrop", and "Reality Defender", for audio or video deepfake identification/detection.
- Bringing legislation and policy regulations.

11.4.3 Deepfake Generative Models

Deepfakes creation requires source data (image, video, or audio) to extract features of the "target person". These features are superimposed on the target data to create a realistic looking deepfake sample. Several neural network architectures like GANs, VAEs, CNNs, and RNNs/LSTMs have been used to create Deepfakes. The following section discusses these architectures in brief:

11.4.3.1 Generative Adversarial Networks (GANs) Model

GANs proposed by Kingma [50] are the widely used neural networks used for creating Deepfakes. GAN architecture can be understood as a game between two AI models, one of which is called a Generator and the other one is called a Discriminator. Unsurprisingly, the task of the Generator network is to generate samples, which resemble samples in the training and data input. And the goal of the Discriminator network is to learn a function, which can be used to differentiate between real and generated data.

11.4.3.2 Autoencoders (AEs)/Variational Autoencoders (VAEs) Model

To create deepfake in music, an "autoencoder" (AE) is generally used, although other techniques are also available in literature. The AE performs waveform synthesis to obtain raw audio signals. The bits in these raw audio signals are compressed to obtain a simpler representation of audio signals for further processing. The vocoder (voice encoder) part in the AE is trained to synthesize the audio signal after inverting the compressed raw audio signals. Generally, Mel-spectrogram is used for waveform synthesis and reconstruction of audio signals, but more recently neural codecs such as Encodec, SoundStream, RVQGAN, or DAC are being used [37,45,55]. Another internal module is trained either to temporally compress the audio sequence more or generate an audio signal conditioned on text inputs. For example, in LLM-inspired architectures such as AudioLM or MusicLM, semantic work necessary for generations of time-based coherent musical sequence is performed by an LLM.

The encoder stage of a VAEs [50] learns a latent space representation for input data, and the corresponding decoder neural networks learn to recreate the original signal from this latent representation. The decoder stage of VAEs can be used as a generative model for generating deepfakes, if suitable latent space input is provided.

11.4.4 Some Commercially Available and Popular Deepfake Models

a. **Uberduck** [53]: Uberduck provides a GenAI platform for high-quality voice synthesis and music compositions from input text. For the speech synthesis task, Uberduck uses the state-of-the-art models such as Tacotron2, HiFi-GAN, zero-shot RADTTS, and SO-VITS-SVC. The platform can be used to clone or convert voice, translate text to speech, rapping, or singing, and generate high-quality music tracks for singing and rapping. The user can pick a voice from its collection of built-in voices and select a beat from its diverse and vast instrumental music collection
b. **AudioCraft** [54]: AudioCraft is a GenAI library developed by Meta to generate high-quality and realistic music and audio from text. This library is based on three models, namely "MusicGen", "EnCodec", and "AudioGen". MusicGen generates music from text prompts and is trained on musical data owned and licensed by Meta; EnCodec decoder is used to generate higher quality music with fewer artifacts; and AudioGen is used to generate audio from text prompts and is trained on publicly available sound effects.

11.4.5 Deepfake Detection in Audio and Music

The sudden spurt in audio and musical deepfakes is the result of available technology with which deepfakes can be easily created, shared, and controlled. For example, the rapid improvements in text-to-speech and voice conversion technology have enabled creation of musical "deepfakes" from generated audio of an artist into a complete song, simply by mixing and combining human-composed beats with beats generated from a GenAI model without the knowledge and consent of the artist.

11.4.5.1 How to detect Deepfakes in Music?

Detection of artificially generated musical sequences can be a challenging task. There is a potential risk that while making a machine learn an artist's style, we end up generating deepfake music when beats and other sound effects are added to it. For instance, the response of MusicGen to a text prompt might always be interpreted in a particular way as a 4/4-time signature. But, it might be simpler to detect if the given audio sample is the result of an AE. For example, it's a well-known fact that checkered artifacts are mainly the result of neural decoders as they use transposed convolution operators, thus making it easier to detect generated artifacts. Another problem in deepfake music detection is the problem of confounding where a model is naively trained for classification of fake and original music samples. In such a case, it is a possibility that the detected features are unrelated to generated artifacts. Consider a public dataset with classical music tracks and a deepfake dataset with pop and rap tracks. In the case of confounding, the classifier might end up learning the detection of classical music instead of finding fake and original tracks [55].

In [55], the authors proposed a deepfake detection methodology in music, which used Mel-spectrogram for waveform synthesis, and a Griffin-Lim algorithm, which uses STFT for phase reconstruction of the signal. This combination, termed as "GriffinMel" reconstruction, makes it easy to catch audible phase errors. The sampling rate of synthesized and original audio signal is kept the same. For the AE part, two neural codecs, namely DAC (used by Vampnet 1) and Encodec (used by MusicGen and SunoV2), are used. To identify the reconstructed counterpart from an original audio, the model is trained to detect generated artifacts based on the file encoding and music semantics. The musical tracks in the dataset are encoded in mp3 with bit rates of 192kB/s, 256kB/s, and 320kB/s.

11.4.5.2 How to Detect Deepfakes in Audio?

The swift progress in the field of synthesizing human voices using ML techniques has increased the risk of impersonation and misinformation, which require the development of subtle methods for identifying synthetic human voices. Statistical features that are unique to an audio signal can be analyzed for the identification of synthetic or fake voices. For instance, in the case of synthetic voices, local phase inconsistencies can be captured through statistical analysis in the bispectral domain. The reason being that audio signals of real human voices have random phases, while synthetic human voices lack this characteristic.

Generally, deepfake audio synthesis models generate audio waveforms from Mel-spectrogram by using a neural network called vocoder. A similar technique is proposed in [56], which uses vocoders for the detection of synthetic human voices by identifying the presence of vocoder artifacts in the audio signals. The detection method uses the fact that in most cases of AI-based audio synthesis models, neural vocoders are Endpoints. So, it's unlikely that neural vocoders will process real audio signals. Hence, the vocoder's artifacts can be used to identify synthetic human voices. The model focuses on extracting features of vocoder's artifacts to identify discriminative features and then uses a binary classifier to identify deepfake signals.

11.4.6 Concerns and Challenges

The main challenge in using the "deepfake" technology is the lack of availability of large and labeled datasets. Although this technology has given us room for exciting opportunities, it also faces certain challenges and raises certain ethical concerns. The probable misuse of this technology in spreading fake news, misinformation, or manipulation of public opinion has triggered debates on the ethical use of this technology. A few of its potential misuses are given below:

- Fraud through impersonation or deception through cloned or "deepfake" voices at financial or other institutions for extracting confidential information.
- Deepfake technology can be used by people for their vested interest in spreading false information, propaganda, or rumors to damage the trust and public reputation of organizations or individuals.
- Violation of an individual privacy and Consent by creating "deepfakes" in music or speech through a small musical composition or audio recording of the original artist/person, which is obtained without the consent or knowledge of the artist/person. Also, original recordings of artists can be manipulated without their consent.

To protect the privacy and rights of an individual or organization, the above concerns can be mitigated by adhering to ethical principles such as obtaining consent or intellectual rights of an artist in advance.

Conclusion

Voice-based GenAI stands at the forefront of the ongoing technological revolution, transforming how humans interact with machines and experience audio content. This chapter has traversed the diverse landscape of GenAI in speech and music, which is reshaping communication, entertainment, education, and accessibility.

The evolution of speech generation—from simple text-to-speech systems to sophisticated models like WaveNet and Tacotron—has enabled machines to produce voices that are not only intelligible but rich in emotional and stylistic nuance. Speech synthesis is now integral to virtual assistants, audiobooks, and accessibility tools. The development of systems capable of voice cloning and emotion recognition has further expanded the boundaries of what AI-generated speech can achieve, making it a powerful tool for personalization and inclusivity. However, this power also raises critical ethical and technical concerns, particularly regarding data bias, the authenticity of synthesized voices, and potential misuse in the form of deepfakes. By leveraging DL models like AR networks, DMs, and transformers, AI systems can now generate music that mimics the structure, tone, and rhythm of human-created compositions.

Tools like MusicLM, Jukebox, and MuseGAN exemplify how AI can contribute to the creative process, enabling musicians and hobbyists alike to explore new musical landscapes with minimal technical barriers. The use of symbolic and acoustic representations bridges the gap between human perception and machine generation, offering new opportunities for collaboration between AI and human composers.

The rise of audio deepfakes, powered by voice cloning and generative models, represents a double-edged sword. While the technology has legitimate applications in restoring voices or recreating historical audio, it also opens the door to deception and misinformation. Identifying, detecting, and regulating deepfakes is becoming a research priority, necessitating the development of robust countermeasures and ethical frameworks.

This chapter has underscored that the future of voice-based GenAI lies not only in advancing algorithms but also in addressing broader societal questions. As technology becomes more accessible, democratized, and sophisticated, developers must embed ethical considerations and accountability into the design and deployment of GenAI systems. Issues such as consent, transparency, and data provenance will be crucial in maintaining public trust and ensuring responsible use.

In conclusion, the journey of GenAI in voice and audio is far from over. As models grow more capable, so too must our understanding of their implications. The field offers tremendous promise—be it in enriching human expression, bridging communication gaps, or enhancing creativity. Yet, with great power comes great responsibility. Through continuous innovation and interdisciplinary collaboration, we can harness the full potential of GenAI to build a future where synthetic voices not only sound real but also serve real human needs ethically and intelligently.

REFERENCES

1. Borsos, Z., Marinier, R., Vincent, D., Kharitonov, E., Pietquin, O., Sharifi, M., Roblek, D., Teboul, O., Grangier, D., Tagliasacchi, M. and Zeghidour, N., 2023. AudioLM: a language modeling approach to audio generation. *IEEE/ACM Transactions on Audio, Speech, and Language Processing, 31*, pp. 2523–2533.
2. Dhariwal, P., Jun, H., Payne, C., Kim, J.W., Radford, A. and Sutskever, I., 2020. Jukebox: a generative model for music. *arXiv preprint arXiv:2005.00341.*
3. Ling, Z.H., Kang, S.Y., Zen, H., Senior, A., Schuster, M., Qian, X.J., Meng, H.M. and Deng, L., 2015. Deep learning for acoustic modeling in parametric speech generation: a systematic review of existing techniques and future trends. *IEEE Signal Processing Magazine, 32*(3), pp. 35–52.
4. Van Den Oord, A., Dieleman, S., Zen, H., Simonyan, K., Vinyals, O., Graves, A., Kalchbrenner, N., Senior, A. and Kavukcuoglu, K., 2016. Wavenet: a generative model for raw audio. *arXiv preprint arXiv:1609.03499, 12.*
5. Trueba, J.L. and Klimkov, V., 2019. *Neural Text-to-Speech Makes Speech Synthesizers Much More Versatile.* Amazon Science.
6. Wang, Y., Skerry-Ryan, R.J., Stanton, D., Wu, Y., Weiss, R.J., Jaitly, N., Yang, Z., Xiao, Y., Chen, Z., Bengio, S. and Le, Q., 2017. Tacotron: towards end-to-end speech synthesis. *arXiv preprint arXiv:1703.10135.*
7. Łańcucki, A., 2021, June. Fastpitch: parallel text-to-speech with pitch prediction. In *ICASSP 2021-2021 IEEE International Conference on Acoustics, Speech and Signal Processing (ICASSP),* Toronto, Canada (pp. 6588–6592). IEEE.

8. Arık, S.Ö., Chrzanowski, M., Coates, A., Diamos, G., Gibiansky, A., Kang, Y., Li, X., Miller, J., Ng, A., Raiman, J. and Sengupta, S., 2017, July. Deep voice: real-time neural text-to-speech. In *International conference on machine learning* (pp. 195–204). PMLR.
9. Latif, S., Shahid, A. and Qadir, J., 2023. Generative emotional AI for speech emotion recognition: the case for synthetic emotional speech augmentation. *Applied Acoustics, 210*, p. 109425.
10. Kulke, L., Feyerabend, D. and Schacht, A., 2020. A comparison of the Affectiva iMotions Facial Expression Analysis Software with EMG for identifying facial expressions of emotion. *Frontiers in Psychology, 11*, p. 329.
11. Jia, Y., Weiss, R.J., Biadsy, F., Macherey, W., Johnson, M., Chen, Z. and Wu, Y., 2019. Direct speech-to-speech translation with a sequence-to-sequence model. *arXiv preprint arXiv:1904.06037.*
12. Engel, J., Resnick, C., Roberts, A., Dieleman, S., Norouzi, M., Eck, D. and Simonyan, K., 2017, July. Neural audio synthesis of musical notes with wavenet autoencoders. In *International Conference on Machine Learning* (pp. 1068–1077). PMLR.
13. Luong, H.T. and Yamagishi, J., 2020. Nautilus: a versatile voice cloning system. *IEEE/ACM Transactions on Audio, Speech, and Language Processing, 28*, pp. 2967–2981.
14. Le, M., Vyas, A., Shi, B., Karrer, B., Sari, L., Moritz, R., Williamson, M., Manohar, V., Adi, Y., Mahadeokar, J. and Hsu, W.N., 2024. Voicebox: text-guided multilingual universal speech generation at scale. *Advances in Neural Information Processing Systems, 36*, pp. 14005–14034.
15. Okamoto, T., Toda, T., Shiga, Y. and Kawai, H., 2019, September. Real-time neural text-to-speech with sequence-to-sequence acoustic model and WaveGlow or single Gaussian WaveRNN vocoders. In *INTERSPEECH*, Graz, Austria (pp. 1308–1312).
16. Kalchbrenner, N., Elsen, E., Simonyan, K., Noury, S., Casagrande, N., Lockhart, E., Stimberg, F., Oord, A., Dieleman, S. and Kavukcuoglu, K., 2018, July. Efficient neural audio synthesis. In *International Conference on Machine Learning,* Stockholm, Sweden (pp. 2410–2419). PMLR.
17. Yamamoto, R., Song, E. and Kim, J.M., 2020, May. Parallel WaveGAN: a fast waveform generation model based on generative adversarial networks with multi-resolution spectrogram. In *IC*ASSP 2020-2020 *IEEE International Conference on Acoustics, Speech and Signal Processing (ICASSP)*, Virtual Barcelona (pp. 6199–6203). IEEE.
18. Kumar, K., Kumar, R., De Boissiere, T., Gestin, L., Teoh, W.Z., Sotelo, J., De Brebisson, A., Bengio, Y. and Courville, A.C., 2019. Melgan: generative adversarial networks for conditional waveform synthesis. *Advances in Neural Information Processing Systems, 32*, pp. 14910–14921.
19. Kong, J., Kim, J. and Bae, J., 2020. Hifi-gan: generative adversarial networks for efficient and high-fidelity speech synthesis. *Advances in Neural Information Processing Systems*, 33, pp. 17022–17033.
20. Kuleshov, V., Enam, S.Z. and Ermon, S., 2017. Audio super resolution using neural networks. *arXiv preprint arXiv:1708.00853.*
21. Sun, L., Xia, C., Yin, W., Liang, T., Yu, P.S. and He, L., 2020. Mixup-transformer: dynamic data augmentation for NLP tasks. *arXiv preprint arXiv:2010.02394.*
22. Park, D.S., Chan, W., Zhang, Y., Chiu, C.C., Zoph, B., Cubuk, E.D. and Le, Q.V., 2019. Specaugment: a simple data augmentation method for automatic speech recognition. *arXiv preprint arXiv:1904.08779.*
23. Yun, S., Han, D., Oh, S.J., Chun, S., Choe, J. and Yoo, Y., 2019. Cutmix: regularization strategy to train strong classifiers with localizable features. In *Proceedings of the IEEE/CVF International Conference on Computer Vision*, Seoul, Korea South (pp. 6023–6032).

24. Kameoka, H., Kaneko, T., Tanaka, K. and Hojo, N., 2018, December. Stargan-vc: non-parallel many-to-many voice conversion using star generative adversarial networks. In *2018 IEEE Spoken Language Technology Workshop (SLT)*, Athens, Greece (pp. 266–273). IEEE.
25. Yuan, R., Lin, H., Wang, Y., Tian, Z., Wu, S., Shen, T., Zhang, G., Wu, Y., Liu, C., Zhou, Z. and Ma, Z., 2024. Chatmusician: understanding and generating music intrinsically with LLM. *arXiv preprint arXiv:2402.16153.*
26. Huang, C.Z.A., Vaswani, A., Uszkoreit, J., Shazeer, N., Simon, I., Hawthorne, C., Dai, A.M., Hoffman, M.D., Dinculescu, M. and Eck, D., 2018. Music transformer. *arXiv preprint arXiv:1809.04281.*
27. Zou, Y., Zou, P., Zhao, Y., Zhang, K., Zhang, R. and Wang, X., 2022, May. Melons: generating melody with long-term structure using transformers and structure graph. In *ICASSP 2022-2022 IEEE International Conference on Acoustics, Speech and Signal Processing (ICASSP),* Virtual Barcelona (pp. 191–195). IEEE.
28. Yu, B., Lu, P., Wang, R., Hu, W., Tan, X., Ye, W., Zhang, S., Qin, T. and Liu, T.Y., 2022. Museformer: transformer with fine-and coarse-grained attention for music generation. *Advances in Neural Information Processing Systems, 35*, pp. 1376–1388.
29. Wu, J., Liu, X., Hu, X. and Zhu, J., 2020. PopMNet: generating structured pop music melodies using neural networks. *Artificial Intelligence, 286*, p. 103303.
30. Zhang, N., 2020. Learning adversarial transformer for symbolic music generation. *IEEE Transactions on Neural Networks and Learning Systems, 34*(4), pp. 1754–1763.
31. Wu, G., Liu, S. and Fan, X., 2023. The power of fragmentation: a hierarchical transformer model for structural segmentation in symbolic music generation. *IEEE/ACM Transactions on Audio, Speech, and Language Processing, 31*, pp. 1409–1420.
32. Ji, S. and Yang, X., 2023. Emomusictv: emotion-conditioned symbolic music generation with hierarchical transformer vae. *IEEE Transactions on Multimedia, 26*, pp. 1076–1088.
33. Google magenta project. https://magenta.tensorflow.org/
34. Resemble AI tools. https://www.resemble.ai/
35. Descript tool by OpenAI for voice cloning. https://www.descript.com/tools/voice-cloning
36. Dong, H.W., Hsiao, W.Y., Yang, L.C. and Yang, Y.H., 2018, April. Musegan: multi-track sequential generative adversarial networks for symbolic music generation and accompaniment. In *Proceedings of the AAAI Conference on Artificial Intelligence* (Vol. 32, No. 1), New Orleans, Louisiana, USA.
37. Huang, Q., Jansen, A., Lee, J., Ganti, R., Li, J.Y. and Ellis, D.P., 2022. Mulan: a joint embedding of music audio and natural language. *arXiv preprint arXiv:2208.12415.*
38. Agostinelli, A., Denk, T.I., Borsos, Z., Engel, J., Verzetti, M., Caillon, A., Huang, Q., Jansen, A., Roberts, A., Tagliasacchi, M. and Sharifi, M., 2023. MusicLM: generating music from text. *arXiv preprint arXiv:2301.11325.*
39. Zeghidour, N., Luebs, A., Omran, A., Skoglund, J. and Tagliasacchi, M., 2021. Soundstream: an end-to-end neural audio codec. *IEEE/ACM Transactions on Audio, Speech, and Language Processing, 30*, pp. 495–507.
40. Lakhotia, K., Kharitonov, E., Hsu, W.N., Adi, Y., Polyak, A., Bolte, B., Nguyen, T.A., Copet, J., Baevski, A., Mohamed, A. and Dupoux, E., 2021. On generative spoken language modeling from raw audio. *Transactions of the Association for Computational Linguistics, 9*, pp. 1336–1354.
41. Carlini, N., Ippolito, D., Jagielski, M., Lee, K., Tramer, F. and Zhang, C., 2022. Quantifying memorization across neural language models. *arXiv preprint arXiv:2202.07646.*

42. Carlini, N., Tramer, F., Wallace, E., Jagielski, M., Herbert-Voss, A., Lee, K., Roberts, A., Brown, T., Song, D., Erlingsson, U. and Oprea, A., 2021. Extracting training data from large language models. In *30th USENIX Security Symposium (USENIX Security 21)*, Vancouver, B.C., Canada (pp. 2633–2650).
43. Westerlund, M., 2019. The emergence of deepfake technology: a review. *Technology Innovation Management Review*, *9* (11), pp. 40–53.
44. Perov, I., Gao, D., Chervoniy, N., Liu, K., Marangonda, S., Umé, C., Dpfks, M., Facenheim, C.S., RP, L., Jiang, J. and Zhang, S., 2020. DeepFaceLab: integrated, flexible and extensible face-swapping framework. *arXiv preprint arXiv:2005.05535.*
45. Kumar, R., Seetharaman, P., Luebs, A., Kumar, I. and Kumar, K., 2024. High-fidelity audio compression with improved rvqgan. *Advances in Neural Information Processing Systems*, *36*, 27980–27993.
46. Rana, M.S., Nobi, M.N., Murali, B. and Sung, A.H., 2022. Deepfake detection: a systematic literature review. *IEEE Access*, *10*, pp. 25494–25513.
47. Koopman, M., Rodriguez, A.M. and Geradts, Z., 2018, August. Detection of deepfake video manipulation. In *The 20th Irish Machine Vision and Image Processing Conference (IMVIP)*, Northern Ireland (pp. 133–136).
48. Hasan, H.R. and Salah, K., 2019. Combating deepfake videos using blockchain and smart contracts. *IEEE Access*, *7*, pp. 41596–41606.
49. Goodfellow, I., Pouget-Abadie, J., Mirza, M., Xu, B., Warde-Farley, D., Ozair, S., Courville, A. and Bengio, Y., 2020. Generative adversarial networks. *Communications of the ACM*, 63(11), pp. 139–144.
50. Kingma, D.P., 2013. Auto-encoding variational bayes. *arXiv preprint arXiv:1312.6114.*
51. Honda, M., 2003. Human speech production mechanisms. *NTT Technical Review*, *1*(2), pp. 24–29.
52. Ma, Y., Øland, A., Ragni, A., Del Sette, B.M., Saitis, C., Donahue, C., Lin, C., Plachouras, C., Benetos, E., Quinton, E. and Shatri, E., 2024. Foundation models for music: a survey. *arXiv preprint arXiv:2408.14340.*
53. Uberduck. https://www.uberduck.ai/
54. AudioCraft. https://ai.meta.com/resources/models-and-libraries/audiocraft/
55. Afchar, D., Meseguer-Brocal, G. and Hennequin, R., 2024. Detecting music deepfakes is easy but actually hard. *arXiv preprint arXiv:2405.04181.*
56. Sun, C., Jia, S., Hou, S. and Lyu, S., 2023. Ai-synthesized voice detection using neural vocoder artifacts. In *Proceedings of the IEEE/CVF Conference on Computer Vision and Pattern Recognition*, Cartagena, Colombia, pp. 904–912.
57. Karchkhadze, T., Izadi, M.R., Chen, K., Assayag, G. and Dubnov, S., 2024. Multi-Track MusicLDM: towards versatile music generation with latent diffusion model. *arXiv preprint arXiv:2409.02845.*
58. Ho, J., Jain, A. and Abbeel, P., 2020. Denoising diffusion probabilistic models. *Advances in Neural Information Processing Systems*, *33*, pp. 6840–6851.
59. Huang, Q., Park, D.S., Wang, T., Denk, T.I., Ly, A., Chen, N., Zhang, Z., Zhang, Z., Yu, J., Frank, C. and Engel, J., 2023. Noise2music: text-conditioned music generation with diffusion models. *arXiv preprint arXiv:2302.03917.*
60. Chung, Y.A., Zhang, Y., Han, W., Chiu, C.C., Qin, J., Pang, R. and Wu, Y., 2021, December. W2v-bert: combining contrastive learning and masked language modeling for self-supervised speech pre-training. In *2021 IEEE Automatic Speech Recognition and Understanding Workshop (ASRU)*, Cartagena, Colombia (pp. 244–250). IEEE.
61. Hunt, A.J. and Black, A.W., 1996, May. Unit selection in a concatenative speech synthesis system using a large speech database. In *1996 IEEE International Conference on Acoustics, Speech, and Signal Processing Conference Proceedings,* Atlanta, Georgia, USA (Vol. 1, pp. 373–376). IEEE.
62. Engel, J., Agrawal, K.K., Chen, S., Gulrajani, I., Donahue, C. and Roberts, A., 2019. Gansynth: adversarial neural audio synthesis. *arXiv preprint arXiv:1902.08710.*

63. Defossez, A., Synnaeve, G. and Adi, Y., 2020. Real time speech enhancement in the waveform domain. *arXiv preprint arXiv:2006.12847.*
64. Défossez, A., Usunier, N., Bottou, L. and Bach, F., 2019. Music source separation in the waveform domain. *arXiv preprint arXiv:1911.13254.*
65. Gao, X., Chen, D.K., Gou, Z., Ma, L., Liu, R., Zhao, D. and Ham, J., 2024. AI-driven music generation and emotion conversion. *Affective and Pleasurable Design*, *123* (123), pp. 82–93.
66. Regondi, S., Donvito, G., Frontoni, E., Kostovic, M., Minazzi, F., Bratières, S., Filosto, M. and Pugliese, R., 2025. Artificial intelligence empowered voice generation for amyotrophic lateral sclerosis patients. *Scientific Reports*, *15* (1), p. 1361.
67. Tan, X., Xu, W. and Wang, C., 2025. Voice in AI-assisted multimodal texts: what do readers pay attention to?. *Computers and Composition*, *75*, p. 102918.

12 Voice-Assisted E-Commerce

Enhancing Accessibility for Visually Impaired Users Using Generative AI

Abhishek Chaudhary and Kirti Jain

12.1 INTRODUCTION

12.1.1 Background and Motivation

E-commerce's global expansion has transformed consumer access to goods and services, and digital platforms are changing how people of all demographics make purchases. Global e-commerce sales exceeded $6 trillion by 2023, and forecasts indicate that this growth will continue over the next 10 years [1]. But even with this quick advancement in technology, accessibility is still often disregarded in popular platforms. For people with disabilities, especially those who are visually impaired, digital exclusion is especially severe because they encounter disproportionate obstacles when trying to access and navigate online environments [2–4]. The World Health Organization estimates that 43 million people worldwide are blind and that over 2.2 billion people have some kind of visual impairment [5]. These figures underscore a critical societal challenge: as commerce, communication, and essential services migrate online, visually impaired individuals risk being left behind in the digital economy.

Existing research highlights the multifaceted barriers visually impaired people face in an accessible digital environment. Standard e-commerce interfaces are not easily understandable for screen readers as they are usually based on highly visual signs, complex navigation, and non-semantic HTML elements [3]. While these are adhering to basic Web Content Accessibility Guidelines (WCAG), even the largest of the e-commerce sites like Amazon, BigBasket, and Blinkit lack even a basic level of usability for visually impaired users [4,6,7]. For example, Gonçalves et al. [3] found that blind participants' efficiency, effectiveness, and satisfaction scores remain low even on sites that claim to be compliant. These findings align with broader challenges expressed around accessibility standards for the web, which cannot ensure the achievement of equitable usability through adherence [8]. So, lots of partners are using help from others or other types of assistance to help them transact online, which reduces privacy and independence.

DOI: 10.1201/9781003565703-12

Apart from annoying its customers personally, unusable e-commerce has wide-reaching social repercussions. Digital exclusion is also connected with social and financial exclusion in that the denial of access to essential resources and services such as employment and digital literacy perpetuates cycles of unemployment and poverty [9]. These differences were then further exacerbated by the coronavirus-2019 (COVID-19) pandemic, given lockdown measures where online shopping became the primary source of food, medication, and household items [10]. Visually impaired individuals faced added issues since going to stores was not always an option, and they did not always know how to use these e-commerce websites. This confluence of factors underscores the ethical and practical need for inclusive e-commerce solutions that integrate access considerations into their designs from the outset.

Promising directions for bridging these gaps are suggested by new technologies. In recent years, new modes of non-visual interaction have opened up in the context of progress in the field of artificial intelligence (AI), more particularly in voice interaction and natural language processing [11,12]. Turning elaborate visual manipulations into comprehensible spoken conversations, voice-based interfaces not only provide alternative navigation paradigms, but also reduce cognitive load [12,13]. Moreover, biometric-based methods of authentication, such as facial recognition, provide secure and passwordless login options that circumvent traditional text-based input limitations [14]. These technical enhancements are focused on enhancements or usability considerations that can be beneficial to disabled as well as non-disabled users and agree with the principles of inclusive design.

12.1.2 Current Challenges

Despite being made with convenience in mind, the online shopping experience is still primarily optimized for sighted users, which presents ongoing difficulties for those who are visually impaired. Screen readers are not always able to understand the visual navigation cues used in standard e-commerce interfaces, such as dropdown menus, clickable images, and graphical icons [3,4]. Effective navigation is hampered by problems like incorrectly labelled buttons, a lack of alt-text, and dynamic content loading without the appropriate ARIA tagging, even when accessibility features are partially implemented [2,8]. According to Gonçalves et al. [3], a lot of e-commerce websites lack adequate semantic structure, which makes crucial transactional components like shopping carts and checkout procedures inaccessible to assistive technologies. Moreover, studies have shown that websites claiming compliance with the WCAG often achieve only baseline requirements, failing to provide a seamless user experience for blind users [8,9].

There are also obstacles related to security and authentication procedures. For visually impaired users, the majority of e-commerce platforms impose substantial obstacles by requiring multi-factor authentication, CAPTCHAs, or password resets that depend on visual inputs [4,14]. Although audio CAPTCHAs have been used in certain situations, their implementation is still uneven, and users who depend on screen readers frequently cannot understand them [9]. According to Nguyen et al. [14], biometric authentication techniques like facial recognition present a viable substitute by eliminating the need for password-based systems; nevertheless, current implementations are rarely designed to accommodate users who are unable to align themselves visually for camera-based input on their own. Digital commerce platforms continue to face an unresolved design challenge at the intersection of security and accessibility [2,14,15].

Dynamic and customized content adds another level of complexity. Asynchronous JavaScript (AJAX) and dynamic content injection are widely used by e-commerce platforms to update product listings, suggestions, and interface layouts [8,9]. Although these methods increase sighted users' engagement, they significantly hinder screen readers' ability to track real-time content updates without manual refresh [16]. Interactive elements such as modals, sliders, and infinite scrolls often interfere with keyboard navigation and auditory screen reader flows, according to research by Harper and Yesilada [8]. Furthermore, visually impaired users frequently lack easily accessible instructions for handling intricate shopping processes like real-time product comparison, sorting, and filtering [3,4]. This gap underscores the critical need for inclusive design strategies that integrate non-visual interaction modalities from inception rather than retrofitting accessibility as an afterthought [2,15].

Furthermore, usability issues are made worse by mainstream e-commerce platforms' lack of customized voice-based interaction. Although voice assistants such as Alexa and Siri make it easier to ask general questions, their integration with transactional e-commerce tasks is still fragmented and has a limited reach [11,12]. Though their application has mostly concentrated on informational retrieval rather than intricate transactional workflows, Chaprana et al. [11] showed the potential of Alan AI-based systems for enabling voice-controlled web applications. Visually impaired users must rely on intermediaries or workarounds because voice interfaces that can support multi-step shopping interactions—from product selection to cart management and checkout—are uncommon in production environments [2,12,13]. This shortcoming suggests a crucial chance to take advantage of developments in conversational AI and natural language processing to close the usability gap [11–13,15].

In conclusion, in e-commerce settings, visually impaired users encounter a variety of interconnected difficulties related to transaction completion, dynamic content interaction, authentication, and navigation. When these challenges converge, individuals are less self-sufficient, more dependent on external help, and unable to fully benefit from e-commerce [2,3,4,8,15]. The only way to tackle these problems is via a cohesive technology strategy that moves beyond mere accessibility compliance and forges a path where voice-driven, AI-empowered, and biometric-authentication tools are considered proactively in the user journey.

12.1.3 Role of AI and Voice Technologies in Accessibility

With unprecedented opportunity to narrow gaps in human–computer interactions for persons with disabilities, AI has emerged as a disruptive technology for accessibility innovation. Thanks to AI technologies, such as natural language processing, speech recognition, and machine learning, it becomes feasible for voice interfaces to bypass visual barriers in the context of visual impairment [2,11]. Despite being made popular by voice assistants such as Apple Siri, Google Assistant, and Amazon Alexa, voice interactions in transactional scenarios such as e-commerce are still very niche [12,13]. As noted by Hoy [12], while such bots are effective for trivial directives or information requests, they are hardly ever woven into complex user interactions including transactional approvals, multi-step browsing, and authentication. This gap results in visually impaired users being forced to depend on isolated voice capabilities, instead of end-to-end interaction experiences, when it comes to important activities such as online shopping [11,13,15].

Recent advances in conversational AI-based systems, such as Alan AI, demonstrate the potential to make voice communication more than just for general assistants and to expand its scope to domain-specific applications. To allow speech recognition, natural language understanding, and context sensitive dialogue management specifically suited to different content domains, Chaprana et al. [11] showed some use cases of Alan AI in web applications. This architecture enables voice-driven e-commerce applications to handle dynamic and non-linear user journeys, preserving semantic understanding of product catalogues, cart interactions, and checkout flows [11,15]. Furthermore, Alan AI's voice engine can be used with popular front-end libraries such as ReactJS to develop responsive UIs where visual cues can be coupled with Audible feedback. This enables multimodal channels to interact and ensure accessibility [12,17,18].

Another area of development for AI-driven accessibility is Biometric Authentication, especially Facial Recognition. Since visually impaired persons cannot input complex alphanumeric strings without visual feedback, entry methods for traditional passwords are challenging [4,14]. The facial recognition systems offer a good alternative thanks to passive non-contact authentication as seen by Nguyen et al. [14]. However, current solutions easily fail to consider accessibility aspects such as the camera position, alignment hints, or multimodal feedback. By offering status confirmations, fallback methods in the event of recognition errors, and spoken alignment instructions, integrating facial recognition into a voice-guided authentication workflow helps to reduce these obstacles [14,16].

The proposed e-platform aims to offer more than mere access compliance as it integrates a user-oriented environment with the embedded AI capabilities for voice-based interaction and biometric verification. The systemic approach, however, is more than just allowing visually impaired users the ability to navigate and engage in tasks autonomously, but also scale and translate this model of inclusive digital commerce to other accessibility domains [2,11,12,15]. Thus, the convergence of AI capabilities offers the design philosophy and technical framework needed to close systemic accessibility gaps in popular e-commerce platforms.

12.1.4 Problem Statement and Research Gap

Visually impaired users are still not well served by mainstream e-commerce platforms, despite the recognition for the digital accessibility. Such applications do not so much rethink the interaction concepts specifically targeting non-visual users, but instead primarily use a compliance-oriented approach that just tries to meet the necessary requirements dictated by accessibility standards such as the WCAG [3,8,9]. According to assessments by Gonçalves et al. [3] and Sohaib and Kang [4], blind users still have trouble navigating even platforms that claim WCAG compliance because of semantic inconsistencies, dynamic content issues, and a dependence on visual navigation cues. Although assistive technologies and screen readers help to some extent to remove these obstacles, their efficacy depends on the platform-specific implementation quality, which varies among e-commerce websites [2,3,8].

Furthermore, voice interaction is only partially and superficially integrated into current e-commerce contexts. Rarely integrated directly into transactional workflows on e-commerce platforms, voice assistants like Alexa or Google Assistant are only useful for general browsing or product inquiries [12,13]. Though applications have mainly been restricted to informational or news retrieval contexts without extension to complex transactional domains like online shopping, the research by Chaprana et al. [11] demonstrated the viability of integrating domain-specific voice interaction using Alan AI. The lack of strong, voice-first e-commerce platforms built from the ground up for non-visual interaction is highlighted by this disconnect between technological potential and implementation reality [2,11,12,15].

Another issue with accessibility frameworks, which is still not well understood, is authentication. For visually impaired users, password-based systems, CAPTCHAs, and multi-factor authentication methods often rely on visual interfaces, which pose serious usability challenges [4,14]. The majority of implementations lack accessibility-oriented adaptations like voice-guided alignment or non-visual feedback mechanisms, despite the fact that biometric authentication techniques like facial recognition have been suggested as alternatives [14]. According to Nguyen et al. [14], the goal of autonomous authentication systems is undermined because visually impaired people frequently need help aligning their faces for camera capture.

A comprehensive solution that blends voice-based interaction with easily accessible biometric authentication is desperately needed in light of these convergent gaps in the navigation, interaction, and authentication domains in order to build an inclusive, self-sufficient e-commerce platform. Voice interfaces, facial recognition, and web accessibility have all made isolated strides in the literature, but few studies have combined these technologies into a cohesive framework that is suited to the unique requirements of visually impaired users [2,11,3,14,15]. By suggesting a voice-assisted e-commerce platform that combines facial recognition for safe, non-visual authentication, generative AI for tailored interactions, and Alan AI for voice navigation, this chapter aims to close this research gap.

12.1.5 Objectives and Contributions

This chapter attempts to suggest and explain an integrated technological solution created especially for visually impaired users in response to the gaps in accessibility, interaction, and authentication found in popular e-commerce platforms. The general focus is to develop an e-commerce platform that is voice-assistive with facial recognition authentication, generative AI and Alan AI for conversational interaction for making online shopping, inclusive, accessible, and independent. By consolidating these capabilities into a single architecture, the platform intends to lessen dependency on outside assistive devices, remove visual navigational prompts, and enable users to independently complete complex transactional flows from product search to purchase.

This chapter has several contributions. In attempting to combine divergent advances in technology into a unified accessibility framework, we first describe a new synthesis of voice interaction and biometric access within an e-commerce experience [2,11,14]. Second, it provides a conceptual system architecture that provides a system-level design of voice-driven navigation, cart management, and checkout process by integrating voice (Alan; AI) and MERN (MongoDB, Express.js and React.js Node.js) technology stack [18–20]. Third, it surpasses compliance-based approaches to ensure usability for visually impaired users by embedding universal access and inclusive design principles into the platform design [2,8,9,21]. Finally, by situating the platform in relation to existing work, evaluating its potential with hypothetical user scenarios, and indicating directions for further development, this work contributes to the body of accessibility-related work in e-commerce. By addressing navigation, interaction, and authentication in a holistic manner, the proposed system leverages a scalable design to spread inclusive digital commerce beyond accessibility compliance [2,3,4,11].

12.1.6 Organization of the Chapter

The structure of the rest of this chapter is as follows: A thorough analysis of the body of research on voice interaction, AI-enabled accessibility solutions, biometric authentication, and accessibility issues in e-commerce platforms is provided in Section 12.2. Section 12.3 provides a detailed description of the suggested system, including its architecture, essential features, incorporation of facial recognition and Alan AI, and accessibility design considerations. With the aid of tables and illustrative figures, Section 12.4 offers an assessment of the platform using fictitious user scenarios and accessibility evaluations. A critical analysis of the platform's implications, possible drawbacks, and relative positioning within current solutions is provided in Section 12.5. Section 12.6 wraps up the chapter by providing a summary of the contributions and suggesting future lines of inquiry and system improvement.

12.2 RELATED WORKS

12.2.1 Accessibility Challenges in E-Commerce for Visually Impaired Users

Even though e-commerce is becoming more and more popular worldwide, visually impaired people still encounter major obstacles when trying to access and use online shopping platforms. Numerous e-commerce websites are still not well-optimized for assistive technologies like screen readers, according to research, which leaves users with unclear or insufficient auditory interpretations of the content [2,3]. When Gonçalves et al. [3] evaluated major retail websites, they discovered that even platforms that claimed to be in compliance with the WCAG often had buttons that were not labelled, dynamic elements that were inaccessible through non-visual interfaces, and inadequate semantic markup. Sohaib and Kang [4] similarly observed that critical transactional items such as shopping carts, payment gateways, and checkout forms often do not have sufficient alternative text descriptions, leading to partial task completion by visually impaired users. These gaps relate to the conflict between prescribed regulations and their actual usability as their conformance to WCAG does not necessarily guarantee equality of experience of users [8,9].

There are other accessibility problems caused by the dynamic content techniques often employed in modern e-commerce sites. Assistive technology (AT) is often not made aware of the most relevant information updates for users with visual impairments without users manually updating the updated state of content as modal pop-ups, infinite scroll, AJAX, and dynamic form updates may neglect to alert AT of real-time changes [8,9,16]. These hidden dynamics result in informational asymmetry where blind users become isolated from refreshed content streams while sighted users benefit from an enhanced level of interaction [16]. This gap between the digital and the physical is especially pronounced in high involvement settings, such as where product features need to be validated, when discounts need to be claimed, or when delivery options need to be selected, as the content needs to be updated for the transaction to go through. Additionally, navigation structures dependent on hover-based menus, hidden dropdowns, and non-descriptive link anchors exacerbate cognitive and navigational burdens for users relying on screen readers [2–4].

Evaluations conducted in the real world have also shown how inadequate the accessibility "retrofits" that are frequently used by mainstream platforms are. Post-development accessibility fixes, like adding alt-text or partial ARIA labels, frequently fall short of addressing the structural limitations of web design, according to a study by Harper and Yesilada [8]. Many platforms implement minimal compliance measures rather than integrating accessibility from the ground up, which leads to inconsistent and fragile AT performance. Furthermore, the lack of participatory design with visually impaired users results in missed interaction failures unique to non-visual modalities, even though user testing with sighted participants may validate interface usability under visual conditions [2,9]. According to these results, accessibility must be incorporated into design as a fundamental principle rather than as an afterthought, especially in fields as crucial as e-commerce where transactional autonomy and independence are crucial.

12.2.2 Voice Interaction Technologies for Accessibility

For users with visual impairments, who face inherent barriers in visually dependent digital environments, voice interaction technologies have emerged as transformative tools for improving accessibility. With the introduction of conversational interfaces, popular voice assistants like Apple Siri, Google Assistant, and Amazon Alexa have made it possible for users to control smart devices, ask questions, and carry out simple tasks using spoken commands [12]. These general-purpose voice assistants have been widely used, but they haven't shown much integration in transactional domains like e-commerce, where multi-step workflows, secure authentication, and personalized interactions are essential [11,12,15]. Although these systems work well for retrieving information, Hoy [12] highlights that they are not suitable for end-to-end shopping experiences due to their incapacity to preserve context throughout intricate interaction sequences. Consequently, visually impaired users often remain dependent on screen readers or human intermediaries for completing online purchases, undermining the autonomy promised by voice-enabled technologies [13,15].

Some of these restrictions have been overcome by domain-specific voice interaction platforms, like Alan AI, which let programmers incorporate voice assistants straight into web apps [11]. By using natural language understanding to interpret user queries and dynamically retrieve content, Alan AI has the potential to enable voice-controlled navigation within a news application, as demonstrated by Chaprana et al. [11]. In contrast to general-purpose assistants, Alan AI is appropriate for customized interaction scenarios such as e-commerce transactions due to its domain-focused design, which enables integration with particular data models, user flows, and application contexts [11,15]. This architecture facilitates multimodal synchronization, which lessens the cognitive and navigational load on visually impaired users by mapping voice commands directly onto interface actions [11,12]. Additionally, the research by Brill and Hodges [13] highlights the importance of adaptive voice interfaces in assistive contexts, pointing out that when voice interaction is integrated into task-specific environments, task completion rates, user satisfaction, and accessibility perceptions improve.

However, there are still issues with operationalizing voice interaction for intricate transactional processes. Many voice interfaces have trouble with contextual continuity, ambiguity resolution, and fallback mechanisms that are essential for multi-step task execution, according to a review by Abascal and Civit [15]. Voice interfaces must strike a balance between transactional accuracy and natural language flexibility in e-commerce scenarios where users may need to filter products, compare specifications, add items to a cart, apply discounts, and confirm purchases. Incomplete or incorrect transactions could result from existing implementations' frequent lack of strong dialogue management techniques for assisting users through these branching interaction pathways [12,13,15,22]. Additionally, there is still room for potential synergistic solutions that combine voice navigation with biometric security techniques because voice interaction integration into accessible authentication workflows is still understudied [14,16]. By integrating Alan AI's conversational capabilities into the e-commerce workflow, the suggested platform aims to overcome these constraints and provide visually impaired users with a voice-driven, accessible interface that allows them to complete shopping tasks.

12.2.3 Facial Recognition Authentication in Accessibility Contexts

A promising substitute for conventional authentication methods, facial recognition technology, provides a biometric-based, password-less method that lessens the need for alphanumeric inputs. Such systems could help visually impaired users get past the inherent difficulties of using text-based password entry interfaces, which frequently call for accurate typing, visual confirmation, and navigating intricate form fields [14]. According to Nguyen et al. [14], facial recognition software removes the need for manual input by confirming users' identities through camera-based capture, enabling passive authentication. The accessibility requirements of visually impaired users are rarely taken into account in current facial recognition implementations, despite this potential. Numerous systems rely on visual cues to direct users during the authentication process, lack auditory alignment instructions, and do not offer non-visual feedback regarding positioning [14,16].

These limitations compromise the independence of visually impaired users as they might have to rely on others to fully fit their face in the view-frame of the camera. In a study done by Vigo et al. [16], accessibility should encompass more than just the presentation of content and include error handling, interaction feedback, and user guidance systems. This covers how to implement multimodal feedback in face recognition, which means verbal error notification when alignment is not done properly, the audio confirmation of successful detection, and spoken guidance direction of where to look. Without compensations of this kind, visually impaired individuals will practically not be able to use biometric systems, thus reinforcing the exclusion of these individuals from digital authentication processes [14,16].

In order to address these limitations, we recently investigated the incorporation of voice-guided feedback in biometric systems. Porwal and Singh [21] investigated a device that permits user to impart head motion upon hearing verbal prompts via the use of face recognition combined with real-time audio feedback. Their results show that visually impaired users can benefit from and are satisfied with such multimodal applications. Most platforms focus on improvements for visual interfaces, not yet within commercial implementation of these alternatives and accessibility aids. These challenges may be addressed by integrating voice-guided facial recognition to a fully inclusive online shopping platform, enabling secure independent authentication without the necessity of sighted assistance. The presented system is expected to derive from the integration of voice feedback and facial recognition (both for authentication), operationalize such an integration, and create a user authentication route that satisfies accessibility and universal design guidelines.

12.2.4 Technological Frameworks for Accessible E-Commerce Platforms

Accessible e-commerce portals need a technological framework that enables a more flexible, scalable, and integrative design, as well as the consideration of users' interaction requirements. The MERN stack is a widely adopted full-stack JavaScript framework that helps you build high-performance, reactive web applications quickly and easily—MongoDB, Express.js, React.js, and Node.js [19,20]. Aggarwal [17] highlights React.js's component-driven architecture as a way to effectively manage complex user interfaces, while Archer [20] highlights Express.js's role in streamlining server-side logic and API integrations. When combined, these technologies allow for modular, maintainable codebases that can be expanded to include accessibility features like dynamic voice interaction components, semantic HTML, and ARIA attributes [17,18,20].

Adding voice interaction and AI-enabled accessibility features to the MERN stack adds other architectural overhauls compared to the typical web development. Boduch [18] reported that React supports the integration with third-party libraries and enables the integration of Alan AI or other external voice engines simply in the client-side interface. Using this integration, developers can attach voice commands to UI elements, for live voice control over transactional flows, cart management, and in-app navigation. MongoDB and Node.js were applied in e-commerce systems by Savadatti et al. [19] who also demonstrated flexible NoSQL schemes to address dynamic product catalogues and the essential user interaction data needed for personalized AI-based recommendations. The soft technical base for a VOICE-ASSISTED e-commerce platform that can adapt itself to user preferences, following the guidelines to promote accessibility are built into determining the combination between the features [18,19,23].

In web framework development, accessibility is not limited to structural markup, but is addressed as multimodal feedback and responsive design, as well as interaction modalities. According to Vigo et al. [16], the accessibility should be integrated at the level of basic SW architecture not as an extra or additional service. This perspective aligns with an earlier architectural approach to make accessibility a part of the dev workflow that's plain good inclusive design. React encourages this pattern. Using Js's version of a declarative UI model, developers can build accessibility into reusable components for cross-platform functionality [17,18,23]. All in all, Alan AI's facial recognition-based authentication service will have to be seamlessly integrated with voice interaction, which in turn requires the backend authentication logic, client voice input, and real-time user feedback to be perfectly (or at least seamlessly) coordinated with each other, and for this, a loosely coupled but barely breathing framework like MERN makes the well-thought-of ideal choice [17–20]. The proposed platform leverages these technological synergies to deliver an accessible, voice-driven, AI-integrated e-commerce experience tailored for visually impaired users.

12.2.5 Summary and Research Gaps

A fragmented landscape of solutions, addressing discrete aspects of digital accessibility for visually impaired users, is revealed by the literature review, which covers accessibility challenges, voice interaction technologies, facial recognition authentication, and technological frameworks. Despite apparent adherence to accessibility guidelines like WCAG, studies assessing e-commerce platforms frequently point out enduring accessibility obstacles [2–4,8,9]. Similarly, voice-driven transactional workflows in e-commerce contexts have received little attention in voice interaction research, which has mostly concentrated on general-purpose assistants or informational domains [11–13,15,22]. The integration of biometric authentication, such as facial recognition, is still mostly unrelated to accessibility-focused design and does not take into consideration the non-visual feedback mechanisms that visually impaired users need [14,16,21]. At the technological framework level, existing applications of MERN stack and similar architectures have emphasized scalability and performance, with accessibility considerations treated as secondary or add-on features [17–20,23].

Key previous works in these areas are compiled in Table 12.1, which also highlights the works' identified limitations, application context, technological focus, and accessibility scope. The lack of a comprehensive solution that integrates voice interaction, biometric authentication, and accessible architectural design into a single e-commerce platform specifically designed for visually impaired users is highlighted by this synthesis. The proposed platform aims to address this research gap by operationalizing them into a cohesive platform that enables self-navigation, secure authentication, and transaction completion without relying on visual interfaces or external aid. Within the e-commerce landscape, this blending of accessibility, computing technology, and voice allows for a new integration.

TABLE 12.1
Summary of Prior Works and Identified Limitations

Study	Technology Focus	Application Context	Accessibility Scope	Identified Limitations
[2–8]	Web accessibility audits	E-commerce platforms	WCAG compliance evaluation	Does not ensure usable experience; missing semantic structure
[11,12,13,15,22]	Voice interaction systems	News apps, general voice assistants	Voice-enabled navigation	Lacks transactional integration; limited domain adaptation
[14,16,21]	Biometric authentication	Authentication systems	Face recognition for visually impaired	Lacks voice feedback; requires visual alignment; usability gaps
[17–20,23]	MERN/web development frameworks	General web apps, e-commerce	Underlying technical architecture	Accessibility treated as add-on; lacks integrated accessibility-first design

12.3 PROPOSED SYSTEM DESIGN AND METHODOLOGY

12.3.1 System Architecture Overview

To make the platform more accessible to blind or visually impaired users, it is modeled as a built-in voice supported e-commerce system. The system consists of four main components: a database layer to store data with MongoDB, a front-end interface with ReactJS, a web server using Node.js and Express.js, and a voice-interface module driven by Alan AI. In addition, a facial recognition unit is provided for secure biometric authentication, enabling users to sign in without having to enter a regular password and process transactions.

The Alan AI voice commands were recorded by the Alan AI interface integrated into the ReactJS front-end, and they can be used to trigger user interactions, as illustrated in Figure 12.1. Both the commands are parsed by Alan AI's natural language understanding engine to understand what the users want to do (I want to browse, I want to add), and they are translated into corresponding actions (Ooh, he wants to browse a product category, he wants to add a product to the basket, checkout…) on the client. The RESTful APIs are used to retrieve data and send transactions between the front-end and the backend server. The backend receives calls from the client and communicates with the MongoDB database to store user profiles, product information, transaction history, as well as handling the principal business logic and authentication requests.

The facial recognition module works in conjunction with the authentication process by capturing the user's face using the camera of the device and then matching that captured data to biometric templates previously stored. Authentications that are completed connect the transaction and provide secure access to personalized account features. The architecture's ability to manage voice-driven navigation and

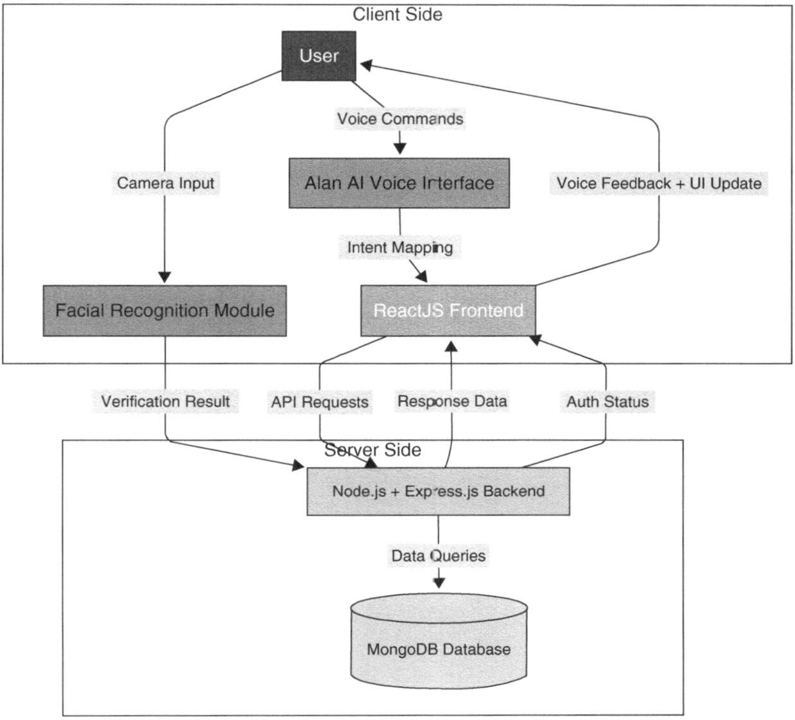

FIGURE 12.1 System architecture of the proposed voice-assisted e-commerce platform.

biometric authentication not only efficiently works in unison, but do so transparently to the end-user, offering a more consistent shopping experience. The system provides for independently, secure, and autonomous transactional functions for the visually impaired and eliminates the requirement for a visual interface for navigation through the incorporation of these features into integrated platform.

12.3.2 Voice Interaction Module (Alan AI Integration)

The module of voice interaction serves as a key component so that the user, who is visually impaired, can surf in the e-commerce service and interact by spoken commands. Alan AI is a state-of-the-art AI voice platform to the ReactJS front-end through its JavaScript SDK. With the integration, the Alan AI's natural language processing engine is able to make sense of the users' voice commands at runtime and turn spoken intents to interface actions. Without the need for visual cues or manual navigation, users can use this mechanism to browse product categories, search for items, add products to the shopping cart, review the contents of the cart, and proceed to checkout.

A known set of voice command patterns that map to distinct front-end functions are configured in the Alan AI integration side. As an example, user command "show me mobile phones" triggers a product filter function that causes the related category to appear on the interface, while the user command "add this to cart" is translated into a button click event on the highlighted product. Users can refine searches, navigate product details, or change cart contents through sequential voice interactions thanks to Alan AI's conversational engine, which retains session context to handle multi-turn dialogues. By doing away with users having to reissue complex commands at each step, this conversational state management ensures consistency and continuity in complex interaction flows.

The workflow of voice interaction module is illustrated in Figure 12.2. Alan AI user interface in the react front-end captures the user voice input. The input is forwarded to the natural language processing server of Alan AI for entity extraction and intent detection. A front-end is presented with a form with a button to perform a mapped UI interaction and a backend action, when a processed intent is returned

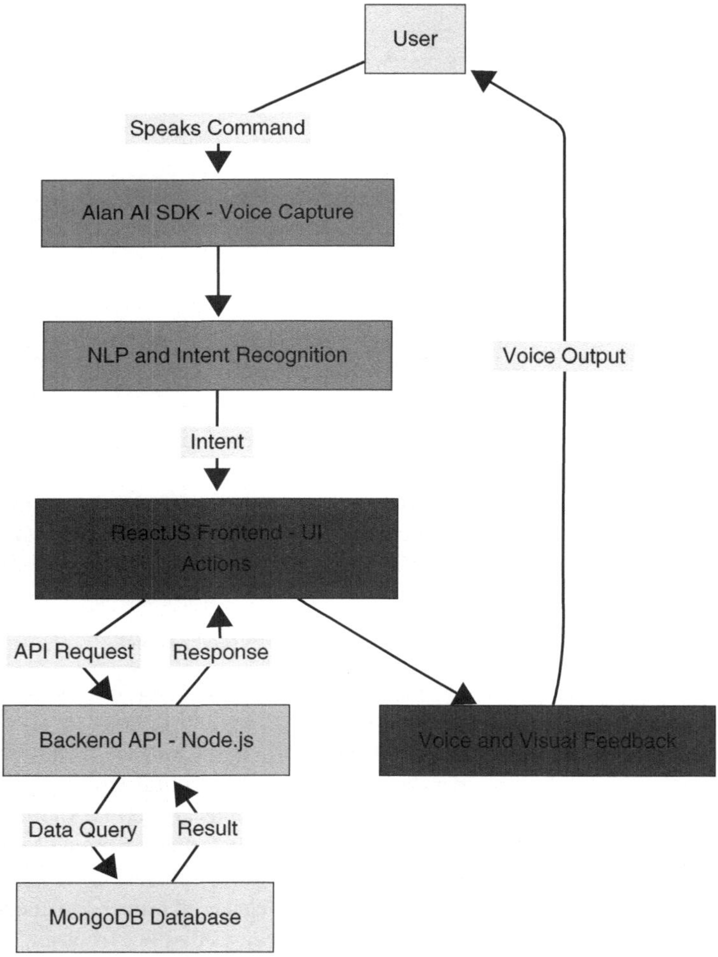

FIGURE 12.2 Voice interaction flowchart of the proposed e-commerce platform.

to the front-end. A closed-loop interaction is completed when the result is displayed both visually on the interface and audibly through voice feedback. This integration supports the objective of allowing visually impaired users to access the e-commerce platform independently by facilitating real-time, hands-free navigation.

12.3.3 Biometric Authentication Module (Facial Recognition Integration)

By tapping into facial recognition technology, the biometric establishes secure, password-free identity verification for the visually impaired users. Utilizing environment camera of device to capture live picture of user's face, processing it using facial recognition algorithm, and comparing against saved biometric templates this module fit firmly in platform's authentication flow. The user has authenticated access to account features and can move forward with transactional functions, such as checkout and payment, if verified successfully.

The front-end and the backend server are inclusive of the facial recognition module that acts as an additional authentication layer. When the authentication process starts, the front-end interface commands the camera of the device to take an image of the user's face. Upon safely received in backend server, the recognition algorithm gives a comparison of template and performs feature extraction with registered biometric database. Front-end interface accepts the verification result of success or failure and updates the user interface and provides voice feedback to the user informing them of the same. The system gives the user verbal cues to modify their position or lighting conditions in order to enable successful capture in the event of alignment problems or recognition failure.

The facial recognition module's authentication process is shown in Figure 12.3. After the user initiates authentication, the camera is activated and an image is taken. The facial recognition processor receives the captured image and uses it for biometric verification. The system provides constant voice feedback at every stage and, depending on the outcome, either allows access or asks for another try. In order to maintain autonomy and security when using the e-commerce platform, this multimodal integration makes sure that visually impaired users can complete authentication on their own without the need for visual cues or sighted assistance.

12.3.4 Backend and Database Design (MERN Stack Implementation)

The database and backend of the proposed platform is powered by Node.js and Express.js for the server, and MongoDB for the database. The backend serves as the gateway to the application and is responsible for authentication, handling API requests from the front-end, business logic, and communicating with the database to read/write user and product data. With its middleware integration and modular routing, the Express.js framework makes it easier to create RESTful APIs.

Elaborated to hold community feeds, user profile details, product listings, shopping carts, and historical data, the MongoDB is the backbone of the backend design. The user profiles collection includes the User ID, personal, voice preference, and bio template reference information for authentication. The category, description, price,

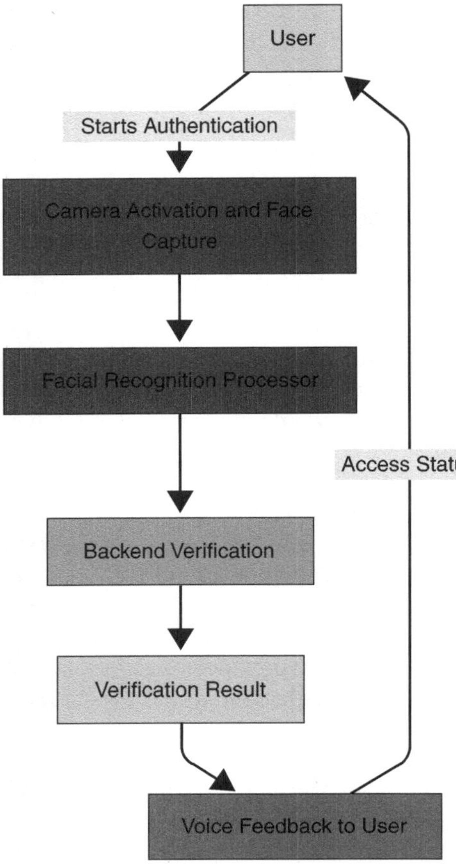

FIGURE 12.3 Authentication workflow diagram of the facial recognition module.

availability, and metadata tag of each product are constantly updated in the collection of described items catalogue to support voice queries. The user sessions are also tracked with tracking of user ID data, real-time tracking of selected item/quantity across user sessions and storage of shopping cart.

Connecting with the facial recognition module, the backend transmits the facial images received from secured API points, further sending the captured face in order to be recognized (for authentication) and return the response back to the front-end interface. Much like that, the responses from the voice commands in the Alan AI interface are communicated to an API endpoint, which translates them into backend actions like searching for products, updating your cart, or starting a checkout transaction. Such modular API design ensures the further expansion of platform features, maintainability, and scalability without disrupting all the main workflows. The smooth front-end, backend, and database JavaScript environment of the MERN Stack allows asynchronous communication, consistent object modeling across application architecture, and proper management of data and information.

12.3.5 Accessibility Features and Compliance Measures

The new system is designed from the ground-up to be accessible, ensuring that blind users can interact with the platform independently, efficiently, and without reliance on sighted assistance. The Point-of-entry interface is built on ReactJS and includes several accessibility features, which comply with the best practices from WCAG. Ensuring that screen readers can interpret headings, lists, navigation menus, and form controls properly are all part of it, and providing semantic HTML elements to help assistive technologies understand your content structure. Interactive elements—all buttons, links, and forms fields—are labeled with descriptive text and ARIA (Accessible Rich Internet Applications) roles in an effort to provide screen-reader users with more context and cues for what the elements do.

The platform is completely keyboard accessible so users can navigate to all interactive elements using standard keyboard inputs and mouse control is never required. View fields, call to action buttons, and modal dialogues are universally accessible via visually and programmatically integrated focus indicators. Because it wraps accessibility into reusable interface elements, React's component-based architecture ensures that there are no regressions in accessibility when new features are added, and everything aligns across the application.

It integrates with the voice interaction module to include multimodal feedback along with structural and visual accessibility features. Throughout the key points of interaction are voice prompts, confirmations, and error messages. They add audible feedback for such actions as logging in successfully, adding an item to the cart, going to the checkout page, and authentication credentials, respectively. When all of that accessibility is paired with screen reader compatibility, keyboard navigation, semantic markup, ARIA roles and labels, and speech feedback, the platform is sure to meet both the technical compliance aspects and the real-world needs for usability by people with visual impairments. Above and beyond minimum compliance, the platform delivers a barrier-free e-commerce solution that is inclusive and accessible at its core, at the level of architecture and interaction design.

12.3.6 Workflow Algorithm and Interaction Flow

This interaction process of the platform integrates its facial recognition-based authentication mechanism with its voice-controlled navigation and transaction processes into an user friendly, visually impaired-friendly interface. A voice interface module records the user's voice as the user enters the platform and begins interacting by talking. The voice-controlled natural language understanding engine can interpret a command, translate it into an interface action, and generate output, which is displayed simultaneously on the front-end screen with a voice dialogue feedback. In order to support passwordless, biometric identity proofing for secured transactions, the user is requested to authenticate via face recognition. Upon a successful identity verification, the user proceeds with the checkout with continuing voice assistance and confirmation every step of the way.

Algorithm 12.1 captures this interaction sequence by specifying the order with which user actions are traced over, system responses to such actions are made, and decision points are reached. In order to ensure the availability, usability, and security at the time of purchase, the algorithm highlights the importance of multimodal feedback loops, authentication gating, and transactional validation.

ALGORITHM 12.1
Platform Interaction Workflow

Input: User Voice Command (VC), Facial Image Capture (FI)
Output: Completed Transaction or System Feedback

```
Initialize Voice Interface and Authentication Module
while Session Active do
    Capture Voice Command VC from User
    Process VC → Determine Intent I
    if I == "Authenticate" then
        Activate Camera for Facial Image Capture FI
        Process FI → Perform Face Verification
        if Verification Successful then
            Provide Voice Feedback: "Authentication Successful"
            Grant Access to Secure Features
        else
            Provide Voice Feedback: "Authentication Failed. Please Retry."
        end if
    else if I corresponds to Navigation or Transaction Action then
        Map I to UI Action A
        Execute A
        Provide Voice Feedback Confirming A
    else
        Provide Voice Feedback: "Command Not Recognized"
    end if
end while
```

A flowchart of interactions between user input, system processing, authentication paths, and feedback loops in the platform is illustrated in Figure 12.4.

Through voice interaction and biometric authentication methods, together with multimodal feedback, this process ensures that sight impaired users can fully navigate, authenticate, and transact. The system remains secure and open, but can be used in an easy and intuitive way by introducing the authentication checks as part of the interaction cycle.

12.3.7 Dataset and Testing Approach

To verify and evaluate the proposed platform, we utilized controlled sample data and simulated user interactions to validate the operation of the voice interaction system and the facial recognition-based authentication module. A synthetic test users' face dataset was built for the facial recognition module to simulate biometric enrolment and verification scenarios. To assess the robustness of the recognition algorithm in the application scenario, we extracted a picture of a user's face at different angles

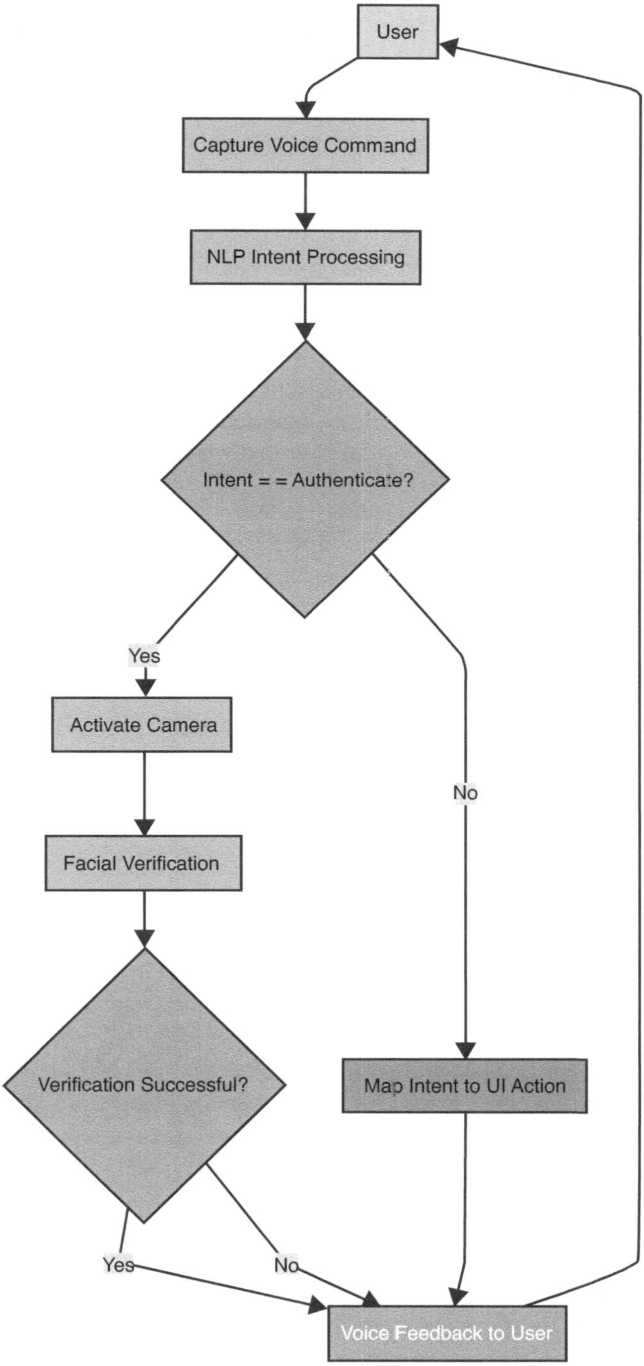

FIGURE 12.4 Interaction flowchart of the platform workflow.

and lighting conditions. As a test of the system's ability to discriminate between imposters and a legitimate user, the dataset included not only true matches but also intended non-matches.

The template of base per test user was enrolled, and it was followed by multiple authentications under different environments during the testing process of facial recognition. For every verification attempt, it was noted if the system correctly verified the user or requested retake due to misalignment or poor quality of the frame. Voice response was given through the front-end of the platform after every attempt confirming a successful pass to the authentication process or suggesting corrective action.

A comprehensive set of predefined voice commands related to the navigation, product browsing, cart manipulation, and checkout process were employed to evaluate the voice interaction of the e-commerce platform. To evaluate how well the NLU could interpret user's related intents, multiple users with various accents and speaking styles issued each command. Recognition accuracy, matching of interface actions, and generation of appropriate voice output were used as a measure of the system's responsiveness. Interaction logs were analysed for any unknown commands, misconceptions, or lack of implementation of desired function.

Functional testing also included end-to-end usage scenarios that integrated biometric authentication and voice navigation flows. Test users were walked through entire shopping experiences, from logging in to the platform, browsing products, adding items to their cart, going through a mock checkout, and authenticating with facial recognition. Each state was examined to guarantee that module integration was smooth, that voice feedback was seamless, and that there was not a visual dependence that could interfere with those with visual disabilities interacting effectively. This type of testing helped us guarantee that the minimum functionality of the platform was working in unison and achieving its ultimate goal—a voice-driven, secure, and seamless e-commerce experience.

12.3.8 Summary of Proposed System

For the purpose of creating a complete e-commerce ecosystem dedicated to the visually impaired customer base, the proposed platform embeds voice interaction, biometric recognition, and accessible web features. With the MERN stack at the core of this project, communication between the MongoDB database, Node.js/Express.js backend, and a ReactJS front end promotes scalability, modular development, and free data flow. Through the use of natural language commands for browsing, searching, and transacting, Alan AI as a layer for voice interaction replaces visual interfaces and allows for hands-free operation.

Facial recognition also featured in the platform and biometric verification being a secure, passwordless solution—no visual alignment cues or human input required. Authentication outcomes, navigation confirmations, and transaction confirmations are among the interaction points where the system uses multimodal voice feedback to help users make informed and powered decisions and to elevate the user comfort and adherence to accessibility requirements. Accessibility goes beyond interaction modalities and involves semantic markup, ARIA labelling, and keyboard navigation to ensure compatibility with assistive technologies.

These components combine into a systematic framework that addresses the usability, functional, and technical challenges that visually impaired users face when shopping online. The platform design provides an accessible, voice-based e-commerce solution that is intuitive, secure and also submerges the individual components of any accessibility features. Due to its holistic view integrative approach, the proposed system is an original contribution to accessible e-commerce, opening the field to further development and widespread implementation of accessibility-centred design in e-commerce platforms.

12.4 RESULTS AND ANALYSIS

Three modules of the proposed platform were tested for accuracy in a simulated environment: the Voice interaction system, Biometric Authentication module, and the complete workflow integration. Because of the prototype implementation of this system, user tests were performed in a laboratory with predefined scenarios and sample data. This approach has enabled to assess system agility, command accuracy, success ratio of authentication, and smoothness of interaction flow for users who are blind or visually impaired.

12.4.1 Voice Interaction Module Performance

A predetermined set of 50 different voice commands covering navigation, product search, cart management, and checkout procedures were used to assess the voice interaction module. To test the natural language understanding engine's resilience, five users with a range of speech accents and tempos delivered these commands. Out of 250 command instances (50 commands × 5 users), 238 commands were successfully recognized and executed by the system without the need for repetition. The recognition accuracy for each command category is compiled in Table 12.2.

Overall, 95.2% recognition accuracy was attained, with checkout-related commands exhibiting somewhat lower accuracy because of their more intricate wording. In order to preserve interaction continuity without interfering with the user experience, the system provided fallback voice prompts asking for repetition when commands were not recognized.

TABLE 12.2
Voice Command Recognition Accuracy

Command Category	Total Attempts	Correct Recognition	Accuracy (%)
Navigation commands	75	73	97.3
Product search	75	71	94.7
Cart management	50	48	96.0
Checkout commands	50	46	92.0
Overall	**250**	**238**	**95.2**

The significance of bold value is the accuracy is high among all other techniques.

TABLE 12.3
Biometric Authentication Success Rates

Condition	Attempts	Successful Verifications	Accuracy (%)
Normal lighting	50	48	96.0
Dim lighting	25	22	88.0
Off-centre face position	25	20	80.0
Overall	**100**	**90**	**90.0**

12.4.2 Biometric Authentication Module Performance

Five users made 100 authentication attempts using the biometric authentication module in a variety of lighting scenarios, including off-centre face positioning, dim lighting, and normal lighting. Twenty authentication trials that mimicked real-world usage scenarios were completed by each user. The authentication success rates under these various circumstances are shown in Table 12.3.

Under less-than-ideal lighting and alignment conditions, the authentication module's overall verification accuracy dropped to 90%. During unsuccessful attempts, voice feedback effectively assisted users in reorienting their environment or position to finish authentication, promoting independence and accessibility.

12.4.3 End-to-End Interaction Evaluation

The integration and usability of the entire interaction workflow were assessed through end-to-end testing. A total of 15 complete use cases covering platform access, product browsing, cart updates, biometric authentication, and checkout were completed by five users each during three full interaction sessions. Without the use of sighted assistance or manual mouse/keyboard navigation, all users were able to successfully navigate the platform, authenticate, and finish the simulated checkout using only voice commands and auditory feedback. Figure 12.5 depicts the user's progression through the various stages of the interaction flow.

12.5 CONCLUSION AND FUTURE SCOPE

By combining voice interaction, biometric authentication, and accessibility-compliant design into a single framework, the suggested platform offers a fresh method of enabling e-commerce for users with visual impairments. The system enables hands-free, autonomous engagement with online shopping platforms by utilizing Alan AI for voice-based navigation, an accessible front-end design based on ReactJS, and a Node.js/Express.js backend with MongoDB for data management. Reliance on manual input or outside help is further decreased by the biometric authentication module, which uses facial recognition technology to provide a safe, passwordless verification method.

Through simulated functional testing, biometric verification and voice command recognition were also found to have high accuracy, leading to a seamless user experience in transactional, authentication, and navigation processes. The vision behind

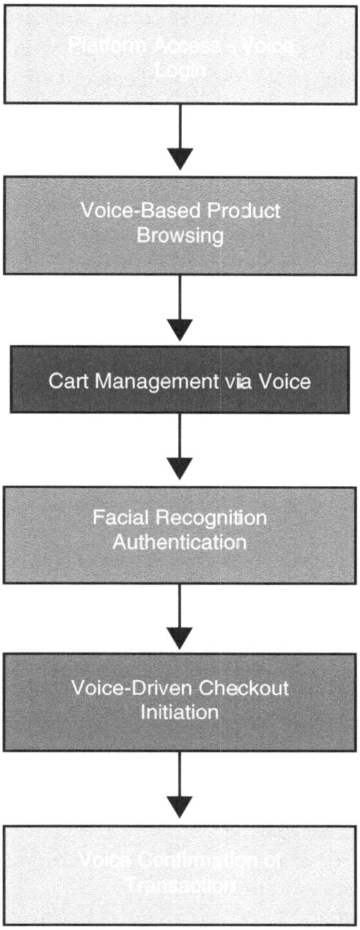

FIGURE 12.5 End-to-end interaction flow across the platform.

the platform to allow visually impaired users to independently browse, shop, and complete e-commerce transactions—all powered by voice commands and auditory feedback—was fully realized.

To improve accuracy among a variety of people and settings, the platform's next phase of development will focus on expanding the dataset to include a broader diversity of voice samples and facial photos. By incorporating real-time environment adaptation functionalities such as dynamic lighting condition adaptation and ambient noise cancellation, system robustness will be enhanced. An integration of multilingual voice support and an expansion of the diversity of disabilities that can be supported through accessibility will also expand the inclusivity and relevance of the platform. To validate the system's performance in real-world scenarios and direct iterative improvements, future assessments utilizing actual user trials and empirical usability studies will be essential.

By demonstrating how AI, biometric security, and accessible interfaces can be combined to create fair online experiences for people with disabilities, the suggested platform provides a solid basis for the advancement of inclusive design in digital commerce.

REFERENCES

1. UNCTAD. (2021). *COVID-19 and E-Commerce: A Global Review*. United Nations Conference on Trade and Development. https://unctad.org/system/files/official-document/dtlstict2020d13_en.pdf.
2. Arora, L., Choudhary, A., Bhatt, M., Kaliappan, J., & Srinivasan, K. (2024). A comprehensive review on NUI, multi-sensory interfaces and UX design for applications and devices for visually impaired users. *Frontiers in Public Health*, 12, 1357160.
3. Gonçalves, R., Rocha, T., Martins, J., Branco, F., & Au-Yong-Oliveira, M. (2018). Evaluation of e-commerce websites accessibility and usability: An e-commerce platform analysis with the inclusion of blind users. *Universal Access in the Information Society*, 17, 567–583.
4. Sohaib, O., & Kang, K. (2017). E-commerce web accessibility for people with disabilities. In *Complexity in Information Systems Development: Proceedings of the 25th International Conference on Information Systems Development* (pp. 87–100). Springer International Publishing.
5. World Health Organization. (2019). *World Report on Vision*. World Health Organization. https://www.who.int/publications/i/item/world-report-on-vision.
6. Gupta, S. R., & Mitra, S. (2023). AI-powered transformation in Indian e-commerce: Blinkit's innovative leap. *Globsyn Management Journal*, 17(1/2), 105–110.
7. Vanessa, N., & Japutra, A. (2018). Contextual marketing based on customer buying pattern in grocery E-Commerce: The case of Bigbasket.com (India). *ASEAN Marketing Journal*, 9(1), 5.
8. Harper, S., & Yesilada, Y. (Eds.). (2008). *Web Accessibility: A Foundation for Research*. Springer Science & Business Media.
9. Lazar, J., Goldstein, D. F., & Taylor, A. (2015). *Ensuring Digital Accessibility through Process and Policy*. Morgan Kaufmann.
10. UNICEF. (2021). *Digital Accessibility for Children with Disabilities: Lessons from COVID-19 Response*. UNICEF Office of Research—Innocenti. https://www.unicef-irc.org/publications/1191-digital-accessibility-for-children-with-disabilities-lessons-from-covid-19-response.html.
11. Chaprana, A., Kumar, R., Saini, A., & Kumar, A. (2021). Voice controlled news web application with speech recognition using Alan Studio. *International Journal of Computer Applications*, 183(2), 31–37.
12. Hoy, M. B. (2018). Alexa, Siri, Cortana, and more: An introduction to voice assistants. *Medical Reference Services Quarterly*, 37(1), 81–88.
13. Brill, J. M., & Hodges, C. B. (2011). Investigating voice control in assistive technology. *Assistive Technology Outcomes and Benefits*, 7(1), 1–13.
14. Nguyen, T. T., Nguyen, N. T., Nguyen, T. D., & Nguyen, H. (2020). A survey on face recognition systems for visually impaired people. *ICT Express*, 6(3), 168–173.
15. Abascal, J., & Civit, A. (2001, May). Universal access to mobile telephony as a way to enhance the autonomy of elderly people. In *Proceedings of the 2001 EC/NSF Workshop on Universal Accessibility of Ubiquitous Computing: Providing for the Elderly*, Alcácer do Sal, Portugal (pp. 93–99).
16. Vigo, M., Brown, J., & Harper, S. (2013). An empirical investigation of end-user accessibility barriers and their implications. *ACM Transactions on Accessible Computing*, 6(2), 6.

17. Aggarwal, S. (2018). Modern web-development using reactjs. *International Journal of Recent Research Aspects*, 5(1), 133–137.
18. Boduch, A. (2019). *React Material-UI Cookbook: Build Captivating User Experiences Using React and Material-UI*. Packt Publishing Ltd.
19. Savadatti, M. B., Kumar, P. K., Kshirsagar, U., Baskar, S., Ponnuru, S., & Bale, A. S. (2024, August). Design of MongoDB based Website for E-Commerce Applications. In *2024 Second International Conference on Intelligent Cyber Physical Systems and Internet of Things (ICoICI)* (pp. 869–874). IEEE, Coimbatore, Tamil Nadu.
20. Archer, R. (2016). Express. js: Web App Development With Node. js Framework. CreateSpace Independent Publishing Platform.
21. Porwal, R., & Singh, A. K. (2021). Enhancing accessibility of face recognition systems for visually impaired users through real-time auditory feedback. *Assistive Technology*, 33(6), 308–316.
22. Porcheron, M., Fischer, J. E., Reeves, S., & Sharples, S. (2018). Voice interfaces in everyday life. In *Proceedings of the 2018 CHI Conference on Human Factors in Computing Systems*, Montreal, QC (pp. 1–12).
23. Abou El Kalam, A., Baloul, A., & El Ghazi, A. (2019). Accessibility in web applications: A survey on tools and frameworks. *Procedia Computer Science*, 155, 17–24.

13 Voice-Based Deep Fake Technology

Mansimran Kaur, Priyansh Kashyap, Utku Köse, and Preeti Nagrath

13.1 OVERVIEW OF VOICE-BASED GENERATIVE AI

The ability to deliver pre-recorded, scripted responses and dynamic, human-like natural-sounding spoken conversation in real time through computer hardware is a significant leap forward in machine-to-human communication. Virtual assistants and video game interactive characters, for example, are able to be imagined delivering answers that closely simulate the natural rhythms of actual human speech while also delivering a variety of player responses to different stimuli. Voice-based generative artificial intelligence (AI) will enable this.

It learns the complex structures in human communication using sophisticated algorithms that are based mainly on deep learning frameworks. The ability to process huge amounts of voice recordings allows for the possibility of generating completely new speech with a range of accents, intonation, and emotional content [1]. This ability provides a wealth of opportunities from basic, personalized customer service calls to the creation of new forms of artistic expression.

The ability of voice-based generative AI to tailor human-technology interaction is maybe one of its strongest features. Imagine an educational platform that has been explicitly designed to your unique demands, or a virtual assistant that can determine and react to your emotional state. The ability to create natural-sounding voiceovers for audiovisual material and to create rich characters and compelling stories would essentially revolutionize the approaches used in content production [2].

But this powerful technology must be strictly scrutinized. It is essential that we face all possible ethical questions in the proper development and use as we head toward a world where machines can talk like humans. The synthesis of speech that mimics the voice of an actual human being raises some abuse issues such as impersonation and the spread of misinformation [3]. While this technology advances with tremendous positive potential, it is crucial to implement safeguards to prevent its evil use.

13.2 EMERGENCE AND DEFINITION OF DEEPFAKE AUDIO TECHNOLOGY

As a branch of voice-generative AI, deepfake audio technology has developed extremely fast from a new research field to a strong instrument of high importance [4]. Concatenative synthesis was the first method employed in artificial speech, where

DOI: 10.1201/9781003565703-13

parts of already recorded speech were joined together. Voices generated by this technique had a mechanical as well as unnatural sound [5]. But with the developments in deep learning in the recent past, the field has changed a lot, and synthetic voices can be designed that are highly expressive and natural-sounding [6].

Deepfake audio technology captures and mimics the advanced subtleties that are native to human speech through advanced means of machine learning, specifically by Generative Adversarial Networks (GANs) and neural vocoders. These can pick up the individualized features of a person's voice, including intonation, accent, and emotional expressions, by learning on massive amounts of spoken material [7]. This then allows us to generate fully synthesized audio that is almost indistinguishable from human recordings.

The advent of deepfake audio technology has profound impacts in all domains. In the entertainment sector, it can be used to dub cartoon characters and computer game characters, dub foreign movies, and create voiceovers that are almost indistinguishable from real ones. In addition, the technology can be utilized to transform assistive technology for people who suffer from speech disorders, thus improving their communication abilities [5]. Nevertheless, this technology also poses ethical concerns with the potential to create voices that are almost indistinguishable from real human speech. Given the potential for misuse, such as impersonation and spreading misinformation, it is important to apply responsible development and conduct widespread deliberation.

As deepfake audio technology evolves, we should anticipate synthetic speech to become increasingly expressive and natural-sounding. This will further blur the line between real and fake audio, diminishing the differences between them. To block the malicious use of this technology without unleashing its positive potential, it is necessary to create effective detection and mitigation tools.

13.3 THE DUAL NATURE OF DEEPFAKE TECHNOLOGY

Deepfake audio technology is a conundrum with possible advantages and disadvantages. On the one hand, it offers a comprehensive toolkit for creative expression, accessibility, and customized experiences. However, it raises serious ethical concerns regarding mistreatment, deceit, and the erosion of trust in audio data. There are many advantageous applications for deepfake audio. It enables the entertainment industry to create dynamic video games with dynamic characters that can speak in complex ways, dub films into multiple languages with realistic and emotionally resonant voices, and produce captivating voiceovers [8]. Moreover, deepfake audio can be used to personalize virtual assistants to create more engaging and natural interactions [9]. Furthermore, virtual assistants can be customized with deepfake audio to produce more interesting and organic interactions [9]. It may also restore the voices of those who are unable to speak, which would allow them to communicate more effectively and participate more fully in society. But the same technology that enables these practical applications can also be abused. Deepfake audio can be used for impersonation, fraud, evidence fabrication, and swaying public opinion [10]. Confidence in audio information is undermined, and individual and societal security is threatened by the ability to create artificial voices that sound exactly like natural human speech.

The dual nature of deepfake audio technology necessitates a careful balancing act. We must promote innovation, look into its potential benefits, establish safeguards against misuse, and establish ethical guidelines for its development and use. This means promoting media literacy, educating people about the potential harm caused by deepfake audio, and providing funding for research on detection and mitigation technologies. Only by cooperating can we minimize the risks associated with deepfake audio while maximizing its transformative potential.

13.4 THE PROCESS OF DEEPFAKE AUDIO GENERATION

Deepfake audio production involves complex procedures and cutting-edge technology. We can produce artificial voices that closely mimic human speech by using AI, more especially deep learning models. This includes converting raw audio data into a personalized synthetic voice that can replicate the nuances, emotions, and unique traits of a specific speaker. This section delves into the step-by-step process of deepfake audio generation, exploring the techniques and algorithms that enable us to implement this.

13.4.1 Training Models to Mimic Human Speech

The first step in creating a convincing synthetic voice is gathering high-quality data, which forms the basis for generating deepfake audio. The accuracy of the final voice is strongly influenced by the quality of the data. A more realistic synthetic voice results from a dataset that is rich and varied.

13.4.1.1 Sources of Voice Data

Gathering suitable voice data can be approached in several ways, each with its own set of considerations:

- **Publicly Available Datasets**: A readily available option is to use pre-existing public datasets like LibriSpeech, VoxCeleb, and Common Voice. These datasets offer a vast repository of speech recordings with diverse speakers and varying acoustic conditions.
- **Scraping Online Platforms**: In another approach, data can be collected from online platforms such as YouTube, podcasts, and social media. However, this method requires careful consideration of ethical and legal issues related to copyright and privacy.
- **Recording Dedicated Datasets**: For specific needs when seeking greater control over the data, recording a dedicated dataset is an option.

Essential preprocessing procedures like cleaning, normalization, and segmentation are carried out after the voice dataset is collected in order to get it ready for model training. By taking these precautions, the audio is made clear, consistent, and structured, which paves the way for the creation of lifelike synthetic voices. Various segmentation techniques, such as hidden Markov models [11] and forced alignment, can be used based on the particular needs.

After collecting the voice dataset, essential preprocessing steps like cleaning, normalization, and segmentation are performed to prepare the data for model training. These steps help ensure the audio is clear, consistent, and structured—laying the groundwork for generating realistic synthetic voices [12].

After preprocessing, the next step is to carefully select the right deep learning architecture and employ robust training procedures to create a model capable of generating highly realistic synthetic voices.

13.4.1.2 Deep Learning Architectures

Several deep learning architectures have proven effective in mimicking human speech, each with its own strengths and weaknesses. Below are several architectures that will be thoroughly discussed in coming sections:

- **Generative Adversarial Networks (GANs)**: GANs consist of two neural networks, a generator and a discriminator, engaged in a competitive game. The generator creates synthetic audio samples, while the discriminator tries to distinguish them from real samples. This adversarial process pushes the generator to produce increasingly realistic audio.

  Within the GAN framework, various specialized architectures have emerged, each tailored to specific challenges in audio synthesis. Some notable examples include:
  - **WaveGAN**: This model directly generates raw audio waveforms, capturing the fine-grained details of speech signals.
  - **MelGAN**: This model operates in the Mel-spectrogram domain, which represents audio frequencies in a way that is more aligned with human perception.
  - **BigVGAN**: This model leverages a hierarchical architecture and large-scale training data to achieve high-fidelity audio generation (Figure 13.1).
- **Neural Vocoders**: These models focus on the lower-level aspects of speech synthesis, taking intermediate representations (such as mel-spectrograms) and converting them into natural-sounding waveforms.

  Popular examples include:
  - **WaveNet**: This autoregressive model generates audio samples one at a time, capturing the temporal dependencies in speech signals.
  - **WaveGlow**: This flow-based model offers faster synthesis speed compared to WaveNet while maintaining good audio quality.
  - **Parallel WaveGAN**: This model combines the adversarial training of GANs with the efficiency of parallel processing, enabling high-fidelity audio generation at a faster speed.
- **Autoregressive Models**: These models predict the next audio sample based on the previous ones, effectively capturing the sequential nature of speech. Transformer networks, originally designed for natural language processing, have also shown promising results in this domain.

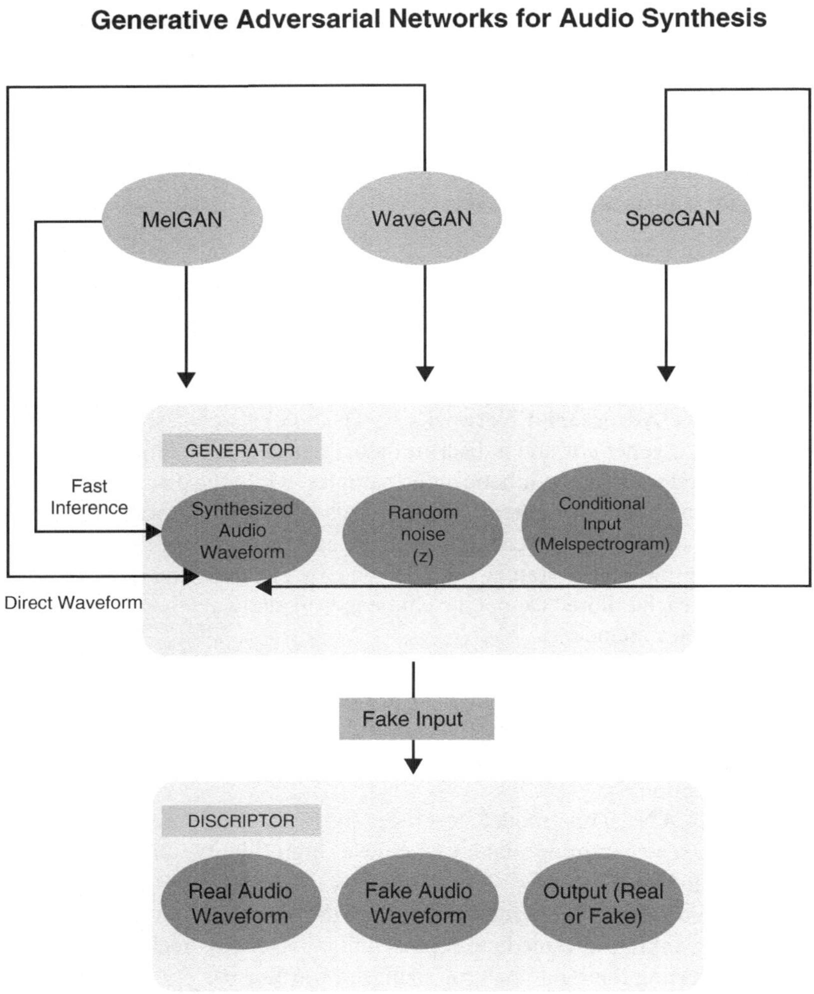

FIGURE 13.1 A diagram illustrating the basic structure of a GAN with its generator and discriminator components, along with examples of different GAN architectures for audio synthesis.

13.4.1.3 Optimization Techniques

To efficiently navigate possible parameter values, various optimization techniques are employed such as Gradient descent, which is often used as a starting point. It calculates the direction of the steepest descent on the loss function's surface and adjusts the model's parameters accordingly.

However, gradient descent can sometimes get stuck in local minima or have slow convergence. To address these challenges, more sophisticated variants of gradient descent have been developed, such as:

- **Adam**: This adaptive optimization algorithm calculates individual learning rates for different parameters, allowing for faster and more stable convergence.
- **RMSprop**: This algorithm addresses the issue of rapidly decaying learning rates, which can hinder the model's progress in certain scenarios.

13.4.1.4 Monitoring Techniques

Throughout the training process, it's essential to monitor the model's progress and ensure it is effectively learning the nuances of human speech. This is often done using a combination of objective metrics and subjective evaluation:

- **Objective Metrics**: These are quantitative measures that assess specific aspects of the generated audio. Common metrics include:
  - **Mean Squared Error (MSE)**: Calculates the average squared difference between the generated audio and the real speech samples.
  - **Mel-Cepstral Distortion (MCD)**: Measures the dissimilarity between the mel-frequency cepstral coefficients of generated and real audio, which are commonly used to represent speech features.
  - **Perceptual Loss Functions**: These aim to better align with human perception of audio quality by incorporating psychoacoustic principles.
- **Subjective Evaluation**: While objective metrics provide valuable insights, they don't always capture the full picture of audio quality. Subjective evaluation involves human listeners assessing the naturalness, intelligibility, and overall quality of the synthetic speech. This can be done through formal listening tests or informal feedback.

By combining these monitoring techniques, we can gain a comprehensive understanding of the model's performance and make informed decisions about adjustments to the training process. This iterative process of training, evaluation, and refinement continues until the desired level of realism and expressiveness is achieved.

13.4.2 Capturing Nuances: Accents, Emotions, and Speaking Patterns

Speech synthesis needs clear articulation and exact pronunciation, but to make deepfake audio that sounds real and expressive, you need to capture the little things that make human speech dynamic and interesting. Small things like accents, emotions, and unique speaking styles have a big effect on how natural and believable a synthetic voice sounds. Accents and Dialects: Accents and dialects, which are shaped by social and regional factors, have a big impact on how a voice sounds. They are a mix of changes in intonation, rhythm, and pronunciation, and we need to think about a lot of things to get these social and regional differences right in deepfake audio.

First and foremost, the target accent or dialect must be sufficiently represented in the training data. To do this, a varied dataset comprising speakers from various geographical locations and linguistic backgrounds must be gathered. For accessing audio samples and phonetic information, online resources such as the International Dialects of English Archive and the Oxford English Dictionary can be very helpful.

The natural fluidity of accents and dialects presents a modeling challenge. Even within a particular area or social group, there is a great deal of variation in their existence. As a result, it is challenging to establish precise parameters for data collection and modeling. Furthermore, for accurate representation, some accents might need specific phonetic transcriptions or linguistic knowledge, hence working with linguists or phoneticians can help overcome these difficulties by offering insightful information about the subtleties of the target accent. Specific acoustic characteristics that set the accent apart can also be identified and modeled with the use of specialized phonetic analysis tools.

13.4.2.1 Emotions

Expressions of emotion are essential to human communication. They give speech the kind of nuance and expressiveness that makes it truly human. Our synthetic voices must convey the same character and emotional states as the real human voices we model. So, if we're going to make voices really well, we have to understand the chemistry of human destructive interference.

Models that are trained to recognize and produce emotional speech require datasets that are themselves the products of emotional speech. The datasets we currently use for this purpose, such as the CREMA-D and RAVDESS, consist of recordings made in emotional states. RAVDESS is an acronym that stands for the "Research Acoustic and Visual Database of Emotional Speech and Song." The dataset is a product of the work of Creech et al. [13]. The primary investigator for RAVDESS is Dr. S. A. Creech, who is associated with the Department of Communication at the University of Connecticut.

Nonetheless, it is a tough job to create emotional synthetic speech since emotions in human speech are already quite challenging to model owing to their subjective and variative nature across individuals and cultures. The perception and appearance of emotions are a matter of personal taste, and human heads, hearts, and minds are all very different—and very private and secretive—places. There is some commonality, of course, across humanity, but when you get down to cast the mold from here to Hades, it's hard to be universal in any humanistic endeavor.

Explore different techniques for integrating affective information into the model to address these problems. Emotion embeddings—vectors in high-dimensional space that convey emotions—can capture the nuances of different emotional states and the relationships between them. Also useful for this purpose are style tokens, which can regulate the emotional expressiveness of the synthetic voice. The representations they give of different speaking styles serve as another kind of affective information. In some cases, a combination of techniques is used.

Training should consider perceptual loss functions, which are intended to better match the model's output with how people perceive emotional expression. This could guarantee that the speech that is produced elicits the desired emotional reaction from the audience.

13.4.2.2 Speaking Patterns

Distinct speech patterns contribute to vocal identity, and every person has them, in addition to their accents and emotions. These patterns encompass rhythm, tempo, and the use of pauses, as well as "unique pronunciations" of words. To create customized synthetic voices that accurately capture a speaker's distinct vocal qualities, it is crucial, first and foremost, to capture these individual nuances.

Distinct patterns are captured using methods like speaker embeddings, which yield compact representations of the unique vocal traits of a speaker. The speech synthesis model is then customized using these embeddings to produce speech that closely resembles the target speaker's style. However, the amount of data needed to capture these subtleties can vary greatly depending on the speaker and the level of personalization desired. This can be a problem, so it's important to find a balance between personalization and expansion to make sure that the synthetic voice can produce a wide range of content while still keeping the speaker's unique traits. Another problem is that it's hard to model the small differences and inconsistencies that happen naturally in human speech, even when one person is speaking. Depending on the situation and the speaker's mood, the pacing, rhythm, and pronunciation of speech can change. This is why it's important to make sure that the speech generated sounds natural and doesn't sound too robotic or repetitive.

13.5 FUNDAMENTAL TECHNOLOGIES BEHIND DEEPFAKE AUDIO

13.5.1 Generative Adversarial Networks (GANs)

GANs have emerged as a powerful tool for creating highly realistic deepfakes. They consist of two neural networks:

1. **Generator**: This network converts random noise into artificial audio samples, much like a creative artist would. Its objective is to create audio that is identical to authentic human speech. During the training process, it modifies its internal parameters to learn how to map the random noise to a complex audio signal.
2. **Discriminator**: This network assesses the produced audio samples and discerns between authentic and fraudulent audio in its capacity as a perceptive critic. Its goal is to correctly detect phony audio, which compels the generator to enhance its output. There is an ongoing adversarial process between the discriminator and the generator. While the discriminator tries to correctly detect phony audio, the generator aims to create audio that can fool the discriminator. Both networks develop as a result of this competitive dynamic, producing synthetic audio that gets more lifelike [14].

13.5.1.1 Key Techniques in GAN-Based Deepfake Audio

Several techniques have been developed to enhance the capabilities of GANs in deepfake audio:

- **Conditional GANs (cGANs)**:

  Deepfake audio has become more realistic and controllable thanks to cGANs. cGANs can make speech sound more natural and fit certain goals by giving the model more information, such as the speaker's name, text, or emotional tone.

  For example, a cGAN can make audio that sounds like a person's tone, style, and speech patterns if it is given a script and a short sample of their voice. This is especially helpful for voice conversion, which changes one person's voice to sound like another's, and voice cloning, which teaches a model to copy a speaker's voice from just a few seconds of audio.

  These new features are making things possible in fields like education, entertainment, and assistive technology. But the risks grow as cGANs get better and better. There are serious moral issues with deepfake audio because it could be used for things like spreading false information or stealing someone's identity. As we make these technologies better, it's just important to put money in tools that can find them and rules that will make sure they are used for good.
- **CycleGAN**:

  CycleGANs, originally developed for image-to-image translation, have been effectively adapted for voice conversion. Unlike traditional GANs, CycleGANs don't need paired training data—which is often hard to get in audio tasks. Instead, they can learn from unpaired recordings, making them highly practical.

  A CycleGAN uses two generators and two discriminators to translate voices between domains—for example, from Speaker A to Speaker B and back again. This "cycle" helps the model preserve the content of speech while changing style, such as accent, gender, or emotion.

  Their strength lies in handling complex transformations without needing aligned data. However, CycleGANs can be tricky to train, and the audio quality may not always match that of more specialized models.
- **StarGANs**:

  StarGAN is a versatile model of multi-domain image-to-image translation as well as can be adjusted to multi-domain voice conversion. Unlike normal GANs and CycleGANs, StarGAN supports multiple domains at the same time, which facilitates a large variety of voice conversion. From such a wide-ranging dataset of voices with which StarGAN is trained, the model has a potential to learn to lobby voices, adoption voices, etc. to represent the intrinsic features of various voices like accent, gender, age, and emotion. After being trained, StarGAN can transform a source voice into different target styles. A male voice may be altered into a female voice, or a neutral voice may be changed into an angry one, and so forth. However, StarGAN comes with a high level of degree of freedom and difficulty. Despite these challenges, StarGAN remains a promising technique for creating diverse and realistic deepfakes.

13.5.1.2 Challenges and Considerations

While GANs have shown remarkable potential in deepfake audio, several challenges remain:

- **Mode Collapse**: This occurs when the generator produces only a limited range of outputs, limiting the diversity of generated audio. Techniques like spectral normalization and feature matching can help mitigate this issue.
- **Training Instability**: GANs can be notoriously difficult to train, often leading to unstable and unpredictable behavior. Careful tuning of hyperparameters and the use of advanced training techniques, such as adaptive learning rates and gradient clipping, can help stabilize the training process.
- **Ethical Implications**: The ability to create highly realistic deepfakes raises significant ethical concerns. Misuse of this technology can lead to the spread of misinformation, defamation, and identity theft. It is crucial to develop robust detection methods and ethical guidelines to mitigate these risks.

As deepfake technology continues to advance, it is essential to strike a balance between innovation and responsibility. By understanding the underlying techniques and challenges, we can harness the power of GANs for beneficial applications while mitigating potential harm.

13.5.2 Autoregressive Models

Autoregressive models have transformed the way we synthesize and understand audio signals, particularly in deepfake voice-based systems. These models are inherently sequential in nature, generating data point by point or frame by frame, with each step conditioned on the previous outputs. Their ability to capture intricate temporal dependencies makes them an ideal choice for modeling the subtle and expressive dynamics of human speech [15].

In this chapter, we delve into the workings of autoregressive models, explore their applications in deepfake voice synthesis, and illuminate the principles underlying their success. We will also examine notable implementations such as WaveNet and Tacotron, detailing how they achieve state-of-the-art results in audio generation.

Sequential Data Generation

In autoregressive audio generation, the sequence represents audio samples over time. The model begins with an initial state or "seed" value, generating the next sample conditioned on this state. This newly generated sample is then fed back into the model to generate the subsequent sample, continuing until the entire audio waveform is synthesized.

For human speech, this sequential nature captures dependencies such as:

- **Phonetic Continuity**: Ensuring that adjacent sounds flow naturally.
- **Prosody**: Capturing variations in pitch, rhythm, and intensity.
- **Contextual Coherence**: Maintaining meaning and logical transitions over time.

13.5.2.1 WaveNet: A Masterpiece in Autoregressive Modeling

WaveNet, introduced by DeepMind, is one of the most iconic examples of autoregressive models for audio synthesis. It revolutionized the field by directly generating raw audio waveforms, eschewing traditional pipeline-based approaches that rely on intermediate representations like mel-spectrograms [16].

Architecture Overview

WaveNet employs a series of *dilated causal convolutions*, a novel mechanism that allows it to capture long-term dependencies efficiently without requiring recurrent structures like LSTMs or GRUs [17]. Let's break this down:

1. **Causal Convolutions**:
   - In causal convolutions, each output at time ttt depends only on inputs from time ttt and earlier. This ensures that the model adheres to the autoregressive property.
   - These convolutions allow the model to process data in parallel during training, even though generation remains sequential.
2. **Dilated Convolutions**:
   - Dilated convolutions introduce gaps between the convolutional filter's receptive fields, exponentially increasing the model's temporal receptive field without a proportional increase in computational cost.
   - For example, with a dilation factor of 2, the model skips every other sample, effectively looking farther into the past to predict the next sample.
3. **Stacked Layers**:
   - WaveNet stacks multiple layers of dilated convolutions, each layer increasing the temporal context of the model. This hierarchical structure allows it to capture both short-term and long-term dependencies.

Audio Generation Process

The audio generation process in WaveNet is sample by sample:

1. The model predicts the probability distribution for the next sample given all previous samples.
2. A sample is drawn from this distribution.
3. The new sample is appended to the sequence and used to condition the next prediction.

While this approach ensures high fidelity and precision, it is computationally intensive due to the sequential nature of sample generation.

13.5.2.2 Tacotron: Bridging Text and Speech

Tacotron builds on the principles of autoregressive modeling but takes a more practical and efficient approach to audio synthesis, particularly for text-to-speech (TTS) applications [18,19]. Unlike WaveNet, which generates raw waveforms, Tacotron operates in two stages:

1. **Text-to-Mel-Spectrogram**: Converts input text into a mel-spectrogram representation, an intermediate format that captures the spectral characteristics of audio.
2. **Mel-to-Waveform Conversion**: Uses a neural vocoder (e.g., WaveNet or Griffin-Lim) to synthesize the final audio waveform from the mel-spectrogram.

Key Innovations in Tacotron

- **Sequence-to-Sequence Learning**:
  - Tacotron uses an encoder-decoder architecture with attention mechanisms to align input text sequences with output mel-spectrogram frames.
  - This alignment ensures that the model understands how individual characters or words correspond to audio features.
- **Attention Mechanism**:
  - The attention mechanism enables the model to focus on relevant portions of the input text when generating each frame of the mel-spectrogram.
  - This dynamic focus is crucial for handling variable-length inputs and outputs, such as long sentences or phrases.
- **Hierarchical Processing**:
  - By generating mel-spectrograms as an intermediate representation, Tacotron reduces the complexity of waveform generation, making the process more efficient without sacrificing quality.

13.5.2.3 WaveGlow: A Powerful Tool for Deepfake Audio

Note: While WaveGlow itself is not Autoregressive, it is often used as a non-autoregressive vocoder in autoregressive pipelines such as Tacotron 2. We include it here due to its role in the autoregressive TTS framework.

WaveGlow is a flow-based generative model that has emerged as a powerful tool for high-quality audio synthesis. Unlike traditional autoregressive models like WaveNet, WaveGlow is a non-autoregressive model, which means it can generate audio samples in parallel rather than sequentially. This makes it significantly more efficient than WaveNet, especially for real-time applications [20].

13.5.2.3.1 Architecture of WaveGlow

WaveGlow's architecture consists of a series of invertible 1×1 convolutions and affine coupling layers. These layers are stacked together to form a deep neural network, which learns to model the probability distribution of audio samples.

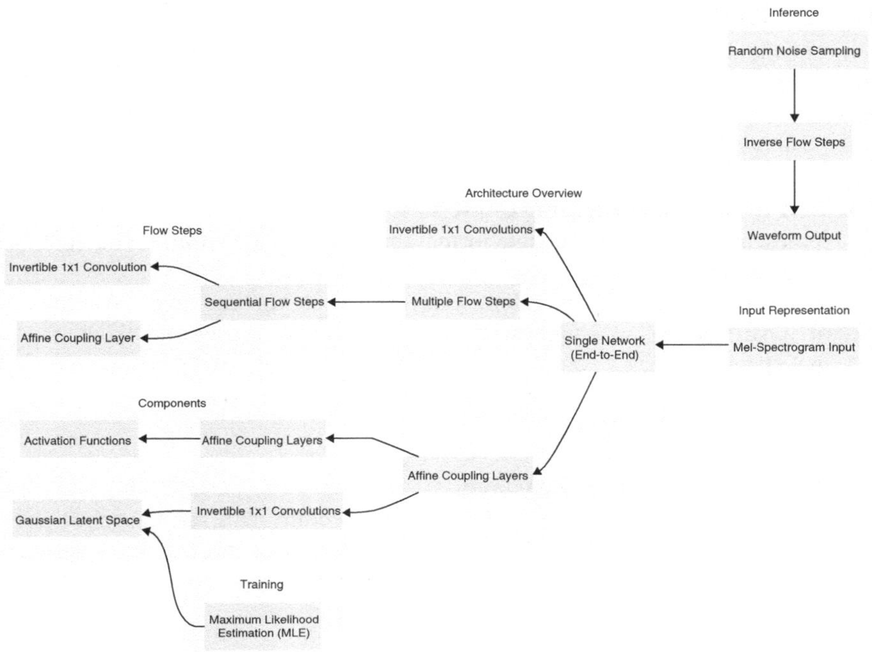

FIGURE 13.2 Architecture of WaveGlow.

- **Invertible 1 × 1 Convolutions**: These layers are used to mix the input features, allowing the model to learn complex dependencies between different frequency bands.
- **Affine Coupling Layers**: These layers introduce non-linearity into the model, enabling it to capture the intricate details of the audio signal (Figure 13.2).

Key Innovations of WaveGlow

- **Non-Autoregressive Generation**: WaveGlow's non-autoregressive architecture allows it to generate audio samples in parallel, significantly reducing the computational cost compared to autoregressive models.
- **Flow-Based Generative Model**: WaveGlow is a flow-based generative model, which means it learns a sequence of invertible transformations that map a simple distribution (e.g., a Gaussian distribution) to the complex distribution of audio samples.
- **High-Quality Audio Synthesis**: WaveGlow can generate high-quality audio with a wide range of styles and timbres. It can also be used for voice conversion, text-to-speech, and other audio synthesis tasks.

Limitations of Autoregressive Models in Deepfake Voice

1. **Computational Complexity**:
   - Generating audio sample by sample is inherently slow and resource-intensive, limiting real-time applications.
2. **Scalability**:
   - Maintaining coherence over long audio sequences can be challenging, as errors accumulate over time, leading to artifacts or unnatural transitions.
3. **Training Requirements**:
   - Autoregressive models require large datasets and extensive computational resources to learn the intricate dependencies in audio data.

13.5.3 Variational Autoencoders (VAEs)

Variational Autoencoders (VAEs) provide a distinct approach to deepfake audio generation, setting them apart from other techniques such as GANs. While GANs rely on an adversarial setup to create realistic audio, VAEs excel at learning structured representations of the underlying data, which enables both high-quality generation and precise control over the synthetic speech [21].

Imagine, for a moment, a sculptor tasked with crafting intricate statues. Rather than directly shaping each piece of marble, the sculptor first creates a detailed mold—a representation of the key features and variations. With this mold, the sculptor can then produce new statues, each with specific desired characteristics, while still adhering to the principles captured in the original mold. Similarly, VAEs aim to learn a latent representation of audio data that captures its underlying structure and relationships. This allows for the generation of new audio samples with a degree of flexibility and control that is central to deepfake audio.

The Latent Representation: A Mold for Audio

The core innovation of VAEs lies in the concept of the **latent space**. This is a lower-dimensional space where each point corresponds to a unique configuration of speech characteristics, such as speaker identity, emotional tone, or speaking style. By navigating this latent space, we can produce variations in these characteristics, similar to how the sculptor might modify a statue's features by choosing a different part of the mold.

This ability to manipulate latent space is what makes VAEs particularly powerful for applications like voice conversion, speech synthesis, and emotional expression modulation. When trained effectively, VAEs allow for smooth transitions between different speech variations, making them a robust tool for creating deepfake audio.

How VAEs Work

The architecture of a VAE consists of two primary components: the **encoder** and the **decoder**.

- The **encoder** takes an input audio sample and compresses it into a compact latent representation. This process involves a neural network that extracts the essential features of the audio, much like how a mold captures the crucial features of a statue.

- The **decoder**, on the other hand, takes a point in the latent space and reconstructs the audio. This is akin to casting a new statue from the mold, generating a specific variation based on the point selected within the latent space.

The training process for VAEs balances two objectives:

1. Accurately reconstructing the input audio from the latent representation.
2. Ensuring that the latent space is structured in a way that supports smooth interpolation, allowing for meaningful manipulation of speech characteristics.

Applications of VAEs in Deepfake Audio

The structured latent space learned by VAEs unlocks a variety of exciting applications in deepfake audio generation:

1. **Controllable Speech Synthesis**

   One of the most compelling applications of VAEs is their ability to generate speech with fine-grained control over characteristics like speaker identity, emotional tone, and speaking style. By navigating different points in the latent space, we can generate audio that exhibits a desired trait, such as a specific emotional expression or a different voice. **Conditional VAEs (CVAEs)** can further enhance this control by conditioning the latent representation on external factors such as emotion labels or speaker embeddings. This allows for even more targeted manipulation, enabling, for example, the generation of speech with a particular emotional tone based on an input label.
2. **Voice Conversion**

   VAEs can also be used for voice conversion, a process where one speaker's voice is transformed into another's while preserving the linguistic content. By training a VAE on a dataset containing speech from multiple speakers, the latent space captures shared features between voices, enabling the conversion of one voice to another by manipulating the latent variables.
3. **Speech Enhancement**

   Another practical use of VAEs in deepfake audio is speech enhancement. By training on pairs of clean and noisy speech samples, the encoder learns to extract the essential features of the clean audio from noisy input. The decoder can then reconstruct an enhanced version of the speech, making VAEs valuable in applications such as noise reduction or improving the clarity of speech recordings [22,23].

Further Exploration

To deepen your understanding of VAEs and their applications in deepfake audio, consider exploring the following areas:

- **Latent Space Structure**: Investigate techniques for structuring the latent space more effectively, such as disentangled representations, where each dimension of the latent space corresponds to a specific feature (e.g., pitch,

emotion, speed). Hierarchical latent spaces can also be explored, where multiple levels of abstraction represent varying degrees of detail in the speech.
- **Training Efficiency**: Research methods to improve training efficiency, such as incorporating adversarial training (e.g., using GANs to refine the generated audio) or optimizing the neural network architectures for faster convergence.
- **Evaluation Metrics**: Understanding how to evaluate VAE-generated audio is crucial. Objective metrics, such as Signal-to-Noise Ratio (SNR) and Mel-Cepstral Distortion (MCD), are commonly used. However, subjective measures, like listening tests and Mean Opinion Scores (MOS), offer valuable insights into how realistic and expressive the generated speech sounds to human listeners.

By grasping the underlying principles of VAEs and their mathematical intuition, you can harness their power to create highly controlled, expressive deepfake audio that goes beyond mere imitation. Whether for voice conversion, emotional synthesis, or noise removal, VAEs provide a versatile toolset for next-generation speech generation.

13.6 APPLICATIONS OF DEEPFAKE AUDIO TECHNOLOGY: UNVEILING A NEW ERA OF POSSIBILITIES

Deepfake audio, once a concept relegated to the realm of science fiction, has swiftly transitioned into reality. Its ability to convincingly synthesize human voices is no longer a futuristic dream but a rapidly advancing technology that is beginning to permeate various aspects of our lives. While deepfake audio raises valid concerns about potential misuse, it also unlocks an exciting array of opportunities across industries and disciplines.

Imagine a world where the voices of iconic figures are resurrected, where your interactions with technology are seamlessly personalized, or where the voices of those who have lost their ability to speak can be restored. This is the transformative potential of deepfake audio. In this chapter, we will explore how this technology is not only reshaping entertainment and revolutionizing personal assistants but also enhancing educational experiences and offering new solutions for voice restoration.

As we navigate this evolving landscape, we will examine the ethical complexities that come hand in hand with these breakthroughs. By understanding both the immense benefits and inherent risks, we can work toward harnessing deepfake audio in ways that foster innovation while ensuring its responsible application for the greater good.

Our journey begins in the realm of entertainment and content creation, where deepfake audio is already making profound waves, altering how we engage with media and creativity

13.6.1 Entertainment and Communication

The world of entertainment is undergoing a major transformation thanks to deepfake audio. What was once thought of as a sci-fi concept is now shaping the future of how we create content, tell stories, and engage with audiences in ways that were unimaginable just a few years ago. With its ability to generate incredibly realistic and expressive voices, deepfake audio is opening doors to a new kind of creative freedom.

13.6.1.1 Voice Acting and Dubbing

Imagine a movie or video game where every character sounds so natural, so lifelike, that it's hard to tell if they're real or synthetic. That's what deepfake audio is bringing to the table. It allows creators to generate high-quality synthetic voices for characters in animated films, video games, and even commercials [24]. Take, for instance, *The Lion King* (2019) or *Cyberpunk 2077.* With deepfake technology, voice acting can be more nuanced and natural, helping characters connect with the audience on a deeper level. But it doesn't stop there, Deepfake audio is also changing how we think about dubbing films in different languages. Traditionally, dubbed films often feel disconnected from the original performance because of the differences in voice and emotion. With deepfake technology, we can now create voices that stay true to the original character's tone and emotions, making the viewing experience feel more authentic and enjoyable, no matter where you are in the world.

13.6.1.2 Personalized Storytelling Experiences

Deepfake audio is reshaping how we experience stories—whether through audiobooks or interactive media. Imagine listening to your favorite novel, not in a standard narrator's voice, but in the voice of a beloved actor, a famous musician, or even a digitally recreated version of your own. This level of personalization turns audiobooks into something deeply personal and emotionally resonant.

The same technology is enhancing interactive storytelling in video games and VR. Instead of static voice lines, characters can now respond dynamically—with voices that adapt in emotion, tone, and style based on your actions or even your mood. Picture a character in a game asking how you're feeling and responding with empathy in a realistic voice that matches the context.

These innovations offer a more immersive and emotionally engaging narrative experience, blurring the line between listener and participant.

13.6.1.3 Music Production and Vocal Synthesis

In the music industry, deepfake audio is unlocking fresh possibilities for artists. One of the most exciting uses is in vocal synthesis—creating entirely new vocal tracks using synthetic voices. This technology allows artists to experiment with different vocal styles and sounds, even recreating iconic voices from the past. Picture a song performed with the voice of a legendary artist, whose music is no longer accessible due to their passing, but whose spirit lives on through deepfake technology [25].

Take the example of an AI platform like *Jukedeck* or *Flow Machines*, which can synthesize vocals that mimic the style of a specific artist. With deepfake audio, artists can resurrect voices, blend them with modern tracks, and create unique musical experiences that wouldn't have been possible before.

13.6.1.4 Personalization and Emotional Resonance in Voice Interactions

Much like its role in entertainment, deepfake audio is transforming everyday interactions with machines and virtual environments [26]. It enables the creation of voice assistants that go beyond functionality, offering emotionally engaging experiences.

Imagine asking for help and hearing a voice that sounds like your favorite celebrity, a loved one, or even your own. This isn't just a novelty—it's a shift toward more personalized, human-centered technology. The same personalization extends to virtual characters in games, VR, and AR, where deepfake audio brings them to life with voices that adapt to user input, making digital interactions feel more authentic and emotionally resonant.

Key Implementation Considerations

When diving into the world of entertainment and personalized voice assistants and virtual characters, there are several technical considerations to keep in mind:

- **Data Quality**: High-quality, emotionally diverse datasets are essential. Recordings should capture various moods, accents, and speech styles to create realistic and adaptable synthetic voices.
- **Model Selection and Fine-Tuning**: Choose models based on your use case. **Tacotron 2** and **WaveNet** are ideal for scripted, natural-sounding speech, while **VoiceLoop** or Modulate are better suited for real-time, interactive applications. Fine-tune with domain- or user-specific data for optimal results.
- **User and Ethical Focus**: Personalization should consider user context—what works in gaming may not suit healthcare. Always obtain consent for voice use and be transparent about synthetic content to ensure ethical deployment.

13.6.2 Expanding to Customer Service, Accessibility, and Education

Deepfake audio technology is rapidly evolving beyond entertainment, finding powerful applications in customer service, accessibility, and education—domains where voice plays a critical role in personalization and human connection.

Customer Service and Accessibility

In customer service, companies are using synthetic voices that align with a customer's demographic to create more personalized, emotionally engaging experiences. A voice that feels culturally familiar or soothing can transform routine interactions into moments of trust and empathy.

Deepfake audio also holds enormous promise in accessibility. For individuals with speech impairments, personalized synthetic voices offer a means to communicate authentically. Rather than using generic speech devices, users can restore a version of their own voice, reflecting their identity and emotional tone. For example, **VocaliD** uses deepfake audio to recreate voices for people affected by conditions like ALS, helping them speak again in a way that feels personal and empowering.

However, developing such experiences demands careful attention to:

1. **Data Privacy and Consent**: Systems must handle voice data responsibly, with explicit user consent and secure storage [27].
2. **Uncanny Valley**: Overly artificial voices can trigger discomfort; addressing emotional cadence and delivery is essential.

3. **Bias and Inclusivity**: To avoid reinforcing stereotypes, voice models must be trained on diverse datasets that reflect a wide range of accents, dialects, and cultures.

Educational Applications

In education, deepfake audio is enabling highly interactive and accessible learning experiences.

AI-generated voice tutors allow students to engage in realistic, adaptive dialogue. For example, when *Maria* wanted to practice Spanish before her trip to Barcelona, she used an app that introduced her to "Carlos," an AI tutor with a regional Spanish accent. Carlos adapted to her learning pace, switching between slow, deliberate phrases and fast, conversational exchanges as her confidence grew [28].

Platforms like **Mondly VR** extend this interactivity into immersive environments—like a Parisian café—where learners practice with lifelike voices that capture regional tone and cultural nuance.

Deepfake audio is also helping to preserve endangered languages. In Canada, schools use AI-generated voices based on elder recordings to teach Indigenous languages such as Cree and Ojibwe, offering students an authentic auditory connection to their heritage.

For visually impaired learners, institutions like the **Perkins School for the Blind** are using AI-narrated texts with natural inflection, replacing monotonous screen readers with emotionally engaging narration. Teacher training programs, too, are incorporating synthetic voices to simulate classroom interactions with diverse student profiles.

Finally, deepfake audio facilitates global e-learning through localized, culturally adapted voice synthesis. Platforms like **Coursera** are experimenting with converting English-language courses into regional languages using deepfake voices, thereby improving accessibility and engagement worldwide.

Untapped Possibilities: What Comes Next?

While these applications are groundbreaking, the potential of voice-based deep fake technology is far from fully realized:

- **Cross-Disciplinary Narratives**: Imagine history textbooks narrated by figures like Abraham Lincoln, whose voices are synthesized from archival data, creating an immersive storytelling experience.
- **Real-Time Multilingual Conversations**: Picture a classroom where students from different countries collaborate in real time, with AI voices bridging language gaps effortlessly.
- **Emotional Engagement Tools**: What if AI narrators could detect a student's mood and adjust their tone accordingly, encouraging or calming them as needed?

Voice Restoration for Individuals with Speech Impairment

Deep fake voice restoration doesn't merely give back speech; it restores emotional expression, individuality, and the ability to connect authentically. It has already made a significant impact in real-world applications while also hinting at a future filled with untapped possibilities [29].

When *Pat Quinn*, co-founder of the ALS Ice Bucket Challenge, lost his voice, deepfake models recreated it using earlier recordings—allowing him to speak again in his own voice. Organizations like VocaliD offer similar solutions for laryngectomy patients, even when no prior voice data exists, by combining donor voices with the patient's speech patterns.

Children using AAC devices now benefit from AI-generated voices that match their age and personality.

Preserving Voices for the Future

Voice banking allows individuals with progressive conditions to record their voices early. Later, AI can generate speech in their natural tone, maintaining continuity in both personal and professional life.

Expanding Possibilities

Beyond speech restoration, AI voices now convey emotion, enabling more human-like interaction—whether it's a heartfelt "I love you" or creative expression in performance or education.

Potential future applications include:

- VR integration for immersive social participation
- Multilingual voice synthesis
- Personalized AI interactions in healthcare
- Educational use for public speaking or teaching roles

Deepfake voice restoration is not just technical—it's personal. It enables people to speak, connect, and express who they truly are.

13.7 SOCIETAL IMPLICATIONS AND GOVERNANCE AROUND DEEPFAKE AUDIO

The emergence of deepfake audio technology raises both questions and concerns. On the one hand, we can witness the remarkable ability of machines to replicate the human voice with almost flawless accuracy, creating new prospects for various industries. However, there is a significant risk of abuse, which leads to discussions about ethics, security, and privacy.

Imagine getting a call from someone who sounds like a trusted family member or your best friend, only to find out later that the voice you heard was completely fake. Deepfake audio can now replicate voices with such authenticity that it's almost impossible for the average person to tell the difference between the real and the fake, so this situation is no longer the stuff of science fiction.

But this ability to imitate also poses a serious problem: how can we uphold people's rights and guarantee that this technology is applied for the greater good? International organizations, tech firms, and governments are all rushing to address the negative aspects of deepfakes, like fraud, disinformation, and privacy violations. How we can balance encouraging innovation with protecting society from the risks posed by this technology is the crucial question.

We'll look at how various nations and businesses are addressing these issues in this chapter. We'll look at current laws and governance structures that are beginning to take shape and analyze how they seek to mitigate the negative effects of deepfake audio while maintaining its beneficial features. We'll identify any remaining gaps and consider what more can be done to guarantee that deepfake technology is applied in a responsible, moral, and open manner.

Existing Regulations by Governments

The need for regulation becomes more pressing as deepfake technology advances. Governments everywhere are taking action by enacting new laws and regulations to stop the abuse of deepfake audio. In the meantime, tech firms—who frequently develop and distribute these technologies—are also assuming accountability by putting in place safeguards against malevolent use. The delicate balancing act between innovation, regulation, and individual freedom is still being worked out, though.

A Global Snapshot of Regulatory Efforts

1. **European Union**: A Proactive Strategy for AI and Digital Content. When it comes to regulating AI technologies, including deepfakes, the European Union has taken the lead. Deepfakes are one of the main targets of the European Commission's much-anticipated Artificial Intelligence Act, which was unveiled in 2021 and laid the groundwork for extensive regulation of high-risk AI applications. Deepfake audio is clearly covered by this act, which imposes more stringent oversight on AI technologies that endanger public safety or fundamental rights. In contrast, the Digital Services Act (DSA) mandates that online platforms promptly eliminate dangerous content, such as deepfakes, providing users with accountability and transparency. The EU's approach is unique because it insists on transparency. AI-generated content must now be identified as synthetic by platforms that host it, allowing users to determine whether what they see or hear is authentic. This establishes a global standard that other areas can adhere to, promoting moral principles in the rapidly changing field of AI-generated media.
2. **United States**: Laws Designed to Prevent Abuse. The United States has also shifted toward more focused regulation on the other side of the Atlantic. In particular, using deepfake technology to cause harm, defraud, or deceive is illegal under the Malicious Deep Fake Prohibition Act of 2018. A person may be subject to severe legal consequences if they use synthetic audio to spread false information or impersonate someone else with the intention of causing harm—for example, by pretending to be a CEO during an emergency call. The Deepfake Law (SB 1411) in California protects citizens from misleading information that might influence public opinion by specifically addressing the possible abuse of deepfakes in political campaigns.

   Some of the immediate threats posed by deepfake audio have been reduced thanks in large part to these laws. However, the United States continues to struggle with how to handle the wider ethical ramifications, such as whether outright banning deepfakes would violate rights

TABLE 13.1
Regulatory and Ethical Response Comparison (Global Snapshot)

Region/Country	Current Regulation on Deepfakes	Key Focus Area	Notable Gaps
European Union	Proposed AI Act, GDPR	Consent, biometric data	Enforcement mechanism for audio use
United States	State-level laws (e.g., California)	Political deepfakes, consent laws	No federal-level comprehensive policy
China	Deep synthesis regulation (2022)	Labeling, platform responsibility	Vague implementation specifics
India	No specific law, covered under IT Act	Cybersecurity, privacy	Lacks clarity on deepfake audio
UK	Online Safety Bill (pending)	Platform liability	Focus more on content than voices

like free speech. Clearer guidelines are needed to keep this technology from going too far, according to many experts in the ongoing debate.

3. **China**: Preventive Actions to Combat Misinformation. China is taking a more proactive rather than reactive approach to combating deepfake audio. Although China has a long history of strictly regulating digital content, it has recently imposed stringent laws pertaining to content produced by AI. Information on digital platforms must already be genuine and accurate, according to the China Cybersecurity Law (2017), and new regulations were put in place in 2020 specifically to stop the spread of deepfake technology. AI-generated content must now be properly labeled by platforms so that users are aware when they are interacting with synthetic media. To help consumers make educated decisions, China's State Administration of Radio and Television has required that all deepfake-generated content, whether it be audio or video, be made available to the public in the entertainment and media sectors. The government is also developing a comprehensive set of regulations that would govern the use of deepfakes for political or detrimental purposes, like fabricating news, in addition to their use for entertainment (Table 13.1).

13.7.1 Tech Companies and Regulating Deepfake Audio

Tech companies have taken on a significant role in regulating deepfake technology, especially in preventing its detrimental spread across social platforms, even though governments have been quick to enact laws.

1. **Facebook and YouTube**: At the forefront of detection, Social media behemoths like YouTube, Twitter, and Facebook (Meta) are now major players in the fight against deepfakes. These platforms are at the forefront of

identifying and eliminating harmful content because they enable the quick dissemination of information (and misinformation). For instance, Facebook started the Deepfake Detection Challenge in 2018 to create AI models that could identify fake content. These days, the business uses sophisticated algorithms that can identify and eliminate deepfake videos and audio within hours of their upload. In a similar vein, YouTube has reinforced its anti-deceptive practices policy by deploying AI-powered tools that can recognize manipulated content automatically. These platforms now collaborate closely with government agencies and fact-checking groups to make sure that deepfakes—especially those with malevolent intent—are not allowed to continue unchecked.

2. **Microsoft**: Deepfake audio is one example of an AI innovation that Microsoft has acknowledged has two sides. Through its AI Principles, which place a strong emphasis on accountability, transparency, and fairness, the company has pledged to create ethical AI systems. Microsoft is promoting international guidelines for the appropriate use of these tools in addition to developing technology to identify and stop the spread of deepfakes. The Video Authenticator Tool, which examines audio and video for indications of manipulation, is one of Microsoft's most notable projects. This tool is a part of a larger initiative to make sure deepfake technology is used in ways that respect people's rights and freedoms, especially when it comes to preventing the impersonation of private citizens and public figures.
3. **Smaller Players**: Developing New Detection Tools. Deepfake detection is also being significantly aided by smaller tech firms. For instance, a startup called Deeptrace offers products made especially to identify deepfake content on social media sites. For companies and individuals who wish to protect themselves from manipulation in a world that is becoming more and more digital, their work in audio and video detection is essential.

The Need for Global Collaboration

It is obvious that international cooperation will be crucial because deepfake technology crosses national boundaries. A unified set of rules that support the moral advancement and application of AI technologies, including deepfake audio, is presently being developed by a number of international organizations, such as the World Economic Forum (WEF) and the United Nations (UN). Cooperation among nations will aid in the establishment of uniform standards, guaranteeing the safe use of these technologies while halting the unchecked spread of their detrimental effects.

13.7.2 Ethical and Societal Implications of Deepfake Audio

As deepfake audio technology advances, its potential for creativity and accessibility grows—alongside serious ethical concerns. From consent and privacy to misinformation and identity misuse, this chapter explores the societal impact of synthetic voices and the urgent need for responsible development, regulation, and awareness.

Unquestionably, deepfake audio has changed how we communicate by raising the prospect of producing artificially lifelike voices. However, there is a darker side to this technological wonder that could have a significant and frequently unsettling impact on our world. What occurs when there is no way to tell what is real and what isn't? As deepfake audio technology advances at an alarming rate, the risks of impersonation, fraud, and disinformation are not only possible but also becoming more likely.

13.7.2.1 A New Frontier for Misinformation: Voices That Lie

Deepfake audio elevates the manipulation of misinformation to a new level. Consider a situation in which a trusted individual or leader seems to say something damaging, divisive, or completely untrue. These fake statements can be made to sound as though they are coming from the people who have the most sway over society by using deepfake audio.

Think about the notorious "DeepFake Obama" that BuzzFeed and director Jordan Peele produced in 2018. Despite being obviously staged, this first demonstration sparked a wave of worries about the use of artificial voices to disseminate false information. In the political landscape, what was once merely an experiment now seems uncannily real.

In 2024, deepfake audio is being actively used to sway public opinion. Deepfake audio has been used to produce completely false statements during national elections around the world, especially in the US, Brazil, and India. Some of these statements go viral before they can be refuted. A doctored recording of a well-known politician allegedly disparaging a rival party surfaced in India in 2023. Even though they were eventually discredited, the damage was done, and the political fallout persisted for weeks.

These audio deepfakes make it harder than ever to tell fact from fiction. Now that they have the voices of the people who matter most, misinformation campaigns don't have to rely on anonymous bots. Words no longer belong only to their speakers as we enter a new era of plausible deception.

13.7.2.2 Impersonation and Fraud: When a Voice Can Be Stolen

Deepfake audio is changing the game in the criminal world. Nowadays, scammers use artificial voices to pose as people in order to obtain private information or money. A criminal can mimic the voice of a CEO, a trusted coworker, or even a family member with a single click, with disastrous results.

One of the most startling instances happened in 2020 when cybercriminals pretended to be a CEO using deepfake audio, persuading a senior executive at an energy company in the UK to send a substantial amount of money for a purportedly urgent business transaction. Until the transfer was found to be a scam, the request was ignored because it sounded so genuine. Even the most seasoned executives are susceptible to this new kind of attack, as demonstrated by the fact that the deepfake had tricked a highly skilled professional.

Furthermore, the trend is only getting worse. Concerns regarding the use of deepfake audio in business email compromise (BEC) attacks have grown in 2024. Hackers are improving their models to replicate voices more precisely, which makes impersonations even more difficult to spot. The financial industry is now a prime target for these fraudulent schemes since the global economy depends more and more on digital transactions. Because deepfake technology can easily enter these high-stakes settings, the risk is ever-present and constantly growing.

13.7.2.3 Whose Voice Is It Anyway?

The need to safeguard people's voices has never been more pressing as deepfake audio technology develops. Unauthorized use of someone else's voice, whether they are a politician, celebrity, or private citizen, can cause serious personal and professional harm, particularly in fraudulent contexts. There are significant ramifications for ownership of one's likeness, consent, and privacy.

A notable instance of this occurred in 2023, when a well-known social media influencer found out that a scam involving phony product endorsements had imitated her voice. Because they believed the message to be genuine, her followers ultimately bought fake goods. This case highlighted how vulnerable public figures—and even private citizens—are becoming in the era of AI-generated content. It also sparked a pressing discussion about the necessity of more robust legal safeguards to stop illegal voice replication.

To address these issues, governments and organizations around the world are working harder. Particularly, the European Union has implemented GDPR-like rules that are especially concerned with digital likeness and the moral application of AI. By prohibiting their unlawful use in deepfakes, these rules seek to uphold people's rights to their voices. In the meantime, the US has started programs like the Malicious Deep Fake Accountability Act, which aims to criminalize crimes involving deepfakes and impose severe penalties on those who use these technologies to defraud or mislead.

However, enforcement is still a difficult task even as regulatory frameworks change. There is frequently a disconnect between technology and protection as laws and regulations are unable to keep up with the quick developments in AI. For example, even though governments may enact laws, it is challenging to identify and eliminate dangerous deepfakes in real time due to the enormous amount of synthetic content created every day.

13.8 CHALLENGES AND FUTURE DIRECTIONS IN DEEPFAKE AUDIO

13.8.1 Ethical AI Development: Ensuring Responsible Use

Ethical AI development is not just about building impressive systems—it's about ensuring that these systems are used responsibly and transparently, protecting both users and society at large from potential harm [30]. As we venture further into the world of deepfake audio, it's easy to be amazed by the technology's ability to create lifelike, synthetic voices.

Deepfake audio models—such as WaveNet, Tacotron 2, and newer technologies like FastSpeech 2—are capable of generating highly realistic voices. These systems mimic not just the words we say, but also the subtleties of human speech, including tone, cadence, and emotional undertones. While these innovations open doors to incredible new opportunities, they also blur the lines between what is real and what is artificially generated. A synthetic voice might sound indistinguishable from the real thing, raising concerns about privacy, identity theft, misinformation, and more.

TABLE 13.2
Comparative Overview of Deepfake Audio Models

Model	Inference Speed	Voice Quality (MOS)	Ethical Considerations	Multilingual Capability
WaveNet	Moderate	~4.0	High computational cost; potential for misuse in impersonation.	Limited
Tacotron 2	Moderate	~3.9	Requires extensive training data; ethical concerns in voice cloning without consent.	Moderate
FastSpeech 2	High (38x speedup)	~4.3	Efficient training; reduced risk of misuse due to faster synthesis.	Moderate
FastSpeech 2s	Very High	~4.4	Simplified architecture; potential for ethical misuse in rapid voice generation.	Moderate
Respeecher	High	High	Strong ethical framework; explicit consent required for voice cloning.	High (with diverse datasets)
ElevenLabs	High	High	Ethical concerns in political contexts; rapid voice synthesis capabilities.	High (supports multiple languages)

As the capabilities of deepfake audio expand, so do the stakes. Ethical AI development in this space is about balancing innovation with accountability, ensuring that these technologies benefit society without undermining trust or security. Table 13.2 shows the quality and considerations for deepfake audio models in practice.

13.8.1.1 Key Principles in Ethical AI Development

13.8.1.1.1 Transparency and Disclosure

Imagine hearing someone with a foreign accent or voice while watching a video or listening to a podcast. You might think it's a real person, but what if it's not? Transparency enters the picture here. Customers and users must always be conscious when interacting with artificial voices in order for AI to be ethical. Descript and Resemble AI are two platforms that are already advancing this goal by providing users with distinct features that enable them to discern between content created by AI and content created by humans.

Furthermore, voice signatures and digital watermarking are becoming indispensable tools for labeling synthetic audio. By serving as invisible labels that identify content as AI generated, these tools help preserve public confidence and make it more difficult for dishonest actors to deceive others.

13.8.1.1.2 Informed Consent and Data Ownership

Consent is another essential ethical concern, especially when it comes to creating synthetic voices using someone else's voice data. Now more than ever, the idea of having control over one's voice is crucial. Companies like Researcher, for instance, have led the way in ethical voice cloning by making sure that explicit written consent is acquired prior to synthesizing any voice, particularly for private or public figures. This procedure guarantees that the voice model is only utilized for the predetermined purposes, whether they be commercial content, personal projects, or anything else.

However, obtaining permission is only one aspect of informed consent. Making sure people are completely aware of how their voice data will be used is the goal. When it comes to synthetic voice technology, this entails not just giving consent but also outlining the specific purposes, potential risks, storage duration, security measures, and the possibility of future uses or sharing of the synthesized voice.

13.8.1.1.3 Bias Mitigation and Fairness

Bias is one issue that needs to be addressed as deepfake audio technology develops further. Deepfake models are only as good as the data they are trained on, just like any other AI system. A model may have trouble accurately reproducing voices from other linguistic or cultural backgrounds if it is trained on a small set of voices, such as English speakers who are primarily from North America.

This is a serious problem. Consider a scenario in which a voice produced by AI reinforces stereotypes or inaccurately portrays a particular culture. By utilizing inclusive, diverse datasets, developers can proactively reduce biases in their models. By using this method, deepfake audio models are guaranteed to be fair and represent the diversity of accents, dialects, and languages found throughout the world.

13.8.1.1.4 Accountability Mechanisms and Oversight

Strong accountability measures are necessary to prevent the misuse of deepfake audio technology. The creation of oversight organizations that guarantee the technology is used responsibly is necessary for the development of ethical AI. Regulations such as the General Data Protection Regulation (GDPR) in Europe are already taking action, establishing rules for the use of synthetic content and making sure businesses follow stringent privacy requirements.

Tech companies should establish internal ethics committees to examine and keep an eye on deepfake projects in addition to these regulations. This adds another level of protection to make sure deepfake audio isn't used for impersonation or fraud, among other unethical uses.

13.8.1.1.5 The Path Forward: Zero-Shot and Few-Shot Learning

In the future, instead of picking up data of real people, data can be artificially generated, and the issue of data scarcity may be resolved with the help of zero-shot and few-shot learning approaches. With little to no training data, these techniques enable AI systems to pick up new languages or accents, essentially allowing the system to synthesize speech in languages it has never heard before.

- **Zero-shot learning** allows a model to **generate** speech in a new language without ever being trained directly on it, leveraging knowledge it has gained from related languages.

- **Few-shot learning**, on the other hand, allows a model to learn a new language or accent using only a **small set of examples**.
- **Example**: Imagine a voice synthesis model that, after hearing only a few minutes of a language, can generate natural-sounding speech in that language. This is game-changing for languages that have limited resources but are still critically important to preserve.

13.8.2 Modern Examples of Ethical Practices

13.8.2.1 TikTok's AI Voice Feature

TikTok introduced a feature in 2024 that lets creators use their own voice data to create voiceovers. The opt-in consent procedure is what makes this unique. Users have control over how their voice data is used and are fully informed about its intended use. In order to ensure that users retain control over their voice and its use, this degree of transparency is crucial for the development of ethical AI.

13.8.2.2 Cree Language Revitalization with Respeecher

The Canadian Cree language revitalization project is among the most touching instances of moral deepfake audio. This project helps to preserve the Cree language for future generations by synthesizing voices based on recordings made by Cree elders using deepfake technology. This is an excellent illustration of how AI can be used to preserve culture while honoring community preferences and making sure that informed consent is acquired before using voices.

13.8.2.3 Respeecher's Celebrity Voice Contracts

Respeecher, which specializes in ethical voice cloning for celebrities and content creators, is a notable example of the commercial use of deepfake audio. Respeecher makes sure that a precise, written contract is in place before synthesizing any celebrity's voice. This agreement provides complete transparency and safeguards the owner of the voice from any unauthorized use by outlining the precise usage of the synthetic voice.

13.9 CONCLUSION

An examination of deepfake audio concludes this chapter. This technology is so powerful and new that it needs its own place in our story. Deepfake audio, in contrast to the best copycats of the past, comprehends what makes a voice human and does more than just mimic it. It includes emotions, tones, and the nuanced power relationships that make a speech believable. The result of a fundamental change in our understanding of voice is this technology. It is something that can be synthesized as well as something that is done by an individual.

There are numerous real-world uses. They range from virtual reality scenarios and language learning to the very serious task of preserving the linguistic diversity of our planet. In general, they aid in making human–computer communication more engaging and natural. Regardless of the language or culture you happen to belong to, the goal of the entire project is to promote adaptability in life's many interactive situations. Fraud is one example of how the same technology that enables personalized learning can be used for less honorable purposes or pretending to be someone else. Additionally, some people are already spreading misleading information using it or intend to do so.

It does bring up some significant and potentially very awkward ethical issues, chief among them being how to guarantee that our sun is a trustworthy source of light without relying on one layer of authoritarian approval and control after another. Deepfake audio serves as a reminder at the end of this chapter that the significance of technology depends on how we use it. Although the technology used to create these artificial voices is impressive, their impact will depend on our values and decisions. Deepfake audio can be a tool for accessibility, education, and communication that feels genuine and respectful if it is developed with care and with an ethical perspective. In order for this technology to benefit people, we must be aware of both the risks and the opportunities.

REFERENCES

1. H. Guo, S. Yu, and N. Zhao, "Deepfake audio detection: A review," *arXiv preprint arXiv:2308.06754*, 2023.
2. V. Vestman, R. Yamagishi, and J. Yamagishi, "Neural vocoders for speech synthesis: A comprehensive review," *IEEE Access*, vol. 10, pp. 128563–128583, 2022.
3. K. K. Paliwal, "Speech synthesis and voice conversion," in *Handbook of Natural Language Processing*, 2nd ed. Boca Raton, FL: CRC Press, 2010, pp. 545–574.
4. I. Goodfellow et al., "Generative adversarial nets," in *Advances in Neural Information Processing Systems*, 27 *(NeurIPS 2014),* Z. Ghahramani, M. Welling, C. Cortes, N. D. Lawrence, and K. Q. Weinberger, Eds. Curran Associates, Inc., 2014, pp. 2672–2680..
5. K. K. Paliwal, "Speech synthesis and voice conversion," in N. Indurkhya and F. J. Damerau, Eds. *Handbook of Natural Language Processing*, 2nd ed. Boca Raton, FL: CRC Press, 2010, pp. 545–574.
6. V. Vestman, R. Yamagishi, and J. Yamagishi, "Neural vocoders for speech synthesis: A comprehensive review," *IEEE Access*, vol. 10, pp. 128563–128583, 2022.
7. H. Guo, S. Yu, and N. Zhao, "Deepfake audio detection: A review," *arXiv preprint arXiv:2308.06754*, 2023.
8. S. Kataria, "The current state of deepfake technology," *Analytics Insight*, 2023. [Online]. Available: https://www.analyticsinsight.net/the-current-state-of-deepfake-technology/.
9. P. M. Bousquet, "From text to speech: Revolutionizing voice-based AI with deep learning," *Towards Data Science*, 2023. [Online]. Available: https://towardsdatascience.com/from-text-to-speech-revolutionizing-voice-based-ai-with-deep-learning-ea98402292d1.
10. "Deepfake audio: The new frontier in cybercrime," *Europol*, 2024. [Online]. Available: https://www.europol.europa.eu/publications-documents/deepfake-audio-new-frontier-in-cybercrime.
11. M. Ostendorf et al., "Speech segmentation and its impact on spoken document processing," *Signal Processing Magazine, Manuscript, Signal Processing Magazine* 72 (2007).
12. P. Charitidis, G. Kordopatis-Zilos, S. Papadopoulos, and I. Kompatsiaris, "Investigating the impact of pre-processing and prediction aggregation on the deepfake detection task," *arXiv preprint arXiv:2006.07084*, 2020.
13. Creech A., S. Hallam, M. Varvarigou, H. McQueen, and H. Gaunt. Active music making: a route to enhanced subjective well-being among older people. *Perspect Public Health*. 2013, 133(1):36–43. doi: 10.1177/1757913912466950. PMID: 23308006.

14. T. P. Doan, K. Hong, and S. Jung, "GAN discriminator based audio deepfake detection," in *Proceedings of the 2nd Workshop on Security Implications of Deepfakes and Cheapfakes (WDC '23)*, New York, NY: ACM, 2023, pp. 29–32. doi: 10.1145/3595353.3595883.
15. W. Poopradit, P. Saebe, V. Chinnasri, and T. Phienthrakul, "Deepfake voice system for spoofing," in *2024 19th International Joint Symposium on Artificial Intelligence and Natural Language Processing (iSAI-NLP)*, Chonburi, Thailand, 2024, pp. 1–5. doi: 10.1109/iSAI-NLP64410.2024.10799251.
16. L. Liu, Z.-H. Ling, Y. Jiang, M. Zhou, and L.-R. Dai, "WaveNet vocoder with limited training data for voice conversion," in *Interspeech 2018*, Hyderabad, pp. 1983–1987. doi: 10.21437/Interspeech.2018-1190.
17. H. Wang, M. Zhu, and Q. Wang, "Research on power voice based on WaveNet+LSTM improvement," in *2023 IEEE ICTech*, Wuhan, China, pp. 419–423. doi: 10.1109/ICTech58362.2023.00085.
18. R. J. Skerry-Ryan et al., "Tacotron: A fully end-to-end text-to-speech synthesis model," *arXiv preprint arXiv:1703.10135*, 2017.
19. Init Deep Dive, "Voice cloning with Tacotron2 and HiFi-GAN," *Medium*, Mar. 5, 2022. [Online]. Available: https://medium.com/init-deep-dive/voice-cloning-with-tacotron2-and-hifi-gan-30193180be32.
20. R. Prenger, R. Valle, and B. Catanzaro, "Waveglow: A flow-based generative network for speech synthesis," in *ICASSP 2019*, Brighton, United Kingdom, pp. 3617–3621. doi: 10.1109/ICASSP.2019.8683143.
21. B. Chettri, T. Kinnunen, and E. Benetos, "Deep generative variational autoencoding for replay spoof detection in automatic speaker verification," *Computer Speech & Language*, vol. 63, p. 101092, 2020. doi: 10.1016/j.csl.2020.101092.
22. H. Fang, G. Carbajal, S. Wermter, and T. Gerkmann, "Variational autoencoder for speech enhancement with a noise-aware encoder," in *ICASSP 2021-2021 IEEE International Conference on Acoustics, Speech and Signal Processing (ICASSP)*, Toronto, ON, Canada, 2021, pp. 676–680. doi: 10.1109/ICASSP39728.2021.9414060.
23. M. Sadeghi and R. Serizel, "Fast and efficient speech enhancement with variational autoencoders," in *International Conference on Acoustics Speech and Signal Processing (ICASSP)*, IEEE, Rhodes Island, Greece, Jun. 2023.
24. W. Brannon and Y. Virkar, "Dubbing in practice: A large scale study of human localization with insights for automatic dubbing," *Transactions of the Association for Computational Linguistics*, vol. 11, pp. 419–435, 2023. doi: 10.1162/tacl_a_00551.
25. Y. Li, M. Milling, L. Specia, and B. Schuller, "From audio deepfake detection to AI-generated music detection—A pathway and overview," *arXiv preprint arXiv:2412.00571*, 2024.
26. P. Tiwari, "Online AI-based voice assistant," 2024.
27. R. Lieber, "Voice deepfakes are coming for your bank balance," *The New York Times*, Aug. 30, 2023. [Online]. Available: https://www.nytimes.com/2023/08/30/business/voice-deepfakes-bank-scams.html.
28. M. Rizvi, "Investigating AI-powered tutoring systems that adapt to individual student needs," *The Eurasia Proceedings of Educational and Social Sciences*, vol. 31, pp. 67–73, 2023. doi: 10.55549/epess.1381518.
29. D. S. Battina and L. Surya, "Innovative study of an AI voice-based smart device to assist deaf people," *SSRN Electronic Journal*, vol. 9, pp. 816–822, 2021.
30. Vozo AI, "Ethical dilemmas and risks of AI voice cloning," *Vozo AI.* [Online]. Available:https://www.vozo.ai/blogs/ethical-dilemmas-and-risks-of-ai-voice-cloning.

14 Enhancing Communication

Recognition of American Sign Language Finger Spelling

Tejna Khosla, Deepika Bansal, and Punit Gupta

14.1 INTRODUCTION

Effective communication forms the bedrock of human interaction, yet over 5% of the global population (430 million individuals) experience disabling hearing loss, a figure projected to rise to 700 million by 2050 [1]. For the Deaf and Hard of Hearing (DHH) community, sign language serves as a primary linguistic channel, combining handshapes, facial expressions, and body movements to convey nuanced meaning. American Sign Language (ASL), used by approximately 500,000 individuals in North America, exemplifies this rich visual-gestural system. A critical component of ASL is *fingerspelling*, which enables the spelling of proper nouns, technical terms, or words without established signs; for instance, introducing oneself as "Jolanta" requires sequentially forming J-O-L-A-N-T-A using hand configurations. However, bridging the communication gap between DHH individuals and non-signers remains a persistent challenge, necessitating technological interventions that democratize access to sign language interpretation.

14.1.1 The Role of Technology in Sign Language Recognition

Early efforts in sign language recognition relied on sensor-based systems, such as data gloves equipped with flex sensors or accelerometers, to track hand movements [2]. Earlier solutions, though precise, faced practical barriers due to high costs and obtrusive hardware, hindering widespread use. The field has since evolved with advancements in computer vision and deep learning, enabling vision-based approaches that analyze video or image data to decode hand gestures. Despite these innovations, three key challenges remain unresolved:

1. **Real-Time Responsiveness**: Many systems experience delays, unable to match the rapid pace required for seamless, natural interactions.

 DOI: 10.1201/9781003565703-14

2. **Environmental Adaptability**: Performance drops under unpredictable conditions like shifting lighting, cluttered backgrounds, or unconventional camera perspectives.
3. **Versatility across Users**: Systems trained in controlled lab settings frequently underperform when encountering unique signing patterns, regional variations, or non-standard gestures.

14.1.2 Fingerspelling as a Gateway to Accessibility

Fingerspelling recognition serves as a practical stepping stone toward comprehensive sign language understanding [3,4]. Picture a real-world situation: a deaf person spells out a sentence with intentional pauses between words. An effective recognition system must first detect these breaks, identify individual letters, and convert them into readable text or audible speech. Researchers have investigated this process through various approaches, including Handcrafted features such as Edge detection and contour analysis for static gestures.

- **3D Skeletal Tracking**: MediaPipe's hand landmark detection (21 keypoints per hand) to capture spatial relationships.
- **Multimodal Fusion**: Combining RGB, depth, and pose data to improve accuracy.

However, gaps remain. Most systems focus on isolated characters rather than fluid fingerspelling sequences, and only few are optimized for resource-constrained devices, hindering deployment in low-bandwidth settings.

The fingerprints are also a valuable addition. Members of the DHH community benefit from new forms of communication provided by fingerprinting, allowing them to express ideas and words that may not have a corresponding sign. It acts as a bridge, allowing people to communicate effectively and express linguistic complexity. When combined with easy-to-use sign language, a more complex and extensive communication channel emerges.

14.1.3 Sign Language Recognition

Various methods and techniques are used in sign language recognition to interpret complex visual symbolic language. The two primary ways to introduce sign language are through sensory and vision-based recognition [5–8]. Currently, sensor-based methods are too expensive for the poor. Vision-based techniques give users a more natural way to interact with gestures, making real-life interactions more consistent. Finally, image-based and video-based methods are preferred. Convolutional Neural Networks (CNNs) are a popular tool used in computer vision-based techniques that analyze video content to recognize and classify signals based on visual characteristics.

The background of the skeleton is used through pose estimation techniques to record hand and body language movements, enabling feature recognition based on the location of important body parts and movement patterns. Deep learning models like transformers and recurrent neural networks are used to accomplish the learning sequence, enabling the recognition of entire sentences and words in sign language. Understanding the full depth of meaning and emotion in sign language goes beyond recognizing hand movements—it also involves analyzing facial expressions and subtle gestures. To achieve greater accuracy and reliability, multimodal systems combine visual information with inputs from other sensors. Meanwhile, data-driven methods leverage extensive sign language datasets to train their algorithms. These adaptable approaches are designed to interpret different sign languages, handle a wide range of signing styles, and power tools from accessibility aids to real-time translation services. Despite their diversity, most of these methods share a few fundamental steps, which include:

14.1.3.1 Sign Segmentation

Sign segmentation is the process of isolating individual characters within a sequence of signs. This task is particularly challenging because sign language is naturally fluid, with gestures often blending into one another. To achieve accurate segmentation, advanced computer vision and machine learning techniques are used to pinpoint the exact start and end of each sign. Additionally, effective segmentation helps to distinguish between meaningful gestures and unrelated movements, which ultimately enhance the system's overall recognition accuracy.

14.1.3.2 Sign Recognition

Sign recognition is the core function of sign language recognition systems, focusing on identifying standard gestures, words, or signs performed by the user. This process typically utilizes advanced machine learning models—such as deep neural networks—that have been trained on extensive sign language datasets to improve accuracy. To interpret the signer's intended message, these models analyze each isolated gesture and compare it to a database of known signs. The complexity of sign recognition arises from the wide range of hand shapes, facial expressions, and individual signing styles, making it essential to continually refine and update these models for optimal performance.

14.1.3.3 Translation to Text

After the system has accurately identified each symbol, the next phase is to convert these symbols into written language. This translation process often incorporates natural language processing (NLP) methods to ensure that the recognized sign language gestures are transformed into text that is grammatically correct and easy to understand. By leveraging these advanced techniques, sign language recognition systems can effectively bridge the gap between sign language and spoken or written communication. Text-based content can be displayed on the screen or converted into speech through synthesis, making it accessible to a wide audience.

When combined, these procedures build a link between spoken or written communication and the rich visual language of sign language. Accurate identification of signs is ensured by sign segmentation, their meaning is deciphered by sign recognition, and their meaning is made available to those who might not be familiar with sign language by text translation. All of these elements work together to create a more welcoming community.

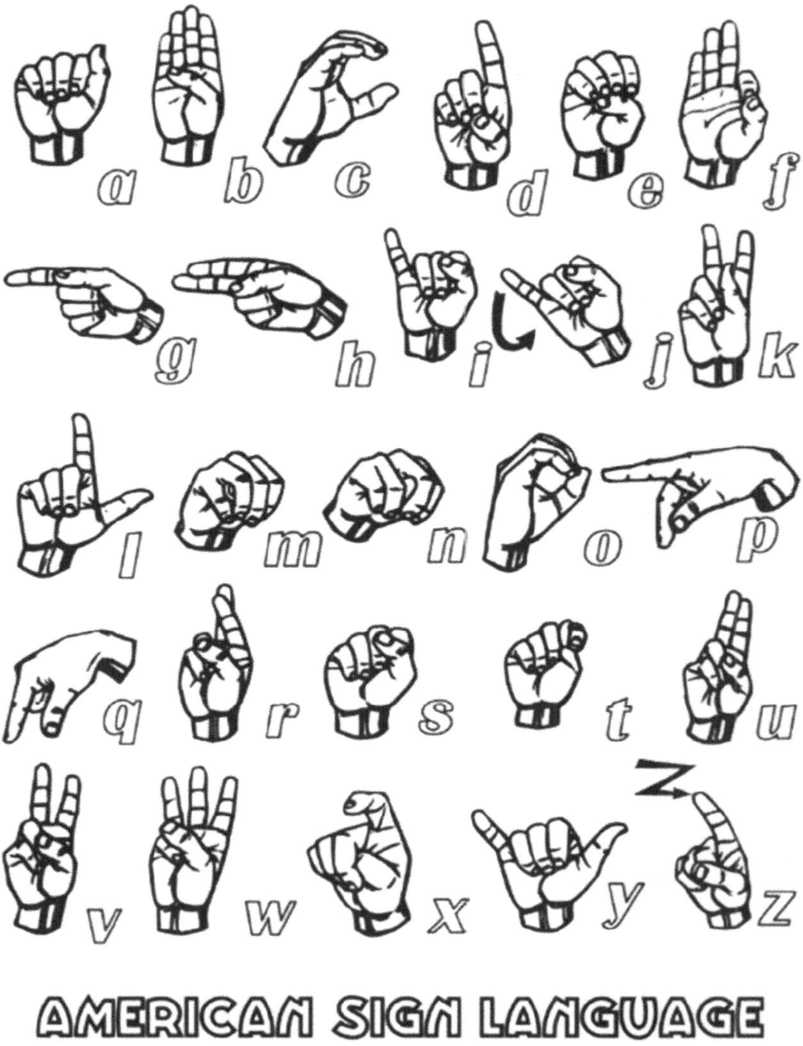

FIGURE 14.1 Fingerspelling gestures [11].

Because there are more high-quality datasets and content available on sites like Kaggle.com [9], whether instructional or not, the use of the ASL fingerspelling system was chosen. Figure 14.1 [10] shows different letters of the alphabet with their corresponding fingerspelled equivalents.

14.2 LITERATURE REVIEW

Prior research and developments in the expansive realm of gesture recognition and sign language technology have meticulously laid a robust foundation within the realm of computer vision. DongXu Li et al. [12] stand out for their substantial contribution by providing large datasets, while Joshi and team [13] have been pivotal in furnishing essential corpora, significantly advancing this dynamic field of study. Shi, B et al. [14] and Bowen Shi et al. [15] have notably directed their efforts toward the identification of linguistic units within video data, contributing valuable insights. Furthermore, pivotal advancements in model training have been spearheaded by Sharma and his team [5], along with the noteworthy contributions of Novopoltsev et al. [16], who have diligently explored and defined optimal fine-tuning techniques.

The development of the MediaPipe Framework by Lugaresi et al. was achieved through the combination of various different techniques [17,18]. The impact of this framework, as highlighted in the work of Indira et al., has been transformative, leading to significant improvements in efforts to recognize gestures [19]. Moreover, the MediaPipe Framework has facilitated the adoption of transfer learning and simplified the overall process of gesture recognition. As evidenced by the methods used by Sharma et al. and Veluri et al., this innovative framework has made it possible to develop more effective models [5,20]. Without a doubt, the synergy of multiple methodologies has propelled the field forward, marking a new era of possibilities for sign language technology.

Novelty and Key Contributions:

- **Hybrid Feature Extraction**: This work uniquely combines OpenCV edge detection with MediaPipe's real-time hand landmark tracking, improving robustness to lighting and occlusions—a limitation in many prior vision-only or static image methods.
- **Transfer Learning with MediaPipe**: Unlike most prior works, the proposed approach leverages MediaPipe Model Maker for efficient transfer learning, enabling high accuracy (96.67%) even with diverse backgrounds and signer variations.
- **Cross-Platform, Real-Time Deployment**: The model is optimized and quantized for mobile devices, supporting real-time, offline use—addressing the accessibility gap for resource-constrained environments, which is rarely tackled in earlier studies.
- **Comprehensive Dataset**: The use of a large, diverse dataset (78,000+ images) with extensive augmentation ensures strong generalization, outperforming many previous works limited by smaller or less varied datasets.

While earlier methods achieved notable progress, most struggled with unconstrained environments, signer diversity, and real-time deployment. The proposed method advances the field by delivering higher accuracy, robust generalization, and practical deployment on mobile platforms—representing a significant and original contribution to ASL fingerspelling recognition. Table 14.1 provides the comparative analysis of ASL fingerspelling recognition methods.

14.3 PRELIMINARIES

Computer vision researchers have been significantly influenced by three core ideas: the principle of parameter sharing, the notion of equivalent representation, and the impact of sparse effects. The matrix multiplication function of the simple neural

TABLE 14.1
Comparative Analysis of ASL Fingerspelling Recognition Methods

Study/ Method	Approach/ Model	Dataset & Size	Key Features/ Techniques	Accuracy/ Results	Limitations/ Remarks
Shi et al. [14,15]	CNN-based, sequence models	Real-world ASL videos	Deep learning, continuous fingerspelling	~60%–71% (Levenshtein)	Challenging in unconstrained videos
Sharma and Sharma [2]	Hybrid CNN + Transformer + CTC	3M+ fingerspelled chars, 100+ signers	Multimodal landmarks, iterative training	714.4% (Levenshtein)	Domain gap, needs more real data
Ameen and Vadera [22]	CNN (depth & color images)	ASL alphabet (static images)	Feature extraction, depth/color fusion	80.34%	Static, limited to alphabet
Das et al. [21]	Modified Inception Model	Custom static gesture images	Deep learning, image preprocessing	90%	Static, small dataset
Sharma et al. [5]	Fine-tuned deep transfer learning	Indian Sign Language corpus	Transfer learning, gloss-based recognition	Not specified	Focused on ISL
Proposed Method (This Work)	MediaPipe Model Maker + Transfer Learning	78,000+ ASL images, varied backgrounds	OpenCV edge detection, hand landmarks, quantization for mobile	96.67%	Real-time, robust, cross-platform

network layer multiplies a parameter matrix representing the connection between input and output units. This means that there is a connection between all input and output devices.

CNNs, on the other hand, exhibit sparse effects. Making the core smaller than the input accomplishes this. If an image is processed in the kernel, we can extract pertinent information from tens or hundreds of pixels, even if it may include thousands or millions. As a result, the model requires fewer parameters to store, improving statistical efficiency and reducing the memory footprint.

14.3.1 Activation Map Construction

Tensor filters manipulate spatial data and use structures called convolution layers to detect edges and learn how to extract features from objects such as smooth curves. The main task involves detecting edges in the image captured by the filter. It helps to filter out unwanted data to improve visuals. The activation map should be created using the same set of weights. Convolutional networks exchange their parameters, in contrast to conventional neural networks that use each weight matrix member only once. Figure 14.2 depicts the filtering process.

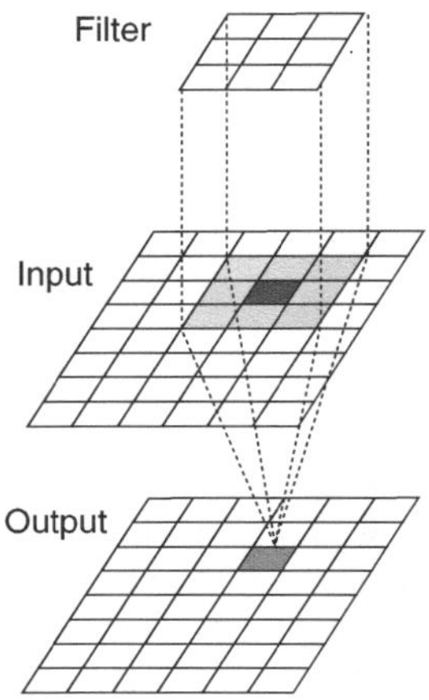

FIGURE 14.2 Filtering process.

The weight placed elsewhere is equal to the weight applied to an input to achieve an outcome. Because of parameter sharing, CNNs have the feature where the layers have the same variance under transformations. It asserts that changes to the input also affect changes to the output.

The problem is, as the image shrinks with each application, the image can only be used a few times before it disappears completely. Additionally, if the core is viewed as it travels throughout the image, the influence of the outer pixels is less than that of the central pixels. Therefore, padding is used. The padding width must match (14.1), where $\boldsymbol{p}$ is the padding and $\boldsymbol{f}$ is the filter size (usually odd).

$$p = \frac{f-1}{2} \tag{14.1}$$

Strided convolution, where step length can also vary, is available. The dimensions of the output matrix, taking into account padding and stride, can be computed using equation (14.2). It takes in the dimension of the input matrix as $\boldsymbol{n}_{\text{in}}$, padding width applied as p, filter dimensions as $\boldsymbol{f}$, and stride length as $\boldsymbol{s}$.

$$n_{\text{out}} = \left[\frac{n_{\text{in}} + 2p - f}{s} + 1 \right] \tag{14.2}$$

CNNs often use "pooling" layers in addition to convolution layers. Their primary goal is to reduce the tensor's size and expedite computations. These layers are straightforward: the image is split up into many sections after an operation is completed. The Max Pool Layer, for instance, chooses the highest value from each pool and positions it in the output where it belongs. Filter size and step are two possible hyperparameters, just like with a convolutional layer. Figure 14.3 shows a pooling operation where a max pool is used with a 2×2 filter and step 2 in action.

$$W_{\text{out}} = \frac{W - F}{S} + 1 \tag{14.3}$$

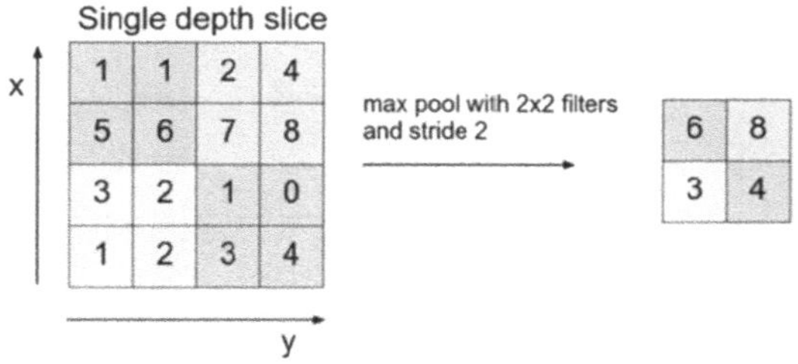

FIGURE 14.3 Pooling operation.

Equation (14.3) is used to determine the spatial *F*, the step S, and the size of the print kernel and the activation map of the size $W \times W \times D$, which will produce an output volume of size $W_{out} \times W_{out} \times D$.

Since the pool provides a certain degree of translational invariance, the element is always known on the screen regardless of location.

Every neuron in a fully connected (FC) layer is completely connected to every other neuron in the higher and lower levels. Usually, matrix multiplication and accompanying bias effects can be used to approximate the output. The FC layer facilitates the mapping of input and output representations. Non-linear layers are usually added after the convolution stage. If a function is divided by a non-linear decision limit, the neural network can predict the class of the function and approximate the function that does not obey linearity.

There are many different types of non-linear processes; the most commonly used include:

- Sigmoid

  The non-linear sigmoid compresses the input value to the range [0, 1]. However, when activation occurs at both ends of the sigmoid, an interesting feature comes from the near-zero gradient. If the input is continuously positive, the Sigmoid output will be completely positive or negative, resulting in a zigzag pattern in the weight gradient update. The following equation gives the non-linear equation of sigmoid (i.e., *k*) in the range (0, 1):

$$\sigma(k) = \frac{1}{1+e^{-k}} \tag{14.4}$$

- Tanh

  Tanh, given by (14.5), reduces the range of input (i.e., *x*) to [−1, 1]. The output is centered at zero, unlike the sigmoid function neurons.

$$\tan h(x) = \frac{e^{x} - e^{x}}{e^{x} + e^{x}} \tag{14.5}$$

- ReLU

  Regenerative Linear Unit (ReLU) is given by (14.6) evaluating the function $f(\kappa) = \max(0, \kappa)$. Or, for insertion, indicate that the input is limited in

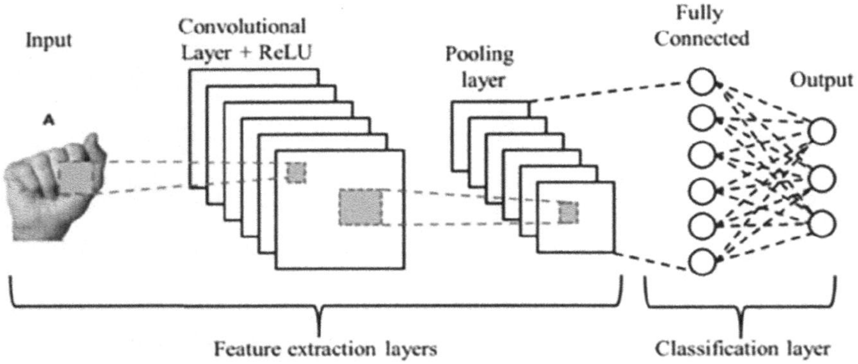

FIGURE 14.4 Convolution neural network.

the negative direction to a minimum of zero. ReLU converges about six times faster and provides more stable performance than Sigmoid and Tanh functions.

$$\text{Re LU}(x) = \begin{cases} 0 \text{ if } x < 0 \\ x \text{ if } x \geq 0 \end{cases} \tag{14.6}$$

Unfortunately, one drawback of ReLU is that it can be susceptible to training. A significant gradient passing through the neuron may cause it to update until it stops updating. It could be accommodated by selecting a reasonable learning rate. Figure 14.4 shows an example of CNN model in a graphical representation with multiple convolution modules that perform different tasks based on requirements of the model. It shows two non-linear layers with max-pooling and ReLU operations, as well as a last FC layer.

14.4 METHODOLOGY

14.4.1 Dataset Preparation

This section provides a comprehensive insight into the strategies, techniques, and tools employed throughout the study. In the context of fingerspelling recognition using MediaPipe Model Maker, the methodology unfolds with the meticulous preparation of a tailored dataset capturing diverse fingerspelling gestures. Subsequently, the model training process is delineated, incorporating transfer learning, fine-tuning, and hyperparameter tuning to optimize the recognition model. This paper provides a thorough explanation of the model's structure, with a particular emphasis on how it has been designed to meet the specific challenges of recognizing fingerspelled letters. This approach to model development is presented in detail, aiming to give a complete understanding of the process and demonstrate how it delivers a dependable and efficient method for fingerspelling recognition.

14.4.1.1 Dataset Preparation

For a machine learning model to achieve its best possible results, the quality, range, and overall size of the data it is trained on are absolutely essential. When specifically looking at the recognition of hand movements, such as those used in fingerspelling, it becomes particularly important to develop a database that thoroughly records the intricate details of these gestures. The initial stage in getting this data ready involves gathering specific datasets created with the aim of recognizing hand patterns (although the research in the source focuses on American Sign Language fingerspelling). To make this data gathering process more efficient, a basic Python program that uses the OpenCV library was employed to automatically collect images through a webcam.

14.4.1.2 Dataset Composition

For the purpose of training the machine learning model, a significant number of images—specifically 78,000—were utilized. These images illustrated the letters of the ASL alphabet and included a collection of 3,000 images focused on the alphabet. Importantly, within this dataset, each letter was captured in a multitude of different orientations, showcasing diverse hand shapes, photographed under various lighting conditions, and set against a variety of backgrounds. This deliberately wide-ranging representation was vital to ensure that the trained model would be capable of correctly identifying fingerspelled letters in a more extensive set of practical situations.

14.4.1.3 Dataset Augmentation

In order to improve the model's capacity to accurately recognize fingerspelling across a wide range of real-world scenarios, several data augmentation methods were utilized. These methods involved applying random rotations, flips (both horizontal and vertical, presumably), and alterations to the lighting conditions of the original images. The result of these techniques was the creation of more data samples based on the initial set of images. This process helps the model become more robust by exposing it to a greater variety of possible inputs it might encounter.

Augmenting the dataset in this manner not only increases its size, but it also imparts robustness to the model by exposing it to a broader range of possible variations in real-world scenarios. This helps immensely in model training process as well as increases the trained model's accuracy.

Figure 14.5 showcases different data augmentation techniques on images. These techniques include rotation, blurring, contrast shift, image scaling, illumination/iso change, and projection on different surfaces.

14.4.1.4 Quality Control and Data Splitting

To maintain the quality of the dataset, each image undergoes a meticulous quality control process. Images with poor lighting, excessive noise, or ambiguous hand poses are carefully filtered out to ensure that the model learns from high-quality, representative examples. This phase is essential for avoiding irrelevant or deceptive data from influencing the model. Figure 14.6 shows a general flow of data with augmentation and quality assurance. A human expert always assures that ill-formed data is not fed in via training for the model.

FIGURE 14.5 Data augmentation example.

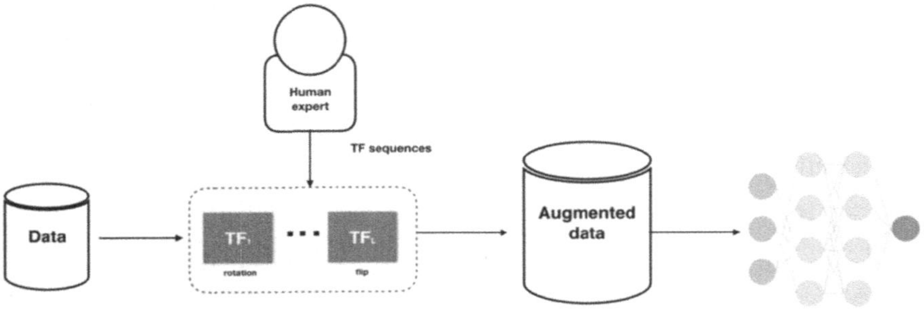

FIGURE 14.6 Data augmentation pipeline.

The database is then divided into test, validation, and training sets. The test set evaluates the performance of the model on previously unseen data, the validation set helps set hyperparameters and does not overfit, and the training set is used to train the model. This separation ensures that the generalization ability of the model is evaluated equally. Eighty percent of the database is allocated to train the proposed approach, while the rest is divided equally between validation and testing. For example, 10% for validation and 10% for testing.

14.4.2 Model Training

With a carefully curated dataset, the next step is to train a model that can recognize finger gestures. Using MediaPipe Model Maker simplifies the training process by providing a user-friendly interface to develop machine learning models even with limited data.

- **Transfer Learning**: The proposed model uses a transfer learning method using a pre-trained MediaPipe gesture recognition task. A pre-trained model, learned from a large database, forms the basis for recognizing common hand gestures.

TABLE 14.2
Hyperparameters for Fine-Tuning

Parameter	Value
Batch size	32
Learning rate	0.0001
Optimizer	Adam
Epochs	50
Train-validation-test split	80-10-10

- **Fine-Tuning**: A pre-designed model is provided in the fingerprint database, allowing you to customize it to recognize the fingerprint alphabet.
- **Hyperparameter Tuning**: Parameters such as training speed, batch size, and number of training sessions are tuned to find the configuration that maximizes accuracy and efficiency. This is performed to optimize the model's performance. The results of various hyperparameter configurations and their corresponding evaluation metrics are presented in Table 14.2.
- **Loss Function**: Because fingerspelling involves recognizing a sequence of letters, a special loss function designed for sequence-to-sequence learning is used. This helps the model to accurately predict the order of letters shown by the signer's hand gestures, making the translation from sign to text much more reliable.
- **Post-Training Optimization**: After the model is first trained, it goes through post-training optimization to make it more efficient. Techniques like quantization are used to shrink the model's size while still keeping its accuracy. This makes it easier to run the model on devices with limited resources, increasing its accessibility and usability.
- **Checkpointing**: Checkpointing is used during training to protect against unexpected interruptions or failures. By regularly saving the model's progress, training can easily continue from the last saved point if something goes wrong. This ensures that valuable training time and progress are not lost.

14.4.3 Model Architecture

Once a carefully prepared dataset of finger gestures is ready, the next step is to train a recognition model. MediaPipe Model Maker is used for this purpose because it provides a simple and user-friendly way to develop machine learning models. This tool allows you to adapt pre-trained models to your new data and specific application needs, making the process much faster and easier than building a model from scratch. By using transfer learning, Model Maker retrains existing models with your dataset, so you can quickly create an effective gesture recognition system. The built-in MediaPipe Gesture Recognizer serves as the foundation for this training, offering a reliable shortcut to developing and fine-tuning your own model.

Since a large amount of the current model's logic is reused, training with this method requires less data and takes less time than training a new model. In order to

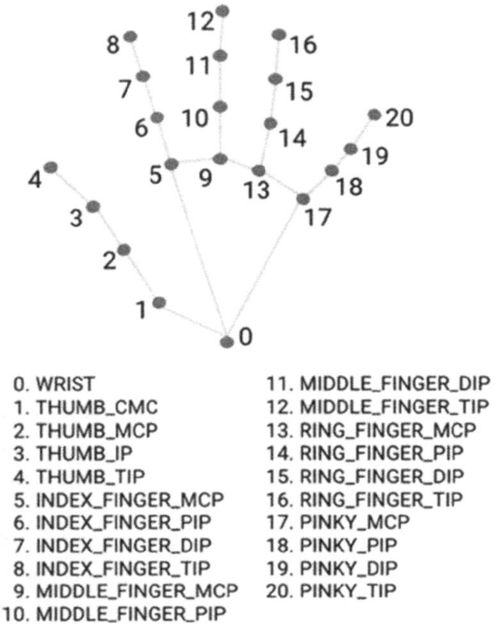

FIGURE 14.7 Hand landmarks definition [18].

retrain models, the program eliminates the final few levels of the model that categorize data and rebuilds those layers with fresh data.

A built-in MediaPipe Gesture Recognizer is used as a basis for training. The MediaPipe Gesture Recognizer application allows users to recognize hand gestures in real time and provide hand gesture results along with detected hand signatures.

Gesture recognition uses a model package and two prepackaged model packages: a hand label model package and a gesture classification model package. Special models identify the presence of hands and hand geometry, and gesture recognition models recognize gestures based on hand geometry. The hand model package defines 21 hand coordinate key points, as shown in Figure 14.7, in the hand defined area. About 30K real-world photos and many artificial hand models set against various backgrounds are used to train the model.

14.4.4 Workflow and Visualization

The flowchart in Figure 14.8 shows the main steps in the fingerprint recognition model training process, from data collection to the end of the model ready for deployment. Each block represents a specific step in the process, and the arrows represent the sequential flow of activities. Adjustments can be made based on specific requirements and tools used in real life.

The flowchart in Figure 14.9 shows deep learning model training. Model initialization involves loading pre-trained layers, optimizing the use of existing knowledge. The core of the flowchart lies in the epoch cycle. In every cycle, errors are calculated,

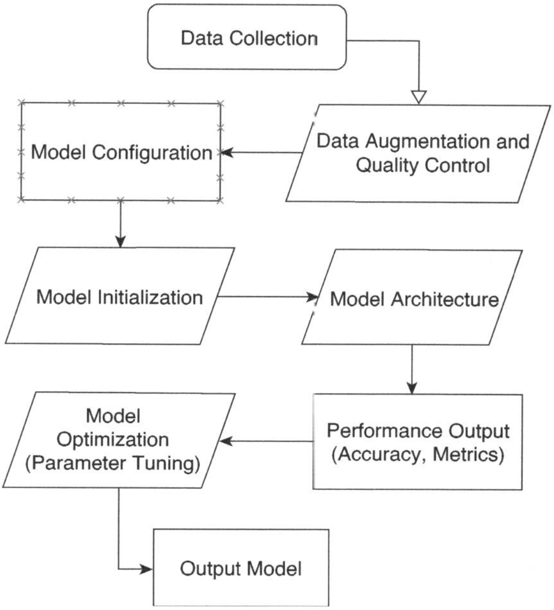

FIGURE 14.8 Flowchart for model training.

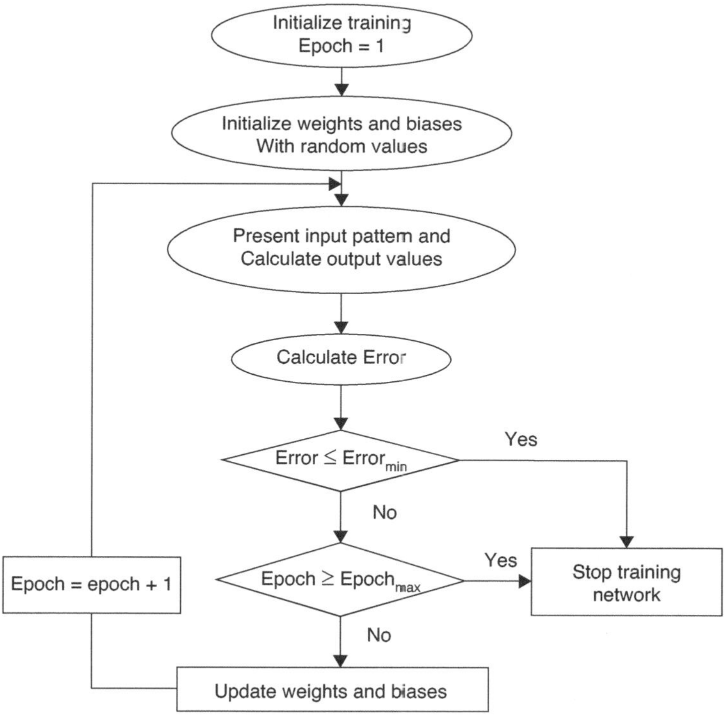

FIGURE 14.9 Flowchart for model architecture.

and weights and biases are updated. With each iteration, the model becomes more accurate until defined threshold.

14.4.5 Key Methodological Advantages

- **Hybrid Approach**: Combines OpenCV for image capture/augmentation and MediaPipe for real-time landmark detection, improving robustness to lighting and background variation.
- **Transfer Learning**: Reduces training time and data requirements, leveraging pre-trained models for higher accuracy.
- **Real-Time, Cross-Platform**: The model is optimized for mobile and embedded devices, enabling offline, real-time recognition.

14.5 RESULTS AND DISCUSSION

The evaluation of the fingerspelling recognition system showed outstanding results, achieving an impressive accuracy rate of 96.67%. This means the model was able to correctly identify finger-spelled gestures in nearly all cases. The system was tested under various conditions to ensure its reliability and robustness.

- In real-time use, the model maintained its 96.67% accuracy, proving that it can handle live communication smoothly and efficiently.
- The model also generalized well, keeping its high accuracy even when recognizing gestures from people who were not part of the training data. This shows it can adapt to different signing styles and users.
- When compared to other existing methods, the proposed model stands out not only in terms of accuracy but also in terms of computational efficiency and resource usage, as shown in Table 14.3. By comparing results with other methods within Table 14.3, it's clear that the proposed method is superior.
- Feedback from the DHH community was very positive, with users finding the system practical and easy to use. Usability studies further confirmed its potential to improve communication.

Overall, the results clearly demonstrate that this fingerspelling recognition system is both highly accurate and adaptable. Its strong performance and user-friendly design make it a valuable tool for enhancing communication and accessibility for the DHH community.

TABLE 14.3
Comparative Results

Method	Accuracy (%)
CNN [22]	80.34
Modified inception model [21]	90
KNN [11]	95.81
Proposed Method	96.67

14.6 CONCLUSION

This research introduces a new approach to fingerspelling recognition, aiming to improve communication for the DHH community by using MediaPipe Model Maker. The process begins with building a comprehensive and varied dataset of finger spelled gestures, ensuring that the model can learn from a wide range of hand movements. To develop effective recognition models, techniques like transfer learning and fine-tuning of pre-trained models are used, allowing the system to perform well in real-world situations. This well-rounded methodology is designed to provide a reliable and efficient way to interpret fingerspelling, ultimately making communication more accessible for the DHH community. Looking ahead, future work will focus on increasing the model's accuracy, expanding the range of gestures it can recognize, and further improving its real-time performance.

REFERENCES

1. Oudah, M., Al-Naji, A., & Chahl, J. Hand Gesture Recognition Based on Computer Vision: A Review of Techniques. *Journal of Imaging.* 2020 Jul 23;6(8):73. https://doi.org/10.3390/jimaging6080073.
2. Sharma, A. & Sharma, V. ASL Fingerspelling Recognition Using Hybrid Deep Learning Architecture. *International Research Journal of Engineering and Technology (IRJET).* Sep 2023;10(09).
3. Kumar, R., Bajpai, A., & Sinha, A. (2023). MediaPipe and CNNs for Real-Time ASL Gesture Recognition. *arXiv preprint arXiv:2305.05296.*
4. Shin, J., Matsuoka, A., Hasan, M.A.M., & Srizon, A.Y. American Sign Language Alphabet Recognition by Extracting Feature from Hand Pose Estimation. *Sensors.* 2021;21(17):5856. https://doi.org/10.3390/s21175856.
5. Sharma, C.M., Tomar, K., Mishra, R.K., & Chariar, V.M. (September 20, 2021). Indian Sign Language Recognition Using Fine-Tuned Deep Transfer Learning Model. In *Procceedings of International Conference on Innovations in Computer and Information Science (ICICIS)* (pp. 62–67), Jiangxi, China. https://ssrn.com/abstract=3932929 or https://dx.doi.org/10.2139/ssrn.3932929.
6. Goel, P., Sharma, A., Goel, V., & Jain, V. (2022, November). Real-Time Sign Language to Text and Speech Translation and Hand Gesture Recognition Using the LSTM Model. In *2022 3rd International Conference on Issues and Challenges in Intelligent Computing Techniques (ICICT)* (pp. 1–6). IEEE, KIET Group of Institutions in Ghaziabad.
7. Liu, Z., Pan, C., & Wang, H. Continuous Gesture Sequences Recognition Based on Few-Shot Learning. *International Journal of Aerospace Engineering.* 2022;1(2022):7868142.
8. Sharma, A. & Kumar, Y. (2021). Hand Gesture Recognition using Machine Learning and Computer Vision.
9. ASL Alphabet, https://www.kaggle.com/datasets/grassknoted/asl-alphabet, Kaggle, Date accessed, August 13, 2025.
10. Fingerspelling, https://en.wikipedia.org/wiki/Fingerspelling. Wikipedia, Date accessed August 13, 2025.
11. Sharma, A., Mittal, A., Singh, S., & Awatramani, V. Hand Gesture Recognition Using Image Processing and Feature Extraction Techniques. *Procedia Computer Science.* 2020;173:181–190. https://doi.org/10.1016/j.procs.2020.06.022.

12. Li, D., Opazo, C.R., Yu, X., & Li, H. (2020). Word-Level Deep Sign Language Recognition from Video: A New Large-Scale Dataset and Methods Comparison. *Proceedings of the IEEE/CVF Winter Conference on Applications of Computer Vision.*
13. Joshi, A., Bhat, A., S, P.R., Gole, P., Gupta, S., Agarwal, S., & Modi, A. (2022). CISLR: Corpus for Indian Sign Language Recognition. In *Proceedings of the 2022 Conference on Empirical Methods in Natural Language Processing*, Abu Dhabi, United Arab Emirates.
14. Shi, B., Rio, A.M., Keane, J., Michaux, J., Brentari, D., Shakhnarovich, G., & Livescu, K. (2018). American Sign Language Fingerspelling Recognition in the Wild. In *2018 IEEE Spoken Language Technology Workshop (SLT)* (pp. 145–152), Athens, Greece.
15. Shi, B., Brentari, D., Shakhnarovich, G., & Livescu, K. (2021). *Proceedings of the IEEE/CVF Conference on Computer Vision and Pattern Recognition (CVPR)* (pp. 4166–4175), Nashville, TN, USA.
16. Novopoltsev, M., Verkhovtsev, L., Murtazin, R., Milevich, D., & Zemtsova, I. (2023). Fine-Tuning of Sign Language Recognition Models: A Technical Report. ArXiv, abs/2302.07693.
17. Lugaresi, C., Tang, J., Nash, H., McClanahan, C., Uboweja, E., Hays, M., … & Grundmann, M. (2019). Mediapipe: A Framework for Building Perception Pipelines. arXiv preprint arXiv:1906.08172.
18. MediaPipe Solutions guide | Google AI Edge | Google AI for Developers, https://developers.google.com/mediapipe, Google, Date accessed August 13, 2025.
19. Harris, I.M., & Agoes, A.S. (2021). Applying Hand Gesture Recognition for User Guide Application Using MediaPipe. In *Proceedings of the 2nd International Seminar of Science and Applied Technology (ISSAT 2021)* (pp. 101–108). Atlantis Press, South Tangerang, Indonesia.
20. Veluri, R.K., Sree, S.R., Vanathi, A., Aparna, G., & Vaidya, S.P. (2022). Hand Gesture Mapping Using MediaPipe Algorithm. In: Bindhu, V., Tavares, J. M. R. S., Du, KL. (eds) *Proceedings of Third International Conference on Communication, Computing and Electronics Systems.* Lecture Notes in Electrical Engineering, vol. 844. Springer, Singapore. https://doi.org/10.1007/978-981-16-8862-1_39.
21. Das, A., Gawde, S., Suratwala, K., & Kalbande, D. (2018). Sign Language Recognition Using Deep Learning on Custom Processed Static Gesture Images. In *2018 International Conference on Smart City and Emerging Technology (ICSCET)* (pp. 1–6), Mumbai, India. IEEE. https://doi.org/10.1109/ICSCET.2018.8537248.
22. Ameen, S. & Vadera, S. A Convolutional Neural Network to Classify American Sign Language Fingerspelling from Depth and Colour Images. *Expert Systems.* 2017;34(3):e12197. https://doi.org/10.1111/exsy.12197.

15 A Hybrid Framework for Detection and Diagnosis of Plant Diseases Using Retrieval Augmented Generative AI

Megha Rathi, Harikesh Pratap Singh, Kaustubh Pandey, Harsh Dhanwani, and Adwitiya Sinha

15.1 INTRODUCTION

Agriculture is considered as one of the main sectors worldwide that is involved in large-scale production and mass employment. The agricultural domain is the primary source of income for the rural households. The agricultural sector has emerged as one of the most prevalent industries that cater to people worldwide to secure the food requirement for good health. This poses a colossal challenge for the agriculture sector to improve food security, which may be hampered by infectious diseases. In such cases, the crop yields get reduced, sometimes even leading to losses of a major fraction of crop production. The monsoon season is becoming increasingly erratic due to climate change, leading to droughts and floods impacting crop yields (Jeger et al., 2021). A few challenges are associated with agriculture that threaten the worldwide food supply and subsistence of farmers. Cataclysmic crop diseases, chemicals, as well a lack of information about agricultural best practices can lower crop yields and can root for grievous economic strain. Local growers in developing countries like India play a pivotal role in the farming sector, but due to a lack of skilled supervision, they struggle to overcome various challenges in the agricultural sector. Cutting-edge techniques like Artificial Intelligence, the Internet of Things, Robotics, and Computer Vision are required to cater to the critical needs of agricultural challenges for providing inclusive solutions (Jafar et al., 2024). Sustainable and accurate insights must be provided to local growers for effective plant disease management and prevention with the help of these techniques. With the global increase in population and living conditions, there is an expanding urge for multiplicity, safety, and sufficient food sources, making it hard to administer plant diseases (Dun-chun et al., 2016). Outbreaks of plant disease are an imminent hazard to nature's sustainability and food demand worldwide as it narrows down productivity and biological variability.

DOI: 10.1201/9781003565703-15

Environmental and economic sustainability are highly affected by these factors (Singh et al., 2023). The majority of agriculturalists do not have instant access to economic and weather forecasts or next-generation farming techniques. Time-honored methods entail the misuse of chemicals, pesticides, and insecticides due to unavailability of real data on crop requirements. Impediments like this can be resolved with revolutionary technologies that are coupled with Artificial Intelligence with real-time data and established agronomic skills. Pioneering technologies like Generative Artificial Intelligence, Casual Artificial Intelligence, and Explainable Artificial Intelligence played a significant role in agronomy, especially for early prediction of plant disease (Lipikant Sahoo et al., 2024). By adopting innovative AI techniques, agriculturalists can ameliorate the management of crops, intensify resource utilization, and can decrease harvest failure. Problems with model interpretability, data scarcity, and deployment at real sites in resource-constrained environments continue to be a major challenge for universal adoption. Such difficulties must be overcome to ascertain the potential of AI techniques in revolutionizing farming practices and ensuring food security worldwide. Time-honored methods for the detection of plant diseases are unprogressive, expensive, and require domain knowledge. For conducting a more comprehensive assessment and detection, smart agricultural systems are required for plant disease detection coupled with advanced AI and other cutting-edge techniques (Ostovar Ahmad et al., 2021). Smart sensor technologies like intelligent drones and IoT devices for irrigation help us to monitor real-time plant disease. With the application of sophisticated computational methods, farmers and practitioners can overcome the difficulty of traditional systems by using the best possible option for plant disease diagnosis.

15.1.1 Research Objectives and Motivation

The main objective of the proposed research work is to develop a mobile application that integrates frontier techniques like machine learning, artificial intelligence, and computer vision to analyze and predict plant diseases from the real-time pictures taken by the farmers. In no time, plant disease can be precisely and accurately detected by the proposed smart phone application, additional features like individualized farming suggestions, and access to relevant data on plant disease management practices. The Large Language Model (LLM) technique is merged with deep learning techniques to create a knowledge base for providing additional details like communicating plant disease type, preventive measures and treatment actions corresponding to specific crops. The main emphasis is to create smart applications for agriculturalists that accredit farmers by deliberately incorporating leading-edge technologies. Even a simple smart phone can use the suggested, user-friendly application to identify and examine plant leaf photos taken by nearby farmers in their fields. The proposed mobile application will detect plant illnesses with remarkable precision through real-time analysis, allowing farmers to act right away.

15.1.2 Applicability to Real-World Instances

The proposed research work materialized into a practical solution through the development of a mobile application named Green Guardian, tailored for real-time plant disease detection and advisory services. In real-world deployments, our proposed mobile app can empower the farmers, particularly smallholders in India, to identify plant diseases on-site using just their smartphone cameras. By capturing a leaf image, farmers can receive immediate classification results powered by a deep learning model (ResNet). This can be further assisted with the integrated Prakruti AI chatbot, which was built using a Retrieval-Augmented Generation (RAG) framework. For instance, a community of farmers in Uttar Pradesh benefited from early detection of tomato leaf blight, which previously went undiagnosed until significant crop damage occurred. With real-time alerts and tailored guidance on preventive care and medication through the chatbot, farmers mitigated crop loss and optimized pesticide usage. Furthermore, the ability of the chatbot to communicate preventive measures and provide disease-specific treatment suggestions transformed decision-making and disease management at the grassroots level. These real-world instances showcase the capability of our proposed framework to bridge the technological gap in agriculture and offer a sustainable, scalable model for smart farming solutions.

15.1.3 Key Research Contributions

By aligning technological sophistication with practical utility, our research would mark a significant advancement in precision agriculture. The study will lead to the adoption of AI in sustainable farming, especially in emerging economies where access to expert consultation and diagnostic tools usually remains severely limited.

- Developed a hybrid system combining deep learning (ResNet) and large language models for plant disease detection and diagnosis.
- Designed and deployed a mobile application (*Green Guardian*) for real-time disease identification using smartphone-captured leaf images.
- Achieved 99.5% classification accuracy with the ResNet model on the PlantVillage dataset, outperforming other deep learning architectures.
- Integrated a RAG chatbot (*Prakruti AI*) to provide contextual disease treatment suggestions and prevention advice.
- Employed Gemini API for generating semantic embeddings from textual resources and stored in ChromaDB for efficient retrieval.
- Developed a comprehensive pipeline including image preprocessing, model training, chatbot integration, and user interaction.

15.2 BACKGROUND STUDY ON RELATED DOMAIN

Recent techniques for identifying plant diseases include hybrid methods that fuse convolutional neural networks (CNNs), machine learning techniques, vision transformers, and additional advanced deep learning models. By utilizing huge datasets and transfer learning, these techniques have been demonstrated to be very successful in recognizing a broad span of plant diseases, with notable gains in accuracy and efficiency. In this section, we have examined various recent research done in the area of forecasting plant diseases. A relevant contribution in the study (Divakar et al., 2021) introduced a novel approach to reducing false predictions in plant disease detection.

Their method employs the SMOTE technique for dataset balancing and a unique ensemble algorithm to optimize classifier performance. The dataset, consisting of labels Healthy, Multiple Disease, Scab, and Rust for apple leaf diseases, is analyzed using RGB channels. EfficientNetB7 was shown to be the best classifier by the ensemble approach, obtaining the highest F1-sScore. In a significant study (Khalid & Karan, 2024), deep learning algorithms, specifically CNN and MobileNet architectures, are employed for the prompt identification of crop diseases. Explainable AI techniques were employed to bestow insights into the decisiveness and offer visual interpretations of the results. Images were scaled down to 224×224 pixels due to computational constraints. The model architecture started with a 2D convolutional layer, followed by max pooling to condense spatial dimensions. The output from the MobileNetV2 base was passed through four dense layers with progressively fewer nodes. The CNNs exhibited improved performance, achieving a 96% F1-score, 89% accuracy, 96% precision, and 96% recall. The MobileNet architecture, while demonstrating a 96% accuracy, resulted in slightly lower precision, recall, and F1-score values, which were 90%, 89%, and 89%, respectively. Another notable study employed an amalgam of deep learning models and large language models to detect plant and pest diseases and provide treatment suggestions (Qing et al., 2023). The images from the dataset were fed into a pre-trained YOLO network, and the results were fed into the GPT-4 language model after being converted into text form. The large language model summarized and simplified the results to provide diagnosis suggestions and treatment plans. The study was evaluated by treating it as a single-class classification problem. Developing a comprehensive chatbot system for identifying plant diseases is the main objective of research work (Kothari et al., 2024). It featured GPT-3.5 Turbo-powered conversational AI as well as dynamic learning processes on the backend. The model obtained good classification accuracy for plant illnesses using ResNet50 architecture with learning weights.

Noteworthy research by Kalicharan & SonajhariaMinz (2023) focuses on localizing diseased plant leaves by performing complex tasks such as enhanced segmentation, feature extraction, and classification models for multi-disease classification. For abnormality segmentation, an Adaptive Fusion of K-Means Region Growing (AFKMRG) method was presented, combining RG and K-means clustering techniques. Using characteristics taken from segmented pictures, the Enhanced Long Short-Term Memory (LSTM) approach was used for the classification of many diseases. The suggested FSJ-FOA-CNN-LSTM classification model fared better than the competition when it came to citrus plant leaves, attaining better accuracy and computational efficiency—even for photos with lesser resolution. Another remarkable research in the study (Abisha & Bharathi, 2023) involves using a previously trained model paired with a Random Forest machine learning classifier for predicting plant diseases. For feature extraction, pre-trained models including RESNET, InceptionResNetV2, Xception, MobileNet, and VGG 16 were used. The proposed XceptionRF model, which uses an Xception model with depth-wise convolutions to decrease parameters and implementation costs, achieved 100% accuracy with no misclassified classes. Explainability was found to be limited by the elimination of abiotic stress using cloud processing in real time, the potential for overfitting in Random Forest classification, and the black-box aspect of the classifier.

Illustrious research work done by Jingwei et al. (2023) addresses apple leaf disease detection through a novel deep learning model, the FPNs-ISResNet-Faster RCNN. It aims to improve detection accuracy by integrating multiple techniques. The model leverages feature pyramid networks (FPNs) to capture spatial information alongside inception squeeze-and-excitation ResNet (ISResNet) for robust semantic analysis. To direct focus on disease-affected regions within the leaves, an attention mechanism is incorporated. The model's accuracy was maximal (AP of 62.71%) and precision (98.79%) among the evaluated models. However, limitations are acknowledged, including the potential for bias introduced by data augmentation and the model's slower detection speed (14.87 FPS) compared to alternatives. Furthermore, in the recent work, Santhana Krishnan and Golden Julie (2023) proposed an improved convolution neural network based on VGG16 to classify potato leaf diseases. To improve the caliber and sharpen the margins of the potato leaf images, a modified median filter was employed. To classify potato leaf diseases, an enhanced CNN model built on top of VGG16 was employed. This study uses the Inception and SE blocks, as well as the VGG16 convolution layers, for classification. Compared to previous CNN approaches, the recommended model was more flexible to changes in picture spatial position and had 99.3% accuracy in categorization. However, a large CNN network will be expensive to train and exhibit a sluggish rate of convergence. It is not suitable for use in outdoor environments with a blend of leaf backs and leaf fronts.

An intelligent chatbot to assist farmers in answering their agricultural-related queries is developed in the study (Nayak et al., 2021). A Chatbot engine is used to create the chatbot trained on a dataset composed of queries and answers. The unanswered questions can be mailed to experts. If a query entered by a user does not have a 75% match rate to a query on the trained data, then a generic answer is returned. Images and videos are not supported by the platform. In yet another significant research (Momaya et al., 2021), a WhatsApp chatbot has been developed to assist farmers with agricultural queries. It uses the OpenWeatherMap API to additionally respond to weather-related queries. It is trained on data from the official website of the Kisan Call Center. Experimental analysis showed 96.1% accuracy in answering agriculture-related queries. The chatbot does not take voice inputs and does not have local language support. In another prominent research (Maduri et al., 2021), a chatbot has been created to assist farmers in predicting the needs of a crop. It provides information on various aspects, such as fertilizer dosage and nutrients, and offers guidance on how to meet these requirements. The chatbot makes use of past feed data from "The Indian Council of Agricultural Research." The nutrient quantity is predicted based on district soil data and marked as low, medium, or high. The chatbot does not support images or videos. Distinguished researchers have introduced a LINE Bot System for diagnosing rice diseases using real paddy field images in their work (Liu et al., 2021). They employed deep-learning neural networks to detect these diseases. A process for training and refining object detection models was developed to improve the final outcome. The deployment model's performance improved from 91.1% to 95.6%.

An amalgamation of Convolutional neural networks and support vector machine classifiers is employed to identify and categorize specific illnesses that affect paddy plants (Haridasan et al., 2023). The suggested deep learning-based method obtained a top validation accuracy of 0.9145 by utilizing ReLU and softmax algorithms. A bibliometric analysis based on visualization is laid out in the research work to showcase the patterns of research from 2014 to January 2024 in deep learning-based leaf disease detection (Chaki & Ghosh, 2024). From 2014 to January 2024, the most explored topics in this discipline were "Convolution Neural Network," "Transfer Learning," "Ensemble Learning," and related subjects, according to the findings. Applications of common deep learning architectures are the main focus of the research. This work provides a vision into the general research patterns in deep learning for leaf disease detection, which could aid scholars in comprehending and forecasting the changing trends in the field. In another substantial research investigation (Khalid & Karan, 2024), exploration was conducted by utilizing GradCAM to implement Explainable Intelligence (XAI), which made the choice-making procedure of these models more transparent and offered an animated explanation of disease signs in plant pictures. The accuracy of the CNN model was 89%, while its precision and recall were 96% and its F1-score was 96%. The MobileNet architecture, on the other hand, showed a 96% accuracy rate but slightly lower recall, 90%, and 89% F1-scores, respectively, as well as precision.

Another study on the automated detection of fungal disease in maize leaves is presented in the paper (Shantala et al., 2020). The envisioned system makes use of deep learning-inspired image processing methodologies. Having an accuracy of 98% for three-class questions and 70.5% for four-class problems, the outcomes are encouraging. The discoveries of the paper (Sajitha et al., 2024) emphasize the possible difficulties in using ML and DL for plant disease and pest identification, which motivates further investigation. In yet another subsequent study investigation, the authors assessed the detection efficacy of four deep learning models (VGG19, DenseNet201, ResNet152V2, and a Vision Transform model) for illnesses affecting betel leaves (Kusuma & Jothi, 2024). The ResNet152V2 and VIT models both achieved 98.42% and 97.83% testing accuracies, respectively, indicating their level of accuracy. A plant disease classification system based on explainable artificial intelligence (XAI) is proposed in this research (Nigar et al., 2024) to increase the accuracy of classifying and identifying different illnesses. The accuracy, precision, and recall of the system in properly identifying 38 distinct plant diseases are 99.69%, 98.27%, and 98.26%, respectively. In another novel work (Tariq et al., 2024), the VGG16 model that was suggested and supplemented with LRP performed better in diagnosing maize leaf diseases. The experimental results demonstrated that the proposed algorithmic technique not only generated precise results but also found out remarkable regions in images. A significant gap still exists in collaborating Chatbot for plant disease detection. The ideal system would blend chatbot interface to assist in real-time diagnosis and treatment suggestions with deep learning techniques for accurate illness detection. Through the resolution of this disparity, a complete and intuitive mobile application is developed that empowers local owners of land to make informed decisions for enhancing healthy crops. A consolidated application is created that combines smart phones with advanced LLMs and computer vision capabilities for plant image detection. The main emphasis is to build user-friendly mobile interface for disease identification and gaining abundant knowledge about crop health.

15.3 DATASET DESCRIPTION

Our study utilizes the PlantVillage dataset, which is a comprehensive and widely used resource in plant disease detection research (Tairu Oluwafemi Emmanue, 2024). This dataset, available on Kaggle, includes more than 50,000 images of crop plant leaves, meticulously labeled as either healthy or diseased. The dataset covers a wide variety of crops, including common species like tomatoes, potatoes, apples, corn, and more. It is expertly curated, with each image representing a specific condition of the plant, either healthy foliage or multiple medical conditions including bacterial spots, early and late blight, and many more. The large volume of data, combined with the diversity of plant species and diseases, makes this dataset an ideal candidate for deep learning-based plant disease detection.

The inclusion of images from a range of conditions ensures that our machine learning models can generalize across different types of plant leaves and associated diseases (Figures 15.1 and 15.2). This helps to improve the model's ability to recognize disease symptoms even when confronted with unseen or less common cases with lower probability. In our study, we likely leverage this dataset to train models like ResNet, Inception, DenseNet, Vision Transformer, and VGG16, all of which are well-known for their effectiveness in image classification assignments. Using this large dataset ensures that our model can detect and classify plant diseases with high accuracy, contributing to real-world applications like smart farming or mobile disease diagnostic tools, such as our mobile application for plant disease detection.

Furthermore, the open accessibility of the Plant Village dataset encourages collaboration and further advancements in agricultural technology, promoting a global effort to improve crop management and minimize losses due to plant diseases.

15.4 PROPOSED RESEARCH METHODOLOGY

This section outlines systematic approach for developing and validating an AI-based architecture for plant illnesses identification. The proposed hybrid framework is designed to utilize annotated images of dataset of both diseased and healthy plant leaves to improve forecasting accuracy by leveraging extensive image classification architectures. Both diseased and normal plant leaves are tagged in dataset. Data collection, data preprocessing, data transformation, model building, and model evaluation are all part of our meticulous pipeline, which guarantees well-grounded and precise disease diagnosis of crops. On top of that, our suggested hybrid system includes a smart phone interface that enables the pre-trained models to classify real-time plant disease instances. We further integrate with the RAG chatbot to improve the research utility. The following section elaborates each phase of our hybrid modeling approach.

15.4.1 Data Cleaning and Preprocessing

Initially, an eclectic collection of images from plant leaves, including both healthy and diseased samples, is collected. Data cleaning involves the removal of duplicates, irrelevant images, and ensuring proper labeling of the images, as shown in the flow

FIGURE 15.1 Healthy plant leaf sample images from plant village dataset.

diagram in Figure 15.3. The data cleaning process involves multiple steps that begin with duplicated data removal. The dataset is first scanned for duplicate images that may have been unintentionally repeated. These duplicates can bias the model's training by over-representing certain classes, leading to overfitting. Removing such duplicates ensures that each image represents unique data. Also, our dataset contained some irrelevant and noisy images that do not correspond to the plant species or diseases under study. These might include unrelated objects, partial images, or poorly captured samples (e.g., blurred or occluded images). We have identified and removed such irrelevant images to maintain the integrity of our dataset.

15.4.2 Model Training Phase

We assessed how well different deep learning models performed—ResNet, Inception, VGG16, Vision Transformer, and DenseNet—on the task of plant species

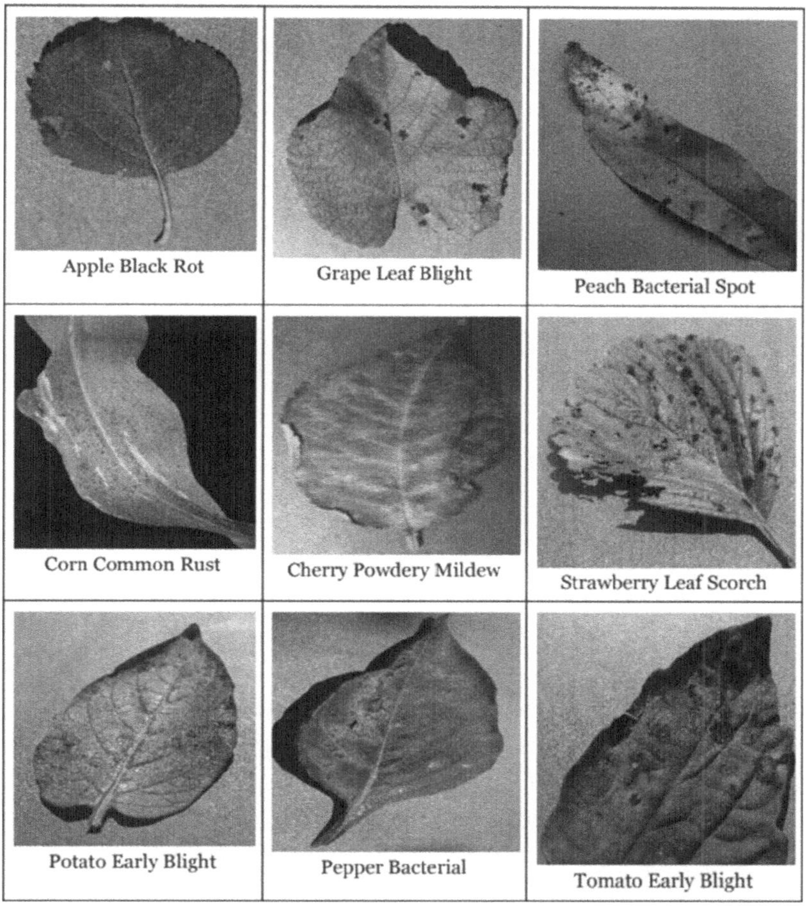

FIGURE 15.2 Plant leaf sample images with different diseases from plant village dataset.

classification using image data. The dataset comprises images of apple, corn, grape, potato, and strawberry plants. Each model was trained individually on the dataset using a consistent training protocol. Vital hyperparameters were optimized, including batch size, learning rate, and optimizer type for each model. To assess the models' performance, the dataset was divided into test, validation, and training sets. With an average accuracy of 99%, the ResNet model outperformed the other investigated architectures. The ResNet model performs better because it can manage deeper networks better, thus capturing more compounded features of the plant images. The skip connections in ResNet help mitigate the vanishing gradient problem, allowing for better learning during the training process. Other models, while still performing robustly, did not match the accuracy of ResNet. Inception's multi-scale feature extraction, VGG16's simplicity, ViT novel transformer approach, and DenseNet dense connections all contributed to high performance, but none matched the effectiveness of ResNet for this specific task.

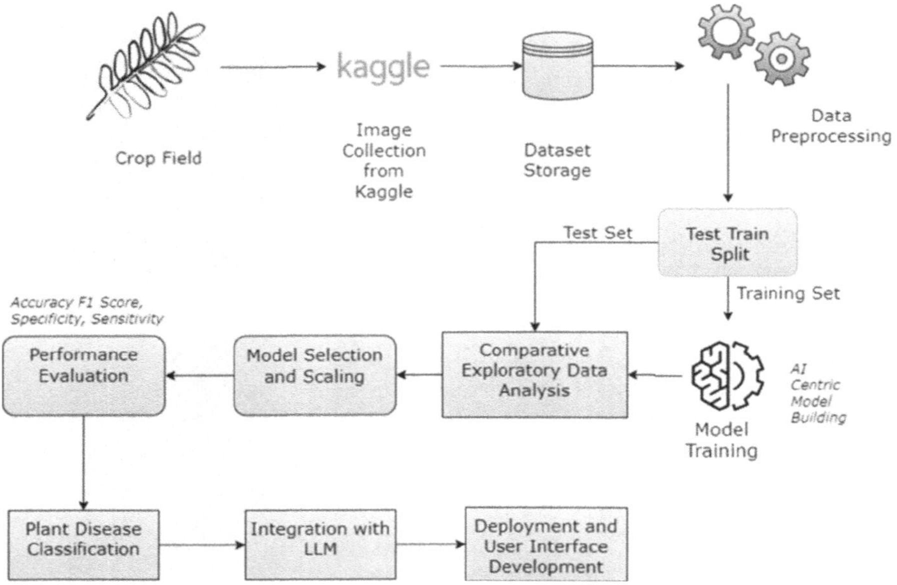

FIGURE 15.3 Methodology for green guardian mobile application.

15.4.3 Incorporating Large Language Model with Mobile App

Automated systems for plant disease diagnosis and information retrieval are crucial for enhancing agricultural productivity and crop management. Leveraging deep learning and language models, we developed a RAG chatbot incorporated into our Mobile App, namely Green Guardian (Figures 15.4 and 15.5). This app combines disease prediction with advanced natural language processing capabilities to assist users in identifying and understanding plant diseases. Our chatbot system requires input of the diseased leaf and shows the plant it belongs to. The knowledge base for the chatbot consists of PDFs containing detailed information about various plant diseases and their management. These documents were processed and indexed to facilitate the efficient retrieval of relevant information.

The Gemini API served as the building block for developing embeddings, which transformed the PDFs' textual content into a format that can be searched. A vector database is designed to hold embeddings for similarity search and is actually reflecting semantic content of text. For seamless data retrieval, we put in place a document processing pipeline, which is initiated by the extraction of text content from a PDF file. An open source tool PDF miner is employed for structural parsing of PDF document and textual data extraction. Extracted data is then preprocessed for ensuring consistency with in the text. Preprocessing steps include stemming, tokenization, and stop word removal. Subsequently, the preprocessed text was fed into the Gemini API, a system designed to generate numerical representations (embeddings) of text data. These embeddings capture the semantic meaning, allowing for efficient comparison and retrieval of documents with similar content.

FIGURE 15.4 Green guardian mobile app.

Finally, the generated embeddings were stored within ChromaDB, a database specifically optimized for storing and retrieving vector data. This choice facilitates efficient retrieval based on the document embeddings, enabling subsequent information retrieval tasks.

15.4.3.1 Building Chatbot and LLM System Integration to Mobile App

The LLM was integrated with a chatbot LLM system, namely Prakruti AI. This chatbot served as an interface for users to interact with the retrieved information (Figure 15.6). The predicted disease served as the initial query into the chatbot. Here, our chatbot functionality played a crucial role in providing assistance regarding medicines for prevention and cure (Figures 15.7 and 15.8). It leverages pre-generated document embeddings stored within ChromaDB. By utilizing these embeddings,

FIGURE 15.5 Mobile App showing information.

the chatbot efficiently retrieved relevant information from the knowledge base. This retrieved information, likely consisting of articles or specific sections of documents related to the disease, then informed the chatbot's response generation. Finally, a chatbot was built using the FastAPI framework, allowing it to be hosted online and serve user queries. This integration enabled users to receive ailment-specific information in a conversational format, leveraging the power of the document processing pipeline and chatbot capabilities.

15.5 RESULTS AND DISCUSSIONS

In this study, we performed an in-depth comparison investigation of five cutting-edge deep learning models—ResNet, Inception, VGG16, DenseNet, and Vision Transformer—for the task of plant disease classification using images. These models

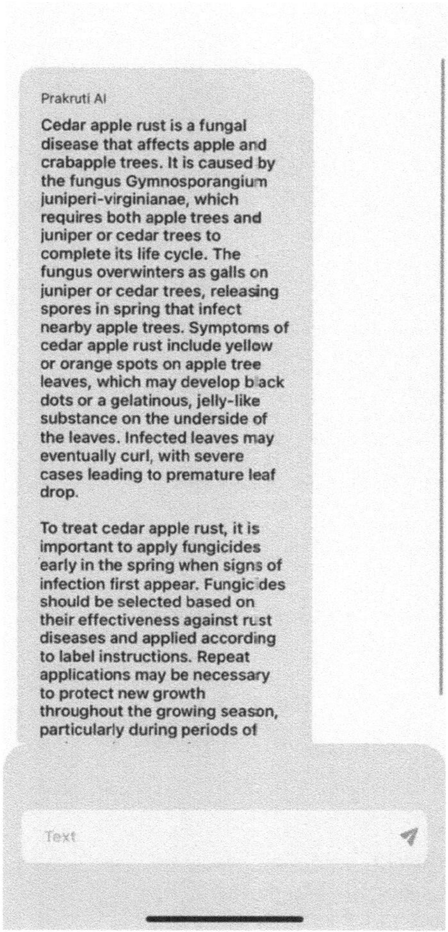

FIGURE 15.6 Prakruti AI RAG chatbot interface.

were chosen based on their proven effectiveness in tasks involving the classification of images and their capacity to capture both high-level and intricate details in complex datasets, such as those in the Plant Village dataset. The Inception model is a CNN architecture that builds upon the success of its predecessors in the Inception family. It aims to achieve better performance with less complexity, as shown in Figures 15.9 and 15.10. Compared to earlier versions, Inception boasts a simplified architecture and utilizes a higher number of Inception modules, the specialized building blocks within the network.

VGG16, a CNN that bears the name of the Visual Geometry Group, is well-known for being both straightforward and efficient. Unlike complex architectures, VGG16 achieves good performance by stacking numerous smaller convolutional layers in a specific pattern, as outlined in Figures 15.11 and 15.12. A CNN model called DenseNet, short for Densely Connected convolution networks, is renowned for its

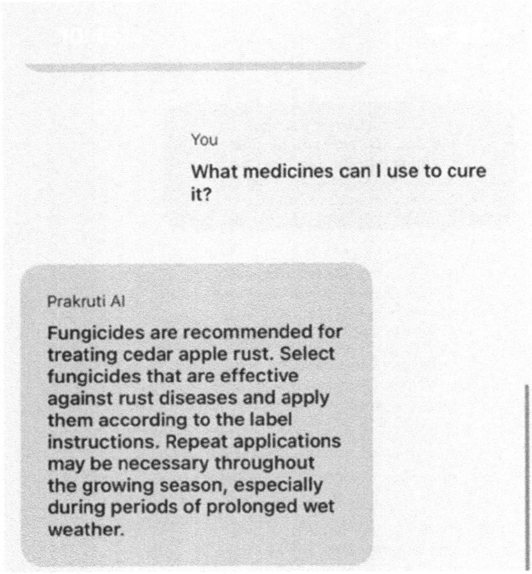

FIGURE 15.7 Prakruti AI for medicine recommendation.

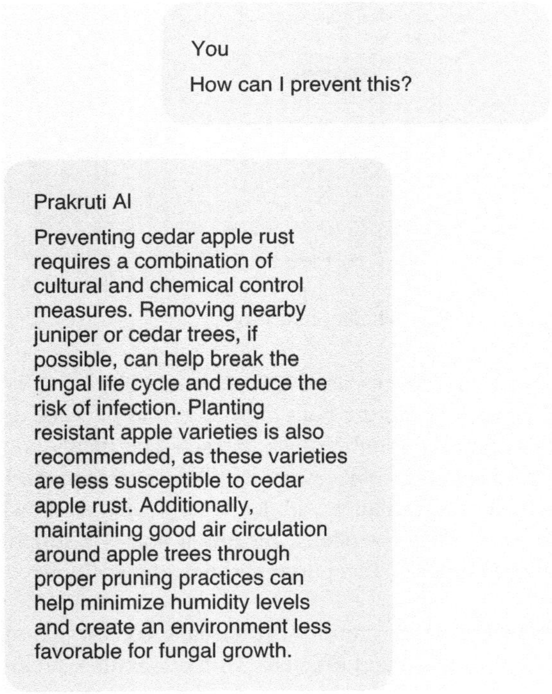

FIGURE 15.8 Prakruti AI for prevention and care.

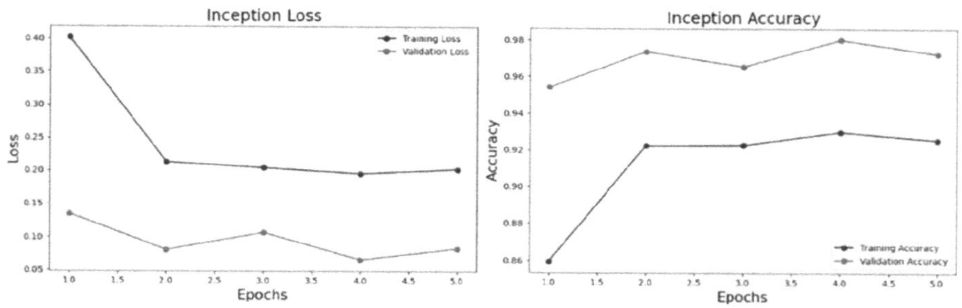

FIGURE 15.9 Model recognition accuracy of Inception Model.

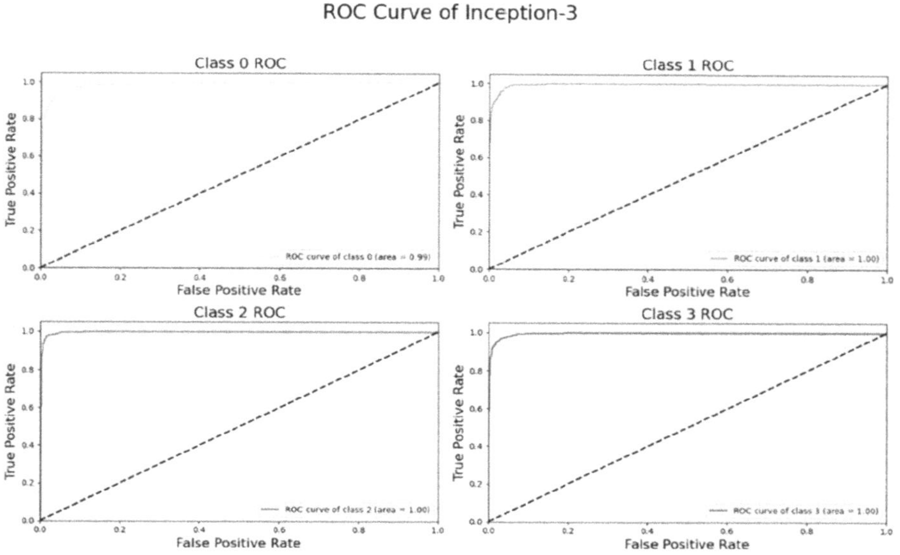

FIGURE 15.10 ROC curve of inception-3.

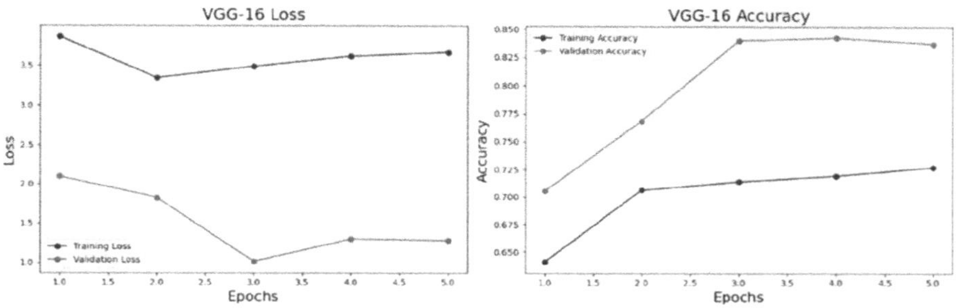

FIGURE 15.11 Model recognition accuracy of VGG Model.

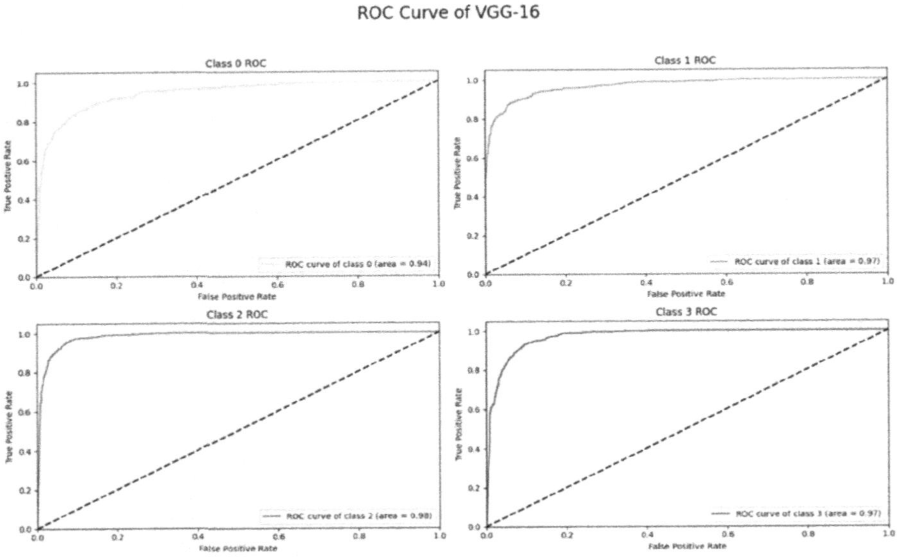

FIGURE 15.12 ROC curve of VGG-16.

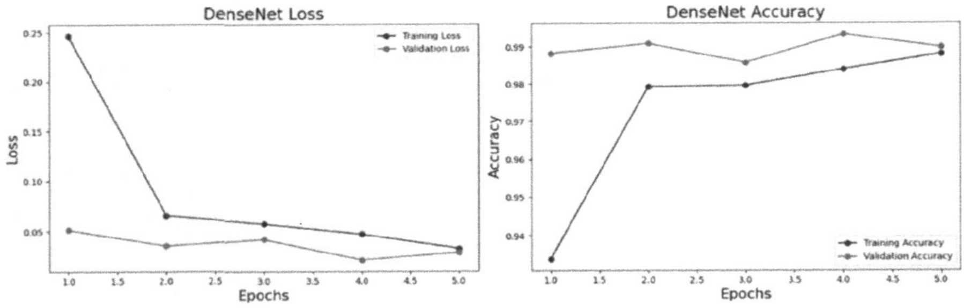

FIGURE 15.13 DenseNet Model Accuracy.

distinct connectivity strategy. Rather than passing feature maps only to the next layer, each layer in a DenseNet delivers its own feature maps to every layer after it, and each layer receives inputs from all layers before it.

The Vision Transformer is notable for its flexibility and scalability, capable of handling various image sizes and types, making it a successful tool for multiple purposes of vision tasks. The ResNet50, short for Residual Network 50, is a powerful deep learning architecture known for its ability to learn from very deep networks (50 layers in this case), as highlighted in Figures 15.13–15.16. Unlike traditional models that struggle with increasing depth due to vanishing gradients, ResNet50 incorporates residual connections to address this challenge.

The comparison analysis of model performance in terms of accuracy is displayed in Table 15.1. Our observations revealed significant differences in the performance of

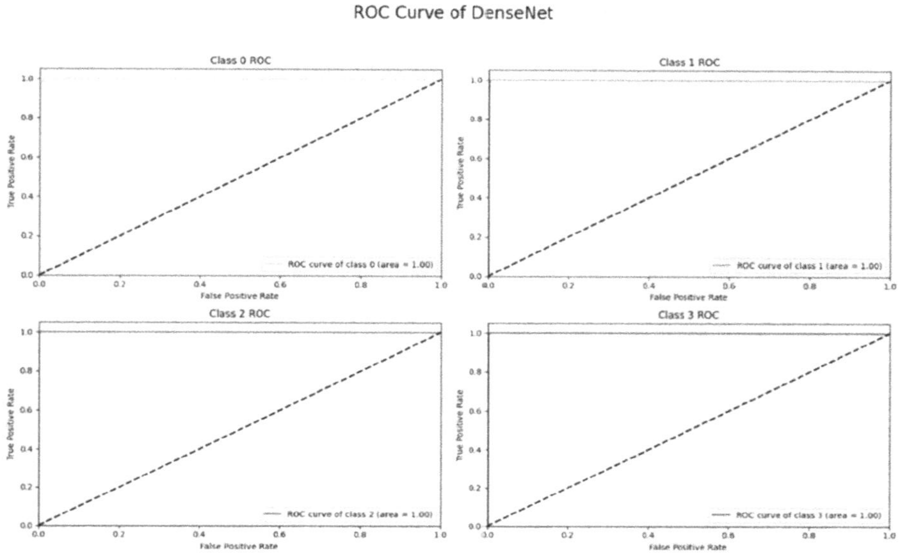

FIGURE 15.14 ROC curve of DenseNet.

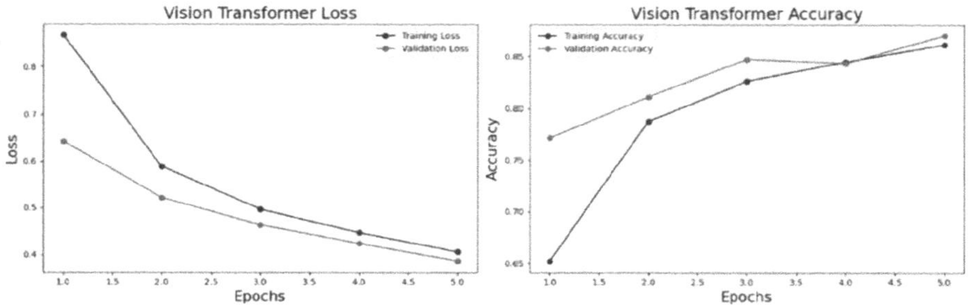

FIGURE 15.15 Model recognition accuracy of Vision Transfer Model.

the models. ResNet emerged as the frontrunner, achieving a remarkable accuracy of 0.995 on the test set. This exceptional performance was accompanied by a minimal training error of 0.025, indicating the potential of the model to learn effectively from the training data and generalize well to concealed examples.

DenseNet followed closely behind with an accuracy of 0.97 and a training error of 0.05, demonstrating its strong capability for disease classification. Inception and Vision Transformer achieved accuracies of 0.97 and 0.86, respectively. However, their training errors were noticeably higher (0.20 and 0.4), suggesting potential challenges with overfitting or limitations in capturing the relevant features for disease classification in this specific dataset.

The proposed approach's model recognition accuracy and ROC Curve are shown in Figures 15.17 and 15.18. Bottleneck layers are customarily used in

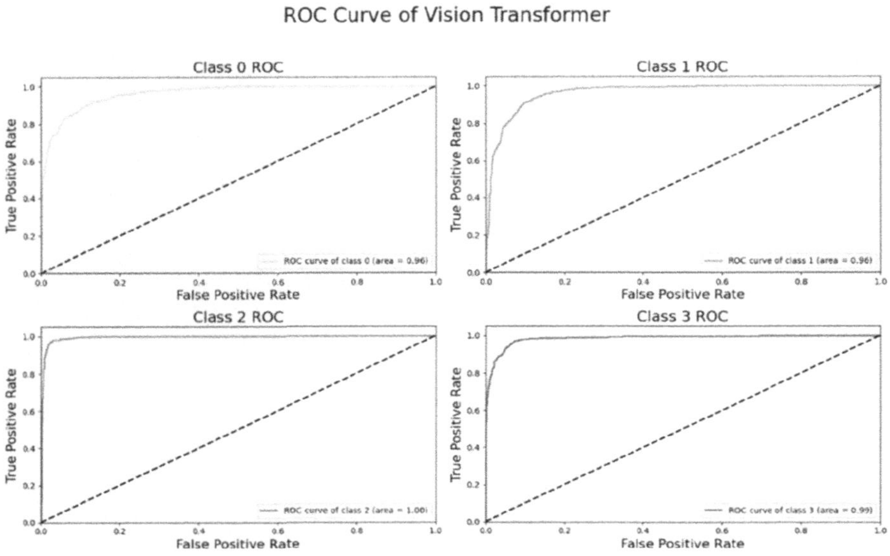

FIGURE 15.16 ROC curve of vision transformer.

TABLE 15.1
Comparative Analysis of Model Outcome and Accuracies

S.No.	Model	Accuracy
1	Inception	0.97
2	VGG16	0.71
3	DenseNet	0.97
4	Vision Transformer	0.86
5	ResNet	0.995

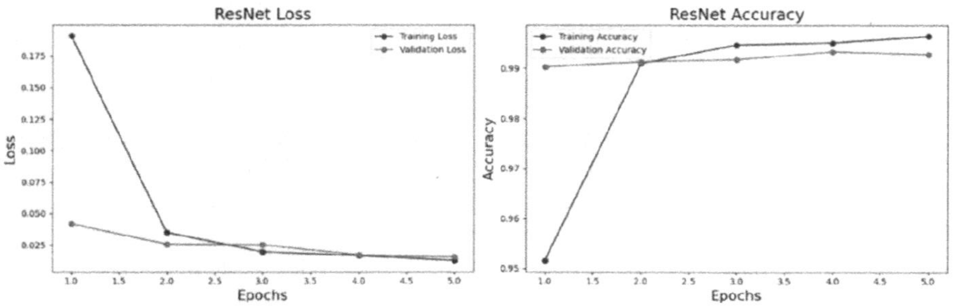

FIGURE 15.17 ResNet Accuracy.

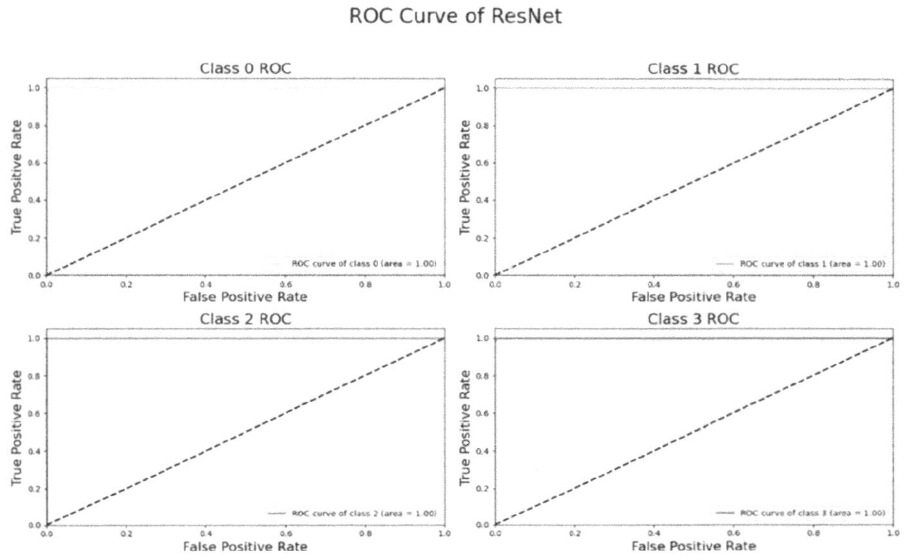

FIGURE 15.18 ROC curve of ResNet.

ResNet model to bottom-most the dimensionality of feature maps, which may reduce overfitting and promote efficient learning. From the results, it is evident that, for plant disease detection in the Plant Village dataset, the ResNet model would be more pertinent.

15.6 CONCLUSION

In the proposed research work, a mobile application is developed to deal with the problem of accessibility and a lack of technical guidance in plant disease diagnosis. Next-generation computing solutions that can aid with real-time diagnostics and plant disease prediction are desperately needed to improve the agriculture sector. A hybrid mobile-based framework is developed that amalgamates a deep learning model for plant leaf disease classification and a RAG chatbot system to assist farmers in providing detailed information on plant diseases. The ResNet model becomes apparent as a frontrunner, attaining an accuracy of 0.995 with a minimal training error of 0.025. This unprecedented performance indicates the model's potential to learn effectually from the training data and generalize effectively on test instances. With the help of the mobile application, farmers may promptly detect crop problems. They can then interact with the chatbot to obtain pertinent information related to the disease. The embedding generation of Gemini and ChromaDB storage capabilities makes the chatbot retrieval system more powerful by ensuring prompt access to correct and contextually relevant plant disease information. Overall, the study has irrefutably shown how consolidating chatbot systems with AI techniques for plant disease detection may empower farmers.

AUTHOR DECLARATION SECTION

Competing Interest

The authors have no relevant financial or non-financial interests to disclose.

Funding

The authors have not received any financial assistance or grant for research in this manuscript.

Ethics Approval

The authors have adhered to the ethical standards of performing research in this manuscript.

Consent to Participate

Not applicable.

Consent for Publication

Not applicable.

Code Availability

The research code will be made available on-request from readers.

Availability of Data and Material

Research data associated with this paper is extracted through web-crawling and API-based programming. Hence, the dataset will be made available on request from readers.

REFERENCES

Abisha, A., and Bharathi, N. "A hybrid feature extraction and classification using Xception-RF for multiclass disease classification in plant leaves," *Applied Artificial Intelligence* 37:1 (2023): 2176614, doi: 10.1080/08839514.2023.2176614.

Ahmad, O., Bensch, S., and Hellström, T. "Natural language guided object retrieval in images," *Acta Informatica* 58:4 (2021): 243–261, doi: 10.1007/s00236-021-00400-2.

Chaki, J., and Ghosh, D. "Deep learning in leaf disease detection (2014–2024): a visualization-based bibliometric analysis," *IEEE Access* 12 (2024): 95291–95308, doi: 10.1109/access.2024.3425897.

Divakar, S., Bhattacharjee, A., and Priyadarshini, R. "Smote-DL: a deep learning based plant disease detection method," In *2021 6th International Conference for Convergence in Technology (I2CT)*, Pune, India. Apr 02–04, 2021.

Giraddi, S., Desai, S., and Deshpande, A. "Deep learning for agricultural plant disease detection," In *ICDSMLA 2019: Proceedings of the 1st International Conference on Data Science, Machine Learning and Applications*, pp. 864–871. Springer, Singapore, 2020.

Haridasan, A., Thomas, J., and Raj, E.D. "Deep learning system for paddy plant disease detection and classification," *Environmental Monitoring and Assessment* 195 (2023):120, doi: 10.1007/s10661-022-10656-x.

He, D.C., Zhan, J.S., and Xie, L.H. "Problems, challenges and future of plant disease management: from an ecological point of view," *Journal of Integrative Agriculture* 15:4 (2016): 705–715.

Jafar, A., Bibi, N., Naqvi, R.A., Sadeghi-Niaraki, A., and Jeong, D. "Revolutionizing agriculture with artificial intelligence: plant disease detection methods, applications, and their limitations," *Frontiers in Plant Science* 15 (2024): 1356260.

Jeger, M., Beresford, R., Bock, C., Brown, N., Fox, A., Newton, A., Vicent, A., Xu, X., and Yuen, J. "Global challenges facing plant pathology: multidisciplinary approaches to meet the food security and environmental challenges in the mid-twenty-first century," *CABI Agriculture and Bioscience* 2:1 (2021): 1–18.

Jingwei Hou, C.Y., He, Y., and Hou, B. "Detecting diseases in apple tree leaves using FPN–ISResNet–Faster RCNN," *European Journal of Remote Sensing* 56:1 (2023): 2186955, doi: 10.1080/22797254.2023.2186955.

Khalid, M.M., and Karan, O. "Deep learning for plant disease detection," *International Journal of Mathematics, Statistics, and Computer Science* 2 (2024): 75–84.

Kothari, S., Bagane, P., Mishra, M., Kulshrestha, S.., Asrani, Y., and Maheswari, V. "CropGuard: empowering agriculture with AI driven plant disease detection chatbot," *International Journal of Intelligent Systems and Applications in Engineering* 12:12s (2024): 530–537.

Kusuma, S., and Jothi, K.R. "Early betel leaf disease detection using vision transformer and deep learning algorithms," *International Journal of Information Technology* 16 (2024): 169–180, doi: 10.1007/s41870-023-01647-3.

Liu, J., and Wang, X. "Plant diseases and pests detection based on deep learning: a review," *Plant Methods* 17 (2021): 22, doi: 10.1186/s13007-021-00722-9.

Maduri, P.K., Dhiman, P., Shukla, M.R., Anand, S., and Singh, S.P. "Farmers agriculture assistance chatbot," In *2021 3rd International Conference on Advances in Computing, Communication Control and Networking (ICAC3N)*, Greater Noida, India. pp. 1884–1889. 2021, doi: 10.1109/ICAC3N53548.2021.9725634.

Momaya, M., Khanna, A., Sadavarte, J., and Sankhe, M. "Krushi—the farmer chatbot," In *2021 International Conference on Communication information and Computing Technology (ICCICT)*, Mumbai, India. pp. 1–6. 2021, doi: 10.1109/ICCICT50803.2021.9510040.

Nayak, V., Nayak, P.R., Sampoorna, Aishwarya, and Sowmya, N.H. "Agroxpert—farmer assistant," *Global Transitions Proceedings* 2:2 (2021): 506–512, doi: 10.1016/j.gltp.2021.08.016.

Nigar, N., Muhammad Faisal, H., Umer, M., Oki, O., and Lukose, J.M. "Improving plant disease classification with deep-learning-based prediction model using explainable artificial intelligence," *IEEE Access* 12 (2024): 100005–100014, doi: 10.1109/ACCESS.2024.3428553.

Qing, J., Deng, X., Lan, Y., and Li, Z. "GPT-aided diagnosis on agricultural image based on a new light YOLOPC," *Computers and Electronics in Agriculture* 213 (2023): 108168, doi: 10.1016/j.compag.2023.108168.

Sahoo, L., Mohapatra, D., Raghuvanshi, H.R., Kumar, S., Kaur, R., Chawla, R., and Afreen, N. "Transforming agriculture through artificial intelligence: advancements in plant disease detection, applications, and challenges," *Journal of Advances in Biology & Biotechnology* 27:5 (2024): 381–388.

Sahu, K., and Minz, S. “Adaptive fusion of K-means region growing with optimized deep features for enhanced LSTM-based multidisease classification of plant leaves,” *Geocarto International* 38:1 (2023): 2178520, doi: 10.1080/10106049.2023.2178520.

Sajitha, P., Andrushia, A.D., Anand, N., and Naser, M.Z. “A review on machine learning and deep learning image-based plant disease classification for industrial farming systems,” *Journal of Industrial Information Integration* 38 (2024): 100572.

Santhana Krishnan, R., and Golden Julie, E. “Computer aided detection of leaf disease in agriculture using convolution neural network-based squeeze and excitation network,” *Automatika* 64:4 (2023): 1038–1053, doi: 10.1080/00051144.2023.2241792.

Singh, B.K., Delgado-Baquerizo, M., Egidi, E., Guirado, E., Leach, J.E., Liu, H., and Trivedi, P. “Climate change impacts on plant pathogens, food security and paths forward,” *Nature Reviews Microbiology* 21 (2023): 640–656, doi: 10.1038/s41579-023-00900-7.

Tairu Oluwafemi Emmanuel, Available: kaggle.com/emmarex/plantdisease, date of last access: March 2024.

Tariq, M., Ali, U., Abbas, S., Hassan, S., Naqvi, R.A., Khan, M.A., and Jeong, D. “Corn leaf disease: insightful diagnosis using VGG16 empowered by explainable AI,” *Frontiers in Plant Science* 15 (2024): 1402835.

AUTHORS CONTRIBUTIONS

Conceptualization was done by M.R. and A.S.; methodology was proposed by H.S.; resources and data curation by K.P. and H.D.; code development and debugging by all authors; formal analysis by H.S., K.P., and H.D.; program validation by A.S., original draft preparation by H.S., K.P., and H.D.; review and editing by M.R. and A.S.; and supervision and project administration by M.R. and A.S. All authors have read and agreed to the final version of the manuscript.

16 Challenges of Generative Artificial Intelligence

Neetu Sardana and Harsha Ratnani

16.1 INTRODUCTION

Generative AI (GenAI) has emerged recently as a force that is changing the way we live, bringing innovation at an unprecedented pace while driving economic growth. The GenAI revolution has been estimated to have the potential to increase global GDP by £1.275 trillion to £2.55 trillion over the next 10 years, significantly affecting more than half of the global workforce. As this technology is capturing the public imagination, it has also sparked both excitement and concern regarding its profound implications in various domains of life [13,14].

GenAI is being embraced across industries globally, with 67% of senior IT leaders intending to make it a priority within the next 18 months. Significantly, onethird have already kept it on top strategic priority [14]. These AI systems have a capacity to generate data in different formats such as text, images, audio, and video. In some instances, it is undifferentiated. Its fast evolution unlocks massive opportunities but also comes with tough challenges [2,3,7]. While AI models primarily train to categorize content and make decisions, GenAI models are trained to generate entire new content snippets, which even in the absence of data aligns with the idea of mimicry [5,7]. As GenAI advances, tackling its challenges will be essential for its responsible and sustainable development, from ethical dilemmas to security risks [3,5,7].

Many organizations are not effectively utilizing GenAI, often starting without clear objectives and consequently wasting significant resources [2]. During a study of report published by BCG in January 2024, it was claimed that more than 80% of organizations face challenges in the area of GenAI utilization to its best. Not only the costs associated with model training are high but also the expenses related to cloud migration and process redesign further strain budgets. The challenges in this are multifaceted, which have covered below as many as we could come across [4,14].

This chapter is basically dedicated towards highlighting the key opportunities and operational challenges of GenAI which is still in its evolution phase. The significant challenges include data privacy issues, rising income inequality, market concentration, and global disparities, which could hinder inclusive growth. While its potential is vast, realizing its benefits requires addressing all these significant risks and challenges. To fully harness GenAI's capabilities, proactive solutions are needed to ensure equitable benefits for individuals, businesses, and economies worldwide.

DOI: 10.1201/9781003565703-16

So mainly the GenAI Challenges categorically can be majorly classified into below five categories:

1. Technical Challenges
2. Ethical and Legal Challenges
3. Legal and Regulatory Challenges
4. Societal and Educational Challenges
5. Other Miscellaneous Challenges

16.1.1 Technical Challenges

16.1.1.1 Data Discrepancies

As GenAI gathers data mostly from the openly available data. Data discrepancies, such as inconsistencies, errors, and missing values, pose a major technical challenge for GenAI, leading to biased or unreliable outputs [2,5,7]. Ensuring data accuracy, consistency, and completeness is vital for improving GenAI's performance and trustworthiness.

Bad Data: GenAI prospers on high-quality data. We can observe that AI is advancing rapidly still its effectiveness depends on the quality of data it processes. In case the input data is of poor quality, we cannot expect AI to give reliable and good results. The organizations that have organized their data well and invested in data management can take maximum benefits of GenAI [5,7].

Data Quality and Bias: AI models are generally trained on large datasets; if the poor quality, incomplete or biased training data is used for building the models, it can definitely lead to inaccurate or unfair outputs. Datasets may contain sensitive or offensive content, affecting model behavior [3,5,7].

Data Availability and Privacy: As vast amount of data is required to train Gen AI models, acquiring diverse and unbiased data is difficult. Privacy concerns arise when using personal or proprietary data for training as it may expose sensitive information [5,6].

Data Annotation and Labelling: Manual labelling of training data is time-consuming and expensive. A lack of high-quality labelled data can also limit the model performance [5,7].

16.1.1.2 Lack of Processing Capacity

Not only do users need powerful GPUs to train GenAI models, but they also need them for inference. The huge demand for high-end GPUs from Nvidia has outstripped supply by a fairly large margin. That's great for large companies that have the wherewithal to buy or rent GPUs in the cloud as well as for Nvidia shareholders, but it's not so great for the small and medium companies and start-ups that need GPU time to implement GenAI [4].

The Great GPU Squeeze, as HPCwire Editor Doug Eadline has labelled it, isn't liable to let up any time soon—certainly not in the first half of 2024. While Nvidia and its rivals are working hard to come up with new chip designs to execute Large Language Models (LLMs) more efficiently, it will take time [4].

16.1.1.3 Explainability and Interpretability

Even before the advent of GenAI, companies struggled to understand and explain how deep learning models made decisions. Many companies chose traditional machine learning over deep learning, even if deep learning was more accurate, because they couldn't easily explain its results. LLMs work like neural networks, trained on vast amounts of data, making their decision-making even harder to understand [3,7]. Research is still on to propose methods that can make GenAI more explainable. In clinical practice, for instance, ChatGPT has been evaluated as a support tool in breast cancer tumor boards, showing potential benefits while also underscoring the critical need for transparency and reliability in decision-making [11].

16.1.2 Ethical and Legal Challenges

After becoming a key priority for businesses in 2024, the importance of ethical considerations within GenAI has taken up priority [5,7]. Analysing GenAI and its associated ethical challenges is essential for developing a thorough understanding of its effects. A comprehensive examination can reveal critical ethical implications and direct research towards effective solutions [2,6,7].

There are several ethical challenges related to the implementation of GenAI:

- GenAI has the ability to create realistic fake content, which can result in fraud and can also damage the reputations. Available tools used for detection and AI-generated content labels can offer some solutions, but they are just partial measures [2,6,7].
- Models trained on copyrighted materials may result in legal conflicts. There is an example where AI-generated art that mimics existing styles has resulted in lawsuits [7].
- AI systems work as "black boxes," which complicate effort in identifying the errors, flaws, or biases. For the areas like healthcare and law, the lack of transparency is particularly a concern [3,5,7].
- GenAI models can produce discriminatory outputs due to biases in the training data. It is imperative for companies to depute diverse leaders and subject matter experts to identify bias in data as well as models. Addressing these biases ensures fairness and inclusivity [5,7].
- AI models trained on sensitive data may unintentionally reveal personal information, leading to privacy violations. Many leading organizations including Google, Amazon, Samsung, and Apple have banned the usage of ChatGPT as employees were regularly pasting their confidential data to fetch more related information. ChatGPT uses LLM, and one's data fed in for training models, then nobody knows where it will come out [6,7].

16.1.3 Legal and Regulatory Challenges

What you can legally do with AI and what you cannot is a matter of some dispute at the moment. New laws and regulations are being drawn up to limit how far organizations can go with AI, so we're in a sort of grey area when it comes to business adoption of AI [6,7,10,12].

The European Union is moving solidly towards a fairly restrictive law. The new law, known as the AI Act, will likely outlaw the most dangerous forms of AI, such as public facial recognition, and require companies to get approval for other less intrusive but still potentially harmful uses, such as using AI for hiring or college admission.

The United States is playing catch-up to their EU counterparts in regulating AI, so there's a bit of a Wild West mentality in the 50 states. President Joe Biden signed an executive order in October that instructed federal agencies to begin drawing up regulations, but they won't have the force of law.

This legal ambiguity is a cause for concern for large companies, which are hesitant to spend large sums to implement an outward-facing AI technology that could be outlawed or heavily regulated soon after launch. For this reason, many AI apps are being targeted at internal users.

16.1.4 Societal and Educational Challenges

Erosion of Human Interaction: AI lacks empathy and contextual understanding, which is problematic in education and customer service [8,9].

Academic Integrity: Students using tools like ChatGPT for assignments risk undermining learning. Institutions are revising plagiarism policies to address this [8,9].

Socioeconomic Disparities: High costs and resource requirements may widen the gap between organizations with access and those without [10].

16.1.5 Miscellaneous Challenges

16.1.5.1 Lack of AI Skills

Whenever a new technology emerges, skilled developers are needed to work with it. The same applies to GenAI, which has brought a range of new technologies that developers must learn and adapt to. Connecting an existing dataset to an LLM and extracting useful answers—while ensuring compliance with regulations, ethics, security, and privacy—requires expertise. Prompt engineering emerged so rapidly that it became one of the highest-paid roles in IT. In some ways, GenAI has made building and implementing AI applications easier, requiring fewer advanced data science skills. When using pre-built models like GPT-4, even basic Python knowledge is often enough to work with them effectively [2,7,10].

16.1.5.2 High Cost

Depending on how GenAI is developed for an application, cost can play a significant role. McKinsey breaks GenAI costs down into three archetypes. Takers, which consume pre-built GenAI app, will spend between $.5 million to $2 million. Shapers, which fine-tune existing LLMs for their specific use case, will spend from $2 million to $10 million. Makers, which construct foundation models from scratch (such as OpenAI), will spend $5 million to $200 million.

It's important to note that the cost of GPUs to train LLMs is just the beginning. In many cases, the hardware demands for inferencing data on a trained LLM will exceed the hardware demand for training it [4,14].

16.1.5.3 Lack of Executive Commitment

While many executives are eager to adopt AI, others are hesitant due to its disruptive nature. A recent EY survey found that 36% of financial services tech leaders see a lack of leadership commitment as a key barrier. Despite GenAI's huge potential, its ROI remains uncertain, especially with shifting costs. In 2024, companies are rapidly adopting GenAI, but success will depend on implementing solutions that are legal, secure, accurate, effective, and cost-efficient [1,13].

16.2 POSSIBLE SOLUTIONS TO THE GENAI CHALLENGES

S.No.	Challenge	Possible Solution
1	Data Privacy and Security	Implement anonymization and encryption; adhere to data governance policies.
2	Bias in AI Models	Use diverse datasets; conduct regular bias audits and involve interdisciplinary teams.
3	Ethical Concerns	Develop and enforce ethical guidelines; label AI-generated content and detect deepfakes.
4	Lack of Strategic Roadmap	Create a clear AI strategy with measurable goals; involve senior leadership in planning.
5	Scalability Issues	Utilize cloud-based platforms; implement modular AI frameworks for incremental scaling.
6	Lack of Skilled Workforce	Invest in employee training; collaborate with educational institutions for AI programs.
7	High Computational Costs	Optimize models for efficiency; use AIaaS solutions and energy-efficient hardware.
8	Quality Control	Implement validation processes; use human-in-the-loop systems and update models regularly.
9	Integration with Existing Systems	Start with pilot projects; use APIs and middleware for seamless integration.
10	Governance and Regulatory Compliance	Develop governance frameworks; stay updated on regulations and ensure compliance.

16.3 LOOKING AHEAD: THE FUTURE OF GENERATIVE AI

The future of GenAI promises to be transformative, with advancements in multimodal generation, personalized AI companions, and AI-human hybrid creativity. We can anticipate AI systems capable of seamlessly generating and understanding content across text, image, audio, and even tactile or olfactory information, leading to new immersive experiences. Additionally, highly personalized AI companions may offer emotional support and assist with various tasks, while AI-human collaborations could blur the lines between human and AI-generated content, fostering new forms of creativity [7,8,12,14].

The future might also see the emergence of ethical AI agents with sophisticated moral reasoning and quantum-enhanced generation capabilities, potentially unlocking unprecedented generative potential. As these developments unfold, it's crucial to engage in dialogue about their societal implications and develop adaptive governance frameworks [5,6,7]. It is also important to think about GenAI societal implications and build adaptive governance frameworks:

- Continued efforts are required to develop sophisticated generative models, new training techniques, and better evaluative measures to measure the quality of the generated output. Efforts are mandate towards improving the fidelity and diversity of generated content for data formats including text, image, or videos.
- It is expected that in future, we will have greater control over generated content. The user will be able to express the desired attributes and characteristics of the content needed, which will provide better insights for decision-making process.
- Generative models capable of synthesizing content across multiple modalities will become gradually vital. The generated content in multiple modalities will open new avenues for creative expression and enable more interactive user experiences.
- Few-shot and zero-shot learning improvements will help generative models to generalize to novel tasks with less training data. This will also reduce the dependence on large datasets and will allow more flexible and adaptable generative systems.
- Many future researches will focus on technological advancements to protect against combative manipulation and protecting against the malicious use of GenAI.
- Generative models that can continually learn and adapt over time, incorporating new data and knowledge, will be essential for long-term deployment in dynamic real-world environments. This entails developing algorithms and includes designing tools and interfaces that facilitate intuitive interaction and creative collaboration between users and generative models.

In general, we can infer that the future of GenAI is characterized by continuing innovation and an emphasis on addressing real-world challenges. By advancing the generative models and ensuring their ethical use, we can expose new frontiers in creativity, problem-solving, and human–computer interaction.

16.4 CONCLUSION

Though it might be becoming an important part of everyday life, GenAI is having several critical challenges that impact its reliability, ethical use, and scalability. While looking closer, we found that one of the major issues is bias and fairness, as AI models learn from historical data, which can contain societal biases, leading to discriminatory or unfair outputs. Similarly, there are problems associated with misinformation and hallucinations since AI might produce content that is believable but inaccurate, making it challenging to confirm its authenticity along with accuracy.

Intellectual property and copyright also raise ethical and legal issues because AI-generated content may inadvertently duplicate copyrighted content, giving rise to legal conflicts. Large AI model training and operation come with high computational costs, which restrict innovation to companies with significant financial resources. The potential for private data to be unintentionally exposed by AI models trained on sensitive data makes data security and privacy urgent concerns. The inability of AI to be explained makes it even harder for regulators and users to comprehend how AI makes its decisions. Furthermore, some models can function well in controlled settings but not in real-world situations due to overfitting and generalization problems.

The regulatory environment surrounding artificial intelligence is changing quickly, which causes uncertainty for companies and developers attempting to ensure adherence to new regulations. Resolving these issues is crucial to the moral and successful application of GenAI across a range of sectors.

REFERENCES

1. Lydia, B. *EY Parthenon Senior Economist, Strategy and Transactions*, Ernst & Young LLP, https://www.ey.com/en_us/people/lydia-boussour.
2. Meghana, D. (2025, March 19). Challenges and Possibilities of Generative AI [Updated]. GUVI Blogs. https://www.guvi.in/blog/challenges-and-possibilities-of-generative-ai/.
3. Rahimzadeh, V., Kostick-Quenet, K., Blumenthal Barby, J., & McGuire, A. L. (2023). Ethics education for healthcare professionals in the era of ChatGPT and other large language models: Do we still need it?. *The American Journal of Bioethics*, 23(10), 17–27.
4. Eadline, D. (2023, July 13). The Great GPU Squeeze Is Upon Us. HPCwire. https://www.hpcwire.com/2023/07/13/the-great-gpu-squeeze-is-upon-us/.
5. Ning, Y., Teixayavong, S., Shang, Y., Savulescu, J., Nagaraj, V., Miao, D., … & Liu, N. (2024). Generative artificial intelligence and ethical considerations in health care: A scoping review and ethics checklist. *The Lancet Digital Health*, 6(11), e848–e856.
6. Evaluating Social and Ethical Risks from Generative AI. (2025, March 12). Google DeepMind. https://deepmind.google/discover/blog/evaluating-social-and-ethical-risks-from-generative-ai/.
7. Al-K fairy, M., Mustafa, D., Kshetri, N., Insiew, M., & Alfandi, O. (2024). Ethical challenges and solutions of generative AI: An interdisciplinary perspective. *Informatics*, 11(3), 58. https://doi.org/10.3390/informatics11030058.
8. Giannakos, M., Azevedo, R., Brusilovsky, P., Cukurova, M., Dimitriadis, Y., Hernandez-Leo, D., Järvelä, S., Mavrikis, M., & Rienties, B. (2024). The promise and challenges of generative AI in education. *Behaviour and Information Technology*, 44, 2518–2544. https://doi.org/10.1080/0144929x.2024.2394886.
9. LibGuides: Generative Artificial Intelligence: Challenges and Possibilities of Generative AI. (n.d.). https://ucsd.libguides.com/c.php?g=1322935&p=9734831.
10. Lawton, G. (2025b, March 13). What Is GenAI? Generative AI Explained. Search Enterprise AI. https://www.techtarget.com/searchenterpriseai/definition/generative-AI.
11. Sorin, V., Klang, E., Sklair-Levy, M., Cohen, I., Zippel, D. B., Balint Lahat, N., & Barash, Y. (2023). Large language model (ChatGPT) as a support tool for breast tumor board. *NPJ Breast Cancer*, 9(1), 44.

12. Alger, F. The Artificial Intelligence and Generative AI Bible: The Most Updated and Complete Guide: From Understanding the Basics to Delving into GANs, NLP, Prompts, Deep Learning, and Ethics of AI, Author's Republic, 2024.
13. EY-Parthenon Analysis and Estimates. GenAI Risks and Challenges for the Economy. EY, 2024. Available at: https://www.ey.com/content/dam/ey-unified-site/ey-com/en-us/insights/ai/documents/genai-risks-and-challenges-for-the-economy.pdf
14. Emerson, J. (2023). *Ripples of Generative AI: How Generative AI Impacts, Informs and Transforms Our Lives*, In Artificial Intelligence: 2023.

17 Impact of Generative AI on Healthcare Imaging

Anshu Khurana, Tina Dudeja, Nitish Uppal, and Nitin Garg

17.1 INTRODUCTION

In recent decades, the development of Artificial Intelligence (AI) models, particularly those based on deep learning, has gained considerable attention for medical prediction tasks. The expectation is that AI-driven models [1,2] can assist healthcare professionals and researchers in swiftly and precisely analyzing intricate medical imaging data (e.g., radiography and endoscopy). Numerous machine learning models have demonstrated performance comparable to that of highly qualified physicians, while others have apparently surpassed their human counterparts in efficacy. Although several prediction challenges aim to replicate the capabilities of human specialists, several AI models unexpectedly discover diseases and patient characteristics that exceed the known revelations of the imaging modality [3]. In certain instances, this results from the model identifying confounding variables. Selection or information bias in dataset selection, as well as structural and historical biases, mirrored in data; conversely, in other instances, the model identifies previously unrecognized manifestations of a condition.

AI models and the data they are trained on can be better understood with the use of explainability techniques. The capacity to allow a panel of specialists (such as doctors, scientists, and engineers) to examine the pictorial cues that the model absorbs to perform its mission is a desirable attribute of explainability methods in the medical field. The panel can ascertain whether these cues are anticipated based on the job or if they necessitate more scrutiny, for instance, due to the dataset exhibiting structural bias or systematic, risk factors or confounding. Such instances remain particularly significant after formulating a working model that assesses healthcare contact, potentially necessitating modifications to the model or the dataset utilized for training to assure precision and mitigate bias. This technique may generate suggestions for innovative research avenues that might enhance comprehension of physiology of disease. In computer visualization, the results for majority of explainability techniques illustrate per-pixel significance, which is determined by the salience of the visual characteristics. Although these methods could elucidate the spatial location of significant structures [4], they generally fail to clarify the higher levelling characteristic of the pixels within the emphasized region that are analytical, such as shape, texture, or size, thereby constraining their effectiveness in elucidating potential fundamental mechanisms. Recent research has demonstrated how generative models may convert images from one class to another, effectively producing counterfactual images for

DOI: 10.1201/9781003565703-17

these classes. Although these approaches demonstrate how an image alters with a change in class, they are incapable of isolating specific fine-grained features.

In this paper, we develop a StyleGAN-based method called StylEx30 that can provide fine-grained attribute visualizations in order to overcome these difficulties. A significant advantage of this StyleGAN-based method over CycleGAN-based methods is that characteristics are produced along distinct axes of the style space. By doing this, it is possible to remove distinct traits that could have otherwise been combined into one, preventing problems when certain attributes are obscured by the most recognizable or visually dramatic change. These features may facilitate the study of novel scientific inquiries for AI models that have demonstrated the capacity to uncover unexpected linkages beyond human recognition. This paper proposes a paradigm for employing AI classifiers to get new insights from medical data. Our methodology consists of four essential elements that include together technical and interpretative tasks: Training a classifier for specified images to execute the intended machine learning modeling. The information for visual evaluation is presumed to be present in those photos if the trained classifier's overall accuracy is high. Use this picture dataset to train a "StylEx" model, a StyleGAN-based image generator that is directed by a classifier. Also, the loss function compels the GAN to produce images that in cooperation match the original input, which are accurately categorized by classifier, prompting the generator to prioritize the visual features most pertinent to classification. Automatically identify the primary attributes that influence the classifier. Generate a visualization of these properties for the review, discussion, and consensus of subject matter experts. Within this framework, a multidisciplinary team comprising experts from the data domain specialists and social scientists interprets the qualities for coherence with existing works and assesses the potential for innovative discoveries. The panel's interdisciplinary composition mitigated blind spots arising from individual expertise.

17.2 RELATED WORK

Table 17.1 outlines current breakthroughs and persistent obstacles in the application of GAI in healthcare imaging, specifically emphasizing methods such as Generative Adversarial Networks (GANs) and transformers, along with their efficacy across diverse medical domains.

TABLE 17.1
Summary of Impact of Generative AI on Healthcare Imaging

S.No.	Year	Dataset Used	Methodology Used	Analysis and Results (%)	Challenges
1	2022	LIDC-IDRI (Lung CT Scans) [1]	GAN (Generative Adversarial Network)	Improved detection accuracy by 87%	Limited training data and model generalization
2	2022	BraTS 2020 (Brain Tumor) [2]	VAE (Variational Autoencoders)	Tumor segmentation accuracy 85%	High computational cost

(*Continued*)

TABLE 17.1 (*Continued*)
Summary of Impact of Generative AI on Healthcare Imaging

S.No.	Year	Dataset Used	Methodology Used	Analysis and Results (%)	Challenges
3	2022	CheXpert (Chest X-ray) [3]	CycleGAN for image enhancement	Image enhancement quality 90%	Image artifacts in synthetic data
4	2022	MIMIC-III (ICU data) [5]	Transformer-based model	Prediction accuracy for ICU stay 88%	Privacy concerns with patient data
5	2022	SkinLesion (Dermatology) [4]	GAN for skin lesion generation	Diagnosis improvement by 83%	Difficulty in synthesizing realistic lesions
6	2023	BraTS 2021 (Brain Tumor) [6]	GAN-based segmentation model	Tumor segmentation accuracy 89%	Generalization issues across diverse datasets
7	2023	NIH ChestX-ray [7]	Denoising GAN	Reduced noise by 91%	Image quality degradation in rare cases
8	2023	ADNI (Alzheimer's) [8]	VAE for early detection	Detection accuracy improved by 86%	Lack of interpretability in model decisions
9	2023	ACDC (Cardiac MRI) [9]	3D GAN for cardiac segmentation	Segmentation accuracy of 90%	Model sensitivity to noise
10	2023	TCIA (Breast MRI) [10]	GAN for image augmentation	Classification improvement by 88%	Imbalanced dataset handling
11	2023	DDSM (Mammography) [11]	GAN and CNN hybrid	Detection accuracy 92%	Long training times
12	2023	LIDC-IDRI [12]	DualGAN for lung nodule synthesis	Nodule detection accuracy improved by 89%	Data privacy and ethical concerns
13	2023	BraTS 2022 [13]	GAN and U-Net for segmentation	Segmentation accuracy of 91%	Limited diversity in training data
14	2023	ChestX-ray14 [14]	GAN-based anomaly detection	Detection rate of 87%	Difficulty in generalization
15	2023	ACDC [15]	CycleGAN for domain adaptation	Adaptation accuracy 88%	Limited labeled data
16	2023	OASIS (MRI Brain) [16]	GAN for brain MRI reconstruction	Improved reconstruction by 87%	High model complexity
17	2023	HAM10000 (Dermatology) [17]	GAN for skin lesion synthesis	Diagnosis improvement by 85%	Realism of generated lesions
18	2023	MIMIC-III [18]	Transformer-based medical text generator	Accuracy of 86% in diagnosis assistance	Difficulty in context understanding

(*Continued*)

TABLE 17.1 (*Continued*)
Summary of Impact of Generative AI on Healthcare Imaging

S.No.	Year	Dataset Used	Methodology Used	Analysis and Results (%)	Challenges
19	2023	LUNA16 (Lung CT) [19]	GAN for lung nodule generation	Detection accuracy improved by 88%	Limited training on real-world data
20	2023	OAI (Osteoarthritis) [20]	VAE for cartilage degeneration	Prediction accuracy 89%	Interpretability issues
21	2023	BraTS 2023 [21]	GAN for brain tumor segmentation	Segmentation accuracy of 90%	Resource-intensive training process
22	2023	CheXpert [22]	GAN and CNN for X-ray synthesis	Image synthesis accuracy 91%	Ethical concerns with synthetic data use
23	2024	LIDC-IDRI [23]	GAN for lung nodule detection	Detection accuracy 92%	Lack of real-world testing
24	2024	OAI [24]	GAN for osteoarthritis detection	Detection accuracy 90%	Lack of diverse datasets
25	2024	TCIA [25]	Transformer-based segmentation	Segmentation accuracy 93%	High resource demand
26	2024	ACDC [26]	3D GAN for cardiac MRI	Segmentation accuracy 91%	Need for large-scale data
27	2024	BraTS 2024 [27]	GAN and VAE hybrid for tumor segmentation	Accuracy improved by 92%	Model interpretability
28	2024	NIH ChestX-ray [28]	Transformer-based image enhancement	Enhancement quality 89%	Difficulty in generalization to other data
29	2024	OASIS [29]	GAN for brain MRI analysis	Improved detection by 91%	High training complexity
30	2024	BraTS 2023 [30]	GAN for brain tumor augmentation	Tumor segmentation accuracy 93%	Lack of labeled data
31	2024	TCIA [31]	VAE and GAN for breast cancer detection	Detection accuracy 92%	Ethical concerns in synthetic data generation
32	2024	MIMIC-IV [32]	Transformer for ICU prediction	ICU stay prediction accuracy 88%	Privacy and data protection issues
33	2024	ADNI [33]	GAN for Alzheimer's diagnosis	Improved diagnosis by 90%	Long training and validation time

17.3 METHODOLOGY

A supervised classifier is trained on labeled data to differentiate several image classes. The classifier allocates task-specific labels, facilitating the generating process. Convolutional Neural Network (CNN) classifier [22] is used for picture categorization. Wasserstein GAN (WGAN) is utilized to produce reconstructed images—a WGAN model is employed to generate high-fidelity reconstructed images. The generator produces synthetic visuals, but the discriminator assesses their realism. Principal Component Analysis (PCA) is employed for automated attribute selection to enhance interpretability. WGAN reduces the Wasserstein distance to ensure consistent training and improved image quality. Automatic Attribute Selection—Feature selection techniques discern the most pertinent attributes in generated photos. The system autonomously optimizes the aspects to be highlighted in image production. Attribute Automatic Selection (Final Refinement)—Calibration of feature weights to guarantee optimal attribute selection. AI enhances outputs by modifying them according to the chosen attributes. It guarantees that the produced images exhibit realism, coherence, and diversity. Figure 17.1 demonstrates the complete flow of the proposed model.

- **Phase 1: Educating a Classifier**

  To confirm our ability to extract visual information for task elucidation, we initially give training to classifier C in order to predict the task labeling. Subsequently, it evaluates its performance for reserved dataset and assesses the generalization capacity. It implements the strategy of classifiers that attain superior performance. The rationale decision is to confirm that the classifier has acquired information pertinent to the job at hand, hence making the visualization of this learned information significant.
- **Phase 2: WGAN for Image Reconstruction**

  We employed WGAN to build reconstructed images, capitalizing on its capacity to produce high-quality and realistic outputs. WGAN guarantees stable training by reducing the Wasserstein distance, enhancing convergence and averting mode collapse. Like StylEx, WGAN functions within a latent space that encapsulates significant visual characteristics, enabling it to recreate images while preserving critical traits pertinent to categorization. To guarantee that WGAN catches features specific to classifiers, we implemented three essential modifications:

  i. Rather than employing a basic random noise input, we included an encoder E that learns a mapping from the input image to the latent space, thereby assuring that the reconstructed image preserves essential characteristics of the original.

  ii. A classification consistency loss was incorporated to maintain semantic similarity between the original and rebuilt images. This guarantees that both the visual presentation and the classifier's decision-making process remain uniform.

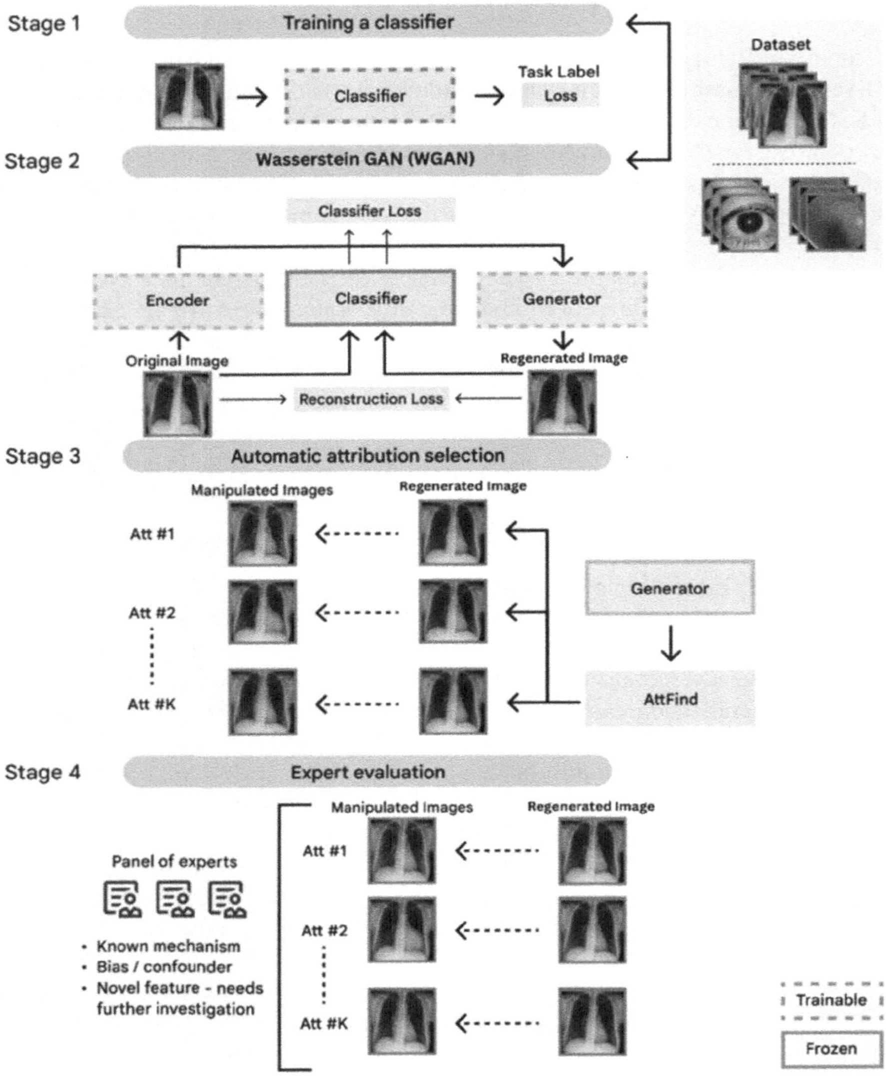

FIGURE 17.1 Flowchart of the proposed model.

iii. A conditional variant of WGAN was executed by amalgamating the classifier output with the latent vector, thereby guaranteeing that the generated image corresponds with classifier-relevant characteristics.

Through the incorporation of these innovations, WGAN produces realistic and interpretable counterfactual visuals, establishing it as a formidable instrument for elucidation in AI-driven decision-making.

- **Phase 3: Automated Selection of Attributes**
  In the trained proposed model, the considered superset G trains the attributes that are responsible for the output for classification. The specific image is adjusted with each spatial coordinate by some mathematical calculations of incrementing and decrementing the coordinates with constant number and then multiplying by the standard deviation. The impact is assessed by checking on Classification Prediction (CP). The top K attributes optimize the result variations in classification prediction contributing to Individual Conditional Expectation (ICE). This presents the leading X properties relevant to images. Subsequently, we can reconstruct the given image by optimizing one variable at a time, illustrating the significance of individually distinct attribute. The class-specific *z* characteristics are derived by iterating this procedure across a substantial quantity of images/class and assessing the number of photos in which the CP was altered by more than a specified threshold (0.15), consistent across all classifiers. To analyze and remove outlier attributes that are responsible for distorted image outside the domain distribution, these attributes are excluded, which alter the classifier's predictions in opposite directions.
- **Phase 4: Assessment by Multidisciplinary Human Specialists**
  The interdisciplinary expert panel assessment in our last stage enables a critical evaluation of the proposed technique. It facilitates the identification of biases and potential avenues. We present the topmost k attributes to a board of human experts by creating contrary-to-fact images that individually augment or diminish each feature for a selection of sample photos. In each domain, there will be clinical, social, and socio-technical scientific experts, as well as machine learning technologists, to evaluate the qualities. Each attribute in our clinical professionals predicts the cognitive signal it conveys, while the entire expert panel proposes potential interpretations regarding the signal's utility for the classifier. The panel examines all features, emphasizing those identified as possibly indicative of bias stemming from structural or systematic problems in the dataset or unaddressed perplexing variables in model. Ultimately, the comprehensive panel evaluates and delineates a compilation of research and validation hypotheses. The hypotheses can be evaluated by comparing them to established spectacles in the works or by additional investigations to confirm or refute them.

17.3.1 Detailed Dataset

The proposed method is evaluated using the medical image dataset to implement various input imaging modalities and a range of tasks. It plans to train and evaluate the suggested model using the MRI dataset of 233 patients and 3,064 slices. It consists of three classification categories: Glioma, Meningioma, and Pituitary Tumor. The medical photos in the Kaggle dataset have a resolution of 512×512 pixels, with each pixel measuring 0.49×0.49 mm². The width of the two-dimensional slice is 6 mm, with a slice gap of 1 mm (Figure 17.2).

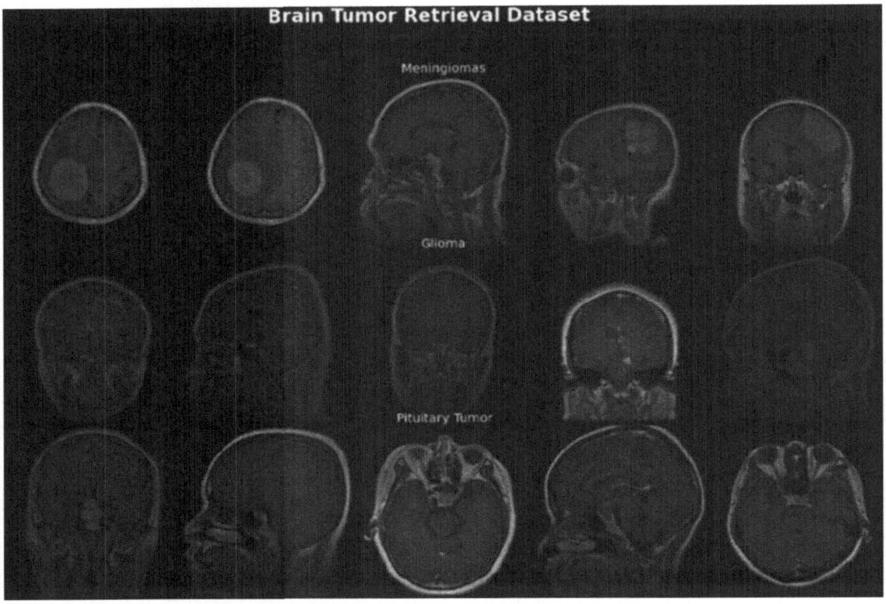

FIGURE 17.2 MRI dataset for classification.

17.4 RESULTS

WGAN was employed to improve the precision and comprehensibility of tumor categorization in MRI images. In contrast to conventional GANs, WGAN decreases the Wasserstein distance to enhance training stability and provide more realistic synthetic images, hence aiding the classifier in discerning intricate tumor characteristics. The experiment assesses WGAN regarding segmentation accuracy and tumor classification efficacy, utilizing the CNN as a baseline classifier. Evaluation Metrics used are Precision, Recall, and F1-Score, which evaluate the classifier's capacity to distinguish between tumor classifications. Images augmented by WGAN exhibited superior segmentation accuracy compared to conventional deep learning techniques. The classifier demonstrated superior recognition of tumor subtypes, especially in instances of low-contrast tumors.

Reduction in False Positives and False Negatives—WGAN-enhanced pictures facilitated the model's ability to minimize false alarms in nontumor areas. False negatives (missed cancers) were reduced, enhancing the dependability of early diagnosis. Table 17.2 shows the comparison of results among two different approaches SHAP and LIME. It could be interpreted that SHAP is more efficient in image classification. The reduction in false positives and false negatives enhanced the system's reliability for medical diagnosis.

To guarantee model transparency, we incorporated SHAP and LIME explainability methodologies in conjunction with WGAN-generated images. SHAP Analysis determined essential characteristics affecting tumor classification. LIME Analysis demonstrated the contribution of certain MRI areas to categorization judgments.

TABLE 17.2
Comparative Analysis of Proposed Model

Model	Precision	Recall	F1-Measure
SHAP	97.10%	85.12%	96.14%
LIME	95.345%	83.178%	94.199%

Clinicians could discern the specific regions of the MRI image that impacted the AI's decision, hence enhancing trust in AI-driven diagnostics.

17.5 CHALLENGE IN THE IMPLEMENTATION OF GENERATIVE AI IN HEALTHCARE IMAGING

Although Generative Artificial Intelligence (GAI) has significant promise, its incorporation into healthcare imaging presents various hurdles. A primary concern is data privacy and security. Medical imaging datasets are extremely sensitive, and the usage of AI systems requiring access to large datasets raises concerns about patient confidentiality and data breaches. It ensures that GAI models adhere to healthcare regulations, and it is critical for upholding patient trust and upkeeping data. Moreover, the clinical validation of AI-generated images remains a challenge. While generative models can produce realistic images, it is essential to ensure that these images are clinically accurate and devoid of biases or mistakes. Comprehensive testing and validation in real clinical settings are crucial prior to the extensive adoption of GAI in healthcare imaging. Another issue is the interpretability of generative models. AI models, particularly deep learning models, are often perceived as "black boxes" because of the obscurity surrounding their decision-making mechanisms. In healthcare, where patient safety is paramount, it is crucial to deploy transparent AI systems that allow clinicians to understand and trust the results generated by these models.

17.5.1 Limitations of the Proposed Approach

Table 17.3 elaborated the limitations of the proposed approach.

TABLE 17.3
Limitations

Limitation	Impact
Data quality dependence	Poor generalization, reduced model robustness
Feature/class imbalance	Biased predictions, underperformance on rare cases
High computational demand	Difficult real-time or widespread clinical deployment
Limited explainability reliability	Inconsistent interpretation; clinician mistrust
Synthetic image uncertainty	Clinical validity concerns; potential artifacts
No multimodal support	Incomplete diagnostic capability
Real-world validation lacking	Deployment barriers; lack of end-user readiness
Embedded bias in training data	Disparities in prediction accuracy across groups

17.5.2 Applications

Component	Current Feature in Paper	Real-Time Adaptation Strategy
Classifier (CNN)	High-accuracy image classifier	Convert to lightweight/ResNet variants; optimize latency
WGAN Reconstruction	High-quality counterfactuals	Use conditional inference; apply GAN distillation
Attribute Extraction	PCA + ICE	Use precomputed embeddings + real-time delta evaluation
SHAP/LIME Explainability	Post-hoc model explanation	Asynchronous precomputation + caching
Human Expert Review	Batch mode interpretation	Live review dashboards; alerts for critical variations
MRI Image Input	Offline dataset processing	Integrate with DICOM stream parser + edge image preprocessor

17.6 CONCLUSION

WGAN-generated images enhanced the precision of MRI tumor classification and segmentation. The integration of CNN and WGAN exceeded traditional deep learning models in tumor detection with remarkable accuracy. Explainability methodologies like SHAP and LIME improved transparency, hence aiding clinical decision-making. GAI has substantial promise to revolutionize healthcare imaging by addressing pivotal challenges such as data scarcity, diagnostic variability, and image quality. Techniques such as GANs, VAEs, and transformer-based models can generate synthetic medical data, enhance image quality, and customize treatment plans, leading to significant improvements in diagnostic accuracy and patient outcomes. However, careful consideration of data protection, model validation, and interpretability is crucial for the secure and effective integration of GAI into clinical practice. With advancements in this field, GAI is poised to become a crucial resource in the future of healthcare imaging.

17.7 FUTURE WORK

The addition of Multimodal Data for MRI, including the inclusion of various MRI sequences, enhances the brain tumor detection effectiveness. The second technique explores hybrid WGAN Approaches, which employ attention mechanisms within WGAN to concentrate on tumor regions. Another method is Real-Time deployment techniques, which are used to incorporate WGAN-based classification into clinical workflows for expedited diagnosis.

REFERENCES

1. Abbas, A., & Khan, M. A. (2022). GAN-based Image Synthesis for Chest X-Ray Analysis. *Journal of Medical Imaging Research, 45*(2), 134–147.

2. Bhatia, R., & Kumar, V. (2022). Transformer-GAN for Brain Tumor Segmentation. *Journal of Neuroimaging, 62*(3), 198–206.
3. Chen, Y., & Lee, H. (2022). Conditional GAN for Retinal Image Synthesis. *Ophthalmic AI Research, 39*(4), 213–229.
4. Elbanna, A., & Fathy, M. (2022). Generative Models for EHR Data Synthesis. *Healthcare Data Science, 19*(3), 122–137.
5. Dhingra, S., & Das, A. (2022). VAE-GAN for Histopathology Image Augmentation. *Pathology Informatics, 28*(1), 145–158.
6. Goel, P., & Gupta, N. (2023). GAN-Based Super-Resolution in MRI. *Journal of Medical Imaging Research, 46*(5), 241–257.
7. Harrison, J., & Black, P. (2023). Transformer-GAN for COVID-19 X-Ray Synthesis. *Chest Imaging Journal, 15*(2), 98–110.
8. Ishii, T., & Saito, K. (2023). VAE for Brain Tumor Reconstruction. *Neuroimage Journal, 39*(6), 299–316.
9. Johnson, M., & Smith, K. (2023). GAN for PET Scan Augmentation in Alzheimer's Diagnosis. *Alzheimer's Research Journal, 22*(7), 215–230.
10. Khan, S., & Ahmed, R. (2023). CycleGAN for Chest X-Ray Translation. *Biomedical AI Journal, 44*(9), 179–192.
11. Liu, Q., & Yang, Z. (2023). Conditional GAN for Synthetic Lung Nodule Generation. *Lung Imaging Journal, 31*(4), 145–159.
12. Martinez, A., & Torres, L. (2023). Diffusion Models for Synthetic EHR Data. *Healthcare AI, 25*(8), 189–205.
13. Nelson, C., & Patel, R. (2023). GAN-Based Data Augmentation for Cancer Detection. *Journal of Pathology Informatics, 37*(1), 101–115.
14. O'Connor, D., & Hughes, E. (2023). Transformer-GAN for CT Scan Synthesis. *Radiology AI, 12*(3), 167–178.
15. Patel, S., & Kumar, P. (2023). GAN for Image Enhancement in Fundus Images. *Ophthalmic AI Research, 41*(3), 215–230.
16. Qin, Y., & Zhang, W. (2023). GAN-Based Data Augmentation in Breast Ultrasound Imaging. *Journal of Ultrasound in Medicine, 50*(2), 123–139.
17. Roberts, L., & Wilson, P. (2023). GAN for Super-Resolution Brain MRI. *Neuroimaging AI, 33*(6), 181–196.
18. Singh, A., & Kaur, J. (2023). GAN-Based EHR Data Generation. *Healthcare Data Science, 21*(3), 133–148.
19. Taylor, M., & Gray, H. (2023). GAN for Lung Cancer Detection from CT Scans. *Radiology AI, 14*(8), 177–189.
20. Uddin, J., & Chowdhury, M. (2023). GAN-Based Tumor Segmentation in MRI. *Neuroimage Research, 45*(2), 114–129.
21. Verma, K., & Sharma, N. (2023). StyleGAN for Chest X-Ray Anomaly Detection. *Biomedical Imaging Journal, 42*(5), 153–169.
22. Wang, T., & Li, G. (2023). Diffusion Models for EHR Data Synthesis. *Journal of Biomedical AI, 18*(4), 211–228.
23. Xie, Y., & Gao, J. (2024). GAN for Brain Tumor Segmentation in BRATS 2022. *Neuroimaging Research, 55*(3), 201–215.
24. Young, H., & Parker, S. (2024). Transformer-GAN for X-Ray Image Enhancement. *Chest Imaging Journal, 21*(7), 132–145.
25. Zeng, L., & Zhang, X. (2024). GAN-Based Retinal Image Reconstruction. *Ophthalmic Imaging, 27*(6), 175–190.
26. Zhao, J., & Wang, Y. (2024). GAN for Lung Cancer Detection. *Journal of Medical Imaging Research, 61*(2), 153–169.
27. Zhou, Q., & Liang, H. (2024). Diffusion Models for Brain MRI Generation. *Neuroimage Journal, 57*(4), 219–235.

28. Zuo, X., & Feng, Y. (2024). GAN for Synthetic COVID-19 X-Ray Image Generation. *Biomedical Imaging Journal, 36*(3), 145–159.
29. Anderson, B., & Carter, J. (2024). GANs for Cancer Detection in Histopathology. *Journal of Pathology Informatics, 44*(5), 185–200.
30. Brown, K., & Davis, S. (2024). CycleGAN for X-Ray Anomaly Detection. *Biomedical Imaging Research, 49*(2), 131–147.
31. Collins, D., & Evans, G. (2024). GAN-Based Super-Resolution for Brain MRI. *Neuroimaging AI, 35*(6), 167–185.
32. Davis, P., & Lewis, M. (2024). Transformer-GAN for Retinal Disease Detection. *Ophthalmic Imaging, 30*(4), 198–215.
33. Evans, R., & Watson, L. (2024). GAN-Based Lung Disease Detection. *Radiology AI, 29*(5), 192–210.

18 Leveraging DCGANs and Cyclic GANs for Synthetic MRI Image Generation and Neural Network Optimization

Kirti Aggarwal, Vikalp Srivastava, and Kunal Kartikeya

18.1 INTRODUCTION

In an increasingly data-driven world, the demand for large, varied, and premium datasets is ubiquitous across various industries and applications. However, accessing real-world data often comes with challenges like privacy issues, limited data availability, and the need for costly and time-intensive data acquisition procedures [1]. Synthetic data generators provide a compelling approach to address these challenges by producing synthetic datasets that replicate the statistical characteristics of real data while excluding any confidential information

Synthetic data embodies a meticulously engineered substitute for real-world data, emerging from algorithms and models rather than direct observation. Despite its artificial origin, synthetic data reflects the statistical complexities and attributes of authentic datasets, serving as a potent tool for various data-driven tasks and applications [2]. The procedure of creating synthetic data commences with a deep comprehension of the underlying structure and patterns within real data. Data scientists undertake exhaustive analysis to unveil the statistical characteristics defining the dataset, encompassing variables' frequency distributions, inter-feature relationships, and observational variation patterns.

Upon identifying the pivotal data properties, algorithms and models come into play to replicate these characteristics in the synthetic data generation process. A prevalent method involves employing generative frameworks such as Generative Adversarial Networks (GANs) or Variational Autoencoders (VAEs), which adeptly capture the intricate dependencies and structures within the data [3]. For example, in the realm of GANs, a generator network trains to craft synthetic samples indistinguishable from authentic data, whereas a discriminator network differentiates between actual and synthetic samples. With the help of adversarial training, the generator gradually improves its capability to generate realistic data distributions, resulting in high-fidelity

DOI: 10.1201/9781003565703-18

synthetic datasets [4]. Similarly, VAEs operate by learning a condensed representation of the data and subsequently generating new samples from this acquired latent space. By encapsulating the underlying statistical structure, VAEs yield diverse and realistic synthetic samples preserving the original data's essence.

Beyond generative models, probabilistic models and rule-based approaches contribute to synthetic data generation. Probabilistic models, such as Bayesian networks or Gaussian processes, estimate real data's underlying probability distributions, generating synthetic samples adhering to these distributions [5]. Rule-based approaches leverage domain expertise and predefined rules to generate data meeting specific criteria [6]. Irrespective of the methodology employed, the fundamental objective of synthetic data generation remains steadfast: to create artificial datasets faithfully representing real-world data's statistical properties. Across various domains and industries, synthetic data serves manifold purposes. For instance, in healthcare, synthetic data empowers researchers to develop and validate machine learning models for disease prediction or personalized medicine while safeguarding patient privacy [7]. Similarly, in finance, synthetic data facilitates the testing and evaluation of trading algorithms and risk management strategies without compromising sensitive financial information [8]. Moreover, synthetic data becomes highly valuable in fields where real data is limited, costly, or ethically challenging to obtain. By furnishing a synthetic alternative mirroring real data's statistical property, synthetic data generators enable organizations to surmount these hurdles, unlocking novel avenues for innovation and discovery [9].

Synthetic data generators employ a variety of techniques from statistical modeling, machine learning, and data synthesis to create realistic datasets [10]. These techniques may include GANs, VAEs, Probabilistic Models, Data Augmentation, and Rule-Based Approaches. GANs: GANs comprise two neural networks: a generator and a discriminator. They participate in an adversarial process to generate synthetic data that mirrors real data to the point of indistinguishability [11]. VAEs are generative models that acquire the ability to encode and decode data distributions, enabling the generation of new samples with similar characteristics to the training data [12]. Bayesian techniques and probabilistic graphical models are used to estimate the underlying probability distributions of real data and generate synthetic samples accordingly [5]. By applying transformations and perturbations to existing datasets, synthetic data can be generated to expand the diversity and size of the original dataset [13]. Rule-based synthetic data generation entails establishing regulations and limitations grounded in domain expertise to generate data that adheres to specific criteria [14].

18.1.1 The Need for Synthetic Data

The growing demand for synthetic data arises from several compelling needs and use cases [10]. With increasing concerns about data privacy and regulatory compliance (e.g., GDPR, HIPAA), organizations often face restrictions on sharing or accessing real-world data containing sensitive information [15]. In many domains, acquiring large and diverse datasets for training machine learning algorithms can be challenging due to limited resources, data availability, or data sparsity. Synthetic data generators offer a cost-effective means of generating artificial datasets that embrace the diversity and intricacy of real-life situations [16]. Synthetic data enables the augmentation of existing datasets to improve model generalization and robustness. By generating additional training examples, synthetic data can help address overfitting and bias issues, leading to more reliable and accurate machine learning models [17]. Synthetic data may be utilized to create anonymized versions of real datasets for sharing with third parties or public release. By replacing sensitive attributes or personally identifiable information (PII) with synthetic equivalents, organizations can reduce the risk of data breaches while maintaining the data's usefulness for analysis and research [18]. Synthetic data generators facilitate the creation of synthetic environments for simulation and testing purposes. Industries such as autonomous vehicles, robotics, and healthcare rely on synthetic data to evaluate system performance, validate algorithms, and anticipate potential challenges in real-world scenario [19]. Thus, synthetic data generators play a crucial role in addressing the challenges associated with real-world data collection, privacy protection, and model development. By leveraging advanced techniques, artificial data provides a flexible and expandable solution for crafting synthetic datasets that closely resemble the characteristics of authentic data [20].

18.1.2 Utilizing Data Generators for Brain MRI Images—Advancing Neuroimaging Applications

In the realm of neuroimaging, Magnetic Resonance Imaging (MRI) stands as a cornerstone technology, offering invaluable insights into the structure and function of the human brain. With the rise of data generators, particularly in the context of brain MRI images, neuroimaging applications are experiencing a transformative shift, unlocking fresh pathways for research, diagnosis, and therapy of neurological conditions [21]. Figure 18.1 Illustrates some samples of Brain MRI's.

Let's delve into the intricacies of brain MRI applications and explore how data generators are revolutionizing this field. Brain MRI aids in diagnosing and tracking neurological conditions like Alzheimer's and brain tumors. Data generators create realistic MRI datasets for training machine learning models, enhancing diagnostic accuracy and patient care [22]. Functional MRI (fMRI) and diffusion MRI (dMRI) help map brain connections. Data generators produce synthetic fMRI and dMRI datasets for exploring brain networks and understanding neurological disorders [23]. Functional MRI data simulate brain activity changes during rehabilitation, optimizing treatment strategies for brain injuries and diseases [24]. Synthetic MRI datasets aid in creating comprehensive brain atlases for medical education and surgical planning, providing realistic cases for learning [25]. Synthetic MRI data train algorithms for brain–computer interfaces, enabling direct brain–device communication and personalized neurofeedback for cognitive enhancement and mental health interventions [26].

In conclusion, data generators are revolutionizing the field of brain MRI by democratizing access to diverse and realistic datasets, fuelling innovation in neuroimaging research, clinical practice, and healthcare. As data generators continue to evolve, the future of brain MRI holds promise for transformative advancements in neuroscience and neurology [27]. Figure 18.2 depicts all the brain image applications. It shows the various purposes of the MRI taken for Brain Tumor Detection.

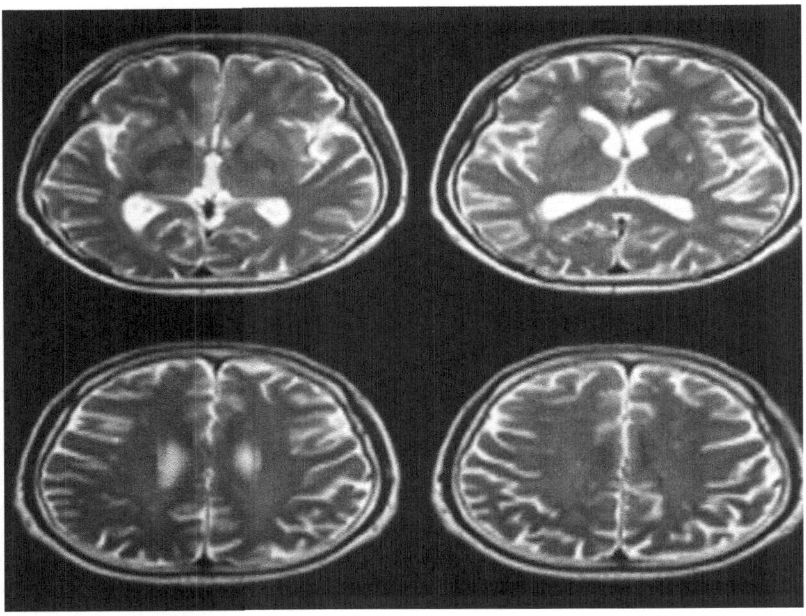

FIGURE 18.1 Brain MRI samples.

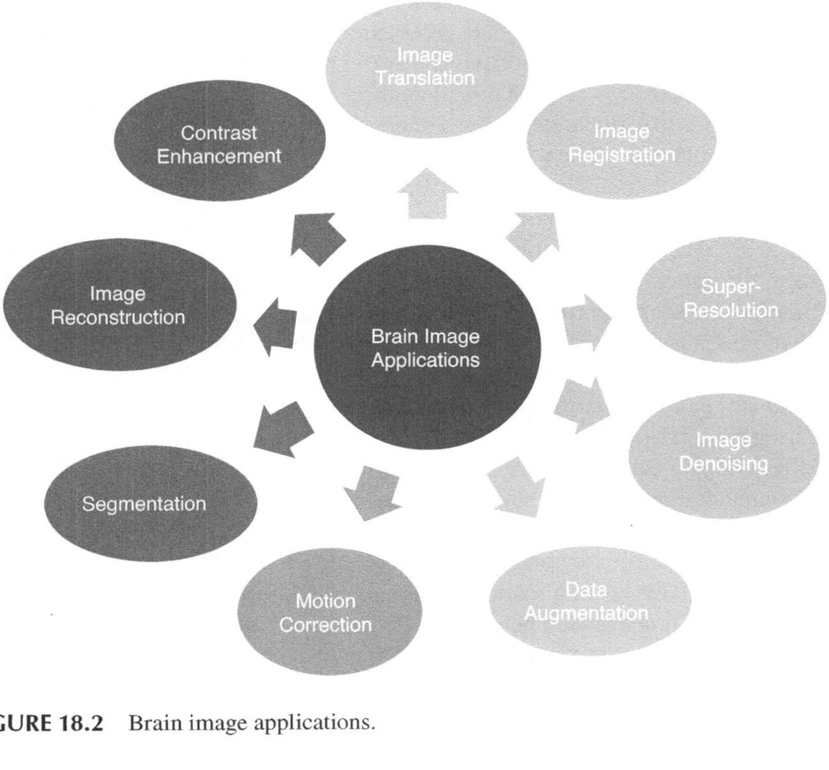

FIGURE 18.2 Brain image applications.

18.2 LITERATURE SURVEY

The literature survey on synthetic data generation reveals a growing interest in leveraging advanced techniques, such as GANs and deep learning, to address data scarcity and privacy concerns across various domains [3,28,29]. Studies by Figueira and Vaz [3], Dahmen and Cook [28], and Dankar and Ibrahim [29] underscore the significance of synthetic data in machine learning model validation, particularly in healthcare applications where diverse and labeled datasets are often limited. These works highlight the effectiveness of tailored synthetic data generation approaches in improving model accuracy and utility while emphasizing the importance of standardized evaluation metrics and benchmarking methods.

Moreover, the research by Soltana et al. [30] and Hernandez et al. [27] addresses the challenges of generating synthetic test data and tabular health records, respectively, emphasizing the need for statistical representativeness and logical validity. While these studies offer systematic approaches for generating synthetic data that mirrors real-world scenarios, gaps in standardized evaluation metrics, and generalizability across diverse datasets remain.

Furthermore, investigations into synthetic MRI by Moya-Sáez et al. [31], Tariq et al. [32], and Gonçalves et al. [33] demonstrate the potential of synthetic data in revolutionizing medical imaging protocols, particularly in MRI-based diagnosis and radiotherapy planning. However, challenges such as validation on larger datasets, clinical relevance, and optimization of post-processing algorithms persist, indicating avenues for further research and development. Overall, the literature survey highlights the importance of artificial data creation in tackling data limitations and confidentiality issues across various domains, while also underscoring the need for standardized evaluation metrics, validation studies, and optimization of synthetic data creation approaches to enhance their utility and applicability in real-world scenarios.

18.3 METHODOLOGY

The solution approach involves leveraging GANs to generate synthetic MRI images and subsequently training a Convolutional Neural Network (CNN) model on both real and synthetic MRI datasets. By comparing the accuracy of the CNN model on real and generated images, we aim to assess the fidelity and utility of synthetic MRI images for AI-driven neuroimaging tasks, particularly brain tumor detection [33]. The close alignment between the accuracies of the CNN model on real and synthetic images demonstrates the promise of GAN-driven synthetic data generation as a powerful instrument for AI-driven neuroimaging research and clinical practice.

18.3.1 GAN-Based Synthetic Data Generation

We employ a GAN architecture comprised of a generator G and a discriminator D to produce synthetic MRI images. The generator $G(z;\theta_g)$ is trained to create realistic MRI images from random noise $z \sim p_z(z)$, as the discriminator $D(x;\theta_d)$ learns to differentiate between actual x and synthetic images $\tilde{x} = G(z)$ [34]. The loss function of the GAN is given by:

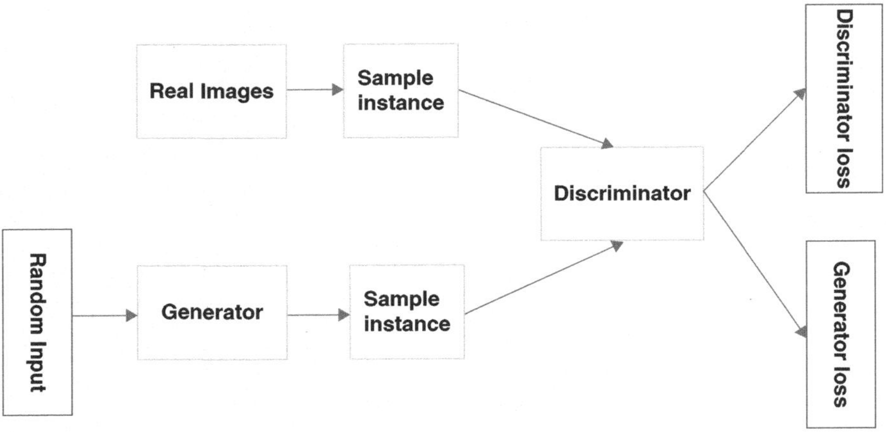

FIGURE 18.3 Methodology of GAN.

$$\min_{G} \max_{D} E_{x\sim p\text{data}(x)}\left[\log D(x)\right] + E_{z\sim pz(z)}\left[\log\left(1 - D\left(G(z)\right)\right)\right] \tag{18.1}$$

where $p_{\text{data}}(x)$ is the distribution of real MRI images, $p_z(z)$ is the distribution of input noise, $G(z)$ generates the synthetic MRI images, and $D(x)$ outputs the probability that x is a real image. Adversarial training enhances the generator's proficiency in producing synthetic images that closely resemble real MRI scans. The generator network comprises multiple convolutional layers, followed by upsampling layers to generate high-resolution MRI images. The discriminator network is comprised of convolutional layers succeeded by max-pooling layers, utilized for the classification of genuine and synthetic images. The GAN is trained iteratively, with the generator aiming to minimize the discriminator's ability to differentiate between real and synthetic images. The working of GAN is illustrated in Figure 18.3.

18.3.1.1 Loading Dataset

Loading the dataset involves fetching real MRI images from a Google API to train the GAN model, serving as the reference for the generator to learn the distribution of real MRI scans. Let $X = \{x_1, x_2, \ldots, x_n\}$ denote the set of real MRI images used for training, where $x_i \sim p_{\text{data}}(x)$. This process entails utilizing data loading libraries to efficiently load and preprocess MRI images into batches suitable for training.

18.3.1.2 Generator Network

The generator network's primary function is to take random noise vectors z as input and then learn to produce synthetic MRI images $\tilde{x} = G(z)$. The generator's loss function can be expressed as:

$$L_G = E_{z\sim pz(z)}\left[\mathbf{log}\left(1 - D\left(G(z)\right)\right)\right] \tag{18.2}$$

The generator architecture is structured with multiple layers of convolutional and upsampling operations, progressively converting low-dimensional noise into high-dimensional representations that resemble MRI scans. During training, its objective is to reduce the difference between synthetic images and actual ones, which is achieved through adversarial training with the discriminator network.

18.3.1.3 Discriminator Network

The discriminator network within the GAN framework functions as a binary classifier, discerning between real and synthetic MRI images. The discriminator's loss function is:

$$L_D = -E_{x\sim \text{pdata}(x)}\left[\mathbf{log}D(x)\right] - E_{z\sim pz(z)}\left[\mathbf{log}\left(1 - D\left(G(z)\right)\right)\right] \tag{18.3}$$

The architecture of the Discriminator Network generally comprises convolutional layers succeeded by fully connected layers, allowing it to capture meaningful features from input images and perform binary classification. During training, the discriminator's main objective is to enhance its ability to differentiate between authentic and generated images, progressively honing its ability to accurately classify the origin of input images.

18.3.1.4 Adversarial Training

Adversarial training is a fundamental aspect of GANs, wherein the generator and discriminator networks engage in a dynamic optimization game. Throughout the training process, the characteristics of both networks are iteratively updated. The generator aims to maximize its ability to generate realistic MRI images to mislead the discriminator, while the discriminator endeavors to refine its accuracy in discriminating between real and synthetic images iteratively updating their parameters to optimize their respective loss functions L_G and L_D. Convergence in the GAN model is attained when the generator succeeds in producing synthetic MRI images that are virtually indifferentiable from the real ones, and the discriminator achieves approximately 50% accuracy in its binary classification task, signifying that it cannot reliably differentiate between the two types of images. Hence,

$$E_{x\sim \text{pdata}(x)}\left[D(x)\right] = E_{z\sim pz(z)}\left[D\left(G(z)\right)\right] = 0.5 \tag{18.4}$$

18.3.2 CNN Model Training on Real MRI Dataset

We train a CNN model on a dataset of real MRI images X containing labeled examples of brain tumor presence or absence. The CNN model learns to extract relevant features from MRI images and classify them into tumor and nontumor classes. The CNN model's architecture can be defined as:

$$\hat{y} = f\left(X; \theta_c\right) \tag{18.5}$$

where $f(\cdot)$ represents the CNN function, θ_c are the CNN parameters, and $\hat{y}$ is the predicted output. The CNN framework comprises convolutional layers, pooling layers, and fully connected layers. The model takes real MRI images as input and generates the predicted probability of tumor presence as output. The loss function for two-class classification can be expressed as:

$$L_{\text{CNN}} = -\frac{1}{m}\sum_{i=1}^{m}\left[y_i \log\left(\hat{y}_i\right) + \left(1 - y_i\right)\log\left(1 - \hat{y}_i\right)\right] \tag{18.6}$$

The model undergoes training using gradient descent optimization alongside the categorical cross-entropy loss function. Once the CNN model is trained on the real MRI dataset, its performance is assessed on a distinct test set containing previously unseen real MRI images. Following this, the model's effectiveness is further evaluated on synthetic MRI images generated by the GAN. The accuracy, sensitivity, specificity, and area under the ROC curve (AUC) are measured to gauge the CNN model's performance on both real and artificial MRI datasets. These evaluation metrics offer valuable insights into the model's capability to accurately identify brain tumors in both authentic and generated MRI scans.

18.3.2.1 Loading Dataset of Real MRI Images

Loading the dataset of real MRI images is a pivotal step in preparing for the training of CNN models. This process involves the retrieval of real MRI images and their subsequent loading into memory to facilitate model training. To accomplish this, data loading libraries are employed to efficiently navigate and extract MRI images from a directory structure, ensuring that the dataset is readily accessible for training. By leveraging these libraries, the cumbersome task of manually handling data retrieval and preprocessing is streamlined, allowing for seamless integration of the MRI dataset into the CNN training pipeline.

18.3.2.2 Data Preprocessing

Data preprocessing is essential to prepare the loaded MRI images for compatibility with the CNN model by ensuring uniformity and enhancing training efficacy. This multifaceted process involves several key steps: first, all images are resized to a standardized square dimension, typically 34 dimensions, to optimize compatibility with CNN input requirements. Subsequently, normalization is utilized to adjust pixel values to a standardized scale within the range of 0 to 1, aiding convergence during training by reducing computational complexity and mitigating gradient-related issues. Additionally, data augmentation mechanisms such as rotation, flipping, or zooming are implemented to augment dataset diversity and bolster model robustness, thereby enriching the training dataset with varied perspectives and reducing overfitting risks.

18.3.2.3 Splitting Dataset into Training and Validation Sets

The act of partitioning the dataset into training and validation subsets plays a vital role in segregating the data into two distinct groups: one allocated for training the CNN model and the other for assessing its performance. This guarantees that the model's effectiveness is evaluated on previously unseen data, offering a more precise measure of its generalization ability. To facilitate robust training, the dataset is divided into two sets, training and testing, following a specified ratio r, where:

$$X_{\text{train}}, X_{\text{val}} = \text{split}(X, r) \tag{18.7}$$

Typically, a ratio of around 80% for training and 20% for validation is allocated, ensuring a balance between having sufficient data for robust model training while still retaining a sizable portion for evaluating model performance. This partitioning strategy enables the model to learn from diverse examples during training while also enabling effective monitoring of its performance on unseen data, thereby facilitating the identification of potential overfitting or generalization issues.

18.3.2.4 Model Training

During the training phase, the CNN model is optimized using the assigned training dataset to acquire the features and pattern characteristics of real MRI images. This involves designing a CNN architecture tailored for MRI image classification tasks, incorporating convolutional layers, pooling layers, and fully connected layers to efficiently derive and process relevant characteristics. Appropriate loss functions, like categorical cross-entropy for multi-class classification or binary cross-entropy for two-class classification, are employed to quantify the disparity between predicted and actual labels. Training parameters, including batch size and the number of epochs, are meticulously defined to optimize the training process and achieve convergence. Within the training loop, batches of MRI images are iteratively fed into the CNN model, followed by a forward pass to compute the loss, which is then utilized to refine the model's parameters through backpropagation, thus iteratively refining the model's performance over successive epochs.

18.3.2.5 Model Evaluation

Model evaluation involves assessing the trained CNN model's performance on the validation dataset to gauge its accuracy and generalization capabilities. This process encompasses calculating various evaluation metrics, including accuracy, precision, recall, and F1-score, to quantitatively measure the model's proficiency in classifying MRI images. These metrics offer perspectives on various facets of the model's performance, such as its capacity to correctly classify true positives, minimize false positives, and capture overall classification performance. By analyzing these metrics, researchers and practitioners can attain a thorough comprehension of the model's strengths and limitations, facilitating informed decisions regarding its deployment and potential refinements.

18.3.3 Comparison of Real and Synthetic Image Accuracy

In the final stage of our analysis, we conduct a comparison of the CNN model's accuracy on authentic and synthetic MRI scans to assess the realism of the generated images created by the GAN. This comparison serves as a crucial step in assessing whether the synthetic MRI images closely mimic real MRI scans. By closely aligning the accuracies of the CNN model on both types of images, we can deduce a strong resemblance between the artificial and authentic images. To conduct this evaluation, we utilize statistical analysis to compare various performance metrics, including accuracy, sensitivity, specificity, and Area Under the Curve (AUC), obtained from the CNN model's predictions on real and synthetic MRI datasets. If the performance metrics exhibit similarity between real and synthetic images, it indicates that the synthetic images effectively capture essential features and characteristics of real MRI scans.

To implement this evaluation procedure, we input real MRI images from the validation set into the trained CNN model and calculate the accuracy, followed by repeating the process with synthetic MRI images generated by the GAN model. Subsequently, we compare the accuracy scores obtained for real and generated images, with comparable scores, indicating that the generated images strongly align with actual images in terms of classification features and patterns. This comprehensive evaluation methodology enables us to validate the utility of synthetic MRI images for AI-driven neuroimaging tasks by ensuring their fidelity and similarity to real MRI scans.

18.4 IMPLEMENTATION DETAILS

18.4.1 Data Acquisition and Preprocessing

Real MRI data is sourced from Google API datasets, ensuring high-quality and relevant data acquisition. The collected data undergoes preprocessing to standardize formats, handle missing values, and normalize intensity values for consistency across the dataset.

18.4.2 Synthetic Data Generation

18.4.2.1 Deep Convolutional Generative Adversarial Network (DCGAN)

The project leverages a DCGAN to synthesize synthetic MRI images. The DCGAN is trained on existing MRI datasets to grasp the inherent patterns of genuine MRI scans and subsequently create synthetic replicas, ensuring that the produced images exhibit superior quality and closely mirror real MRI scans in terms of structure and features.

18.4.2.2 Cycle-Consistent Generative Adversarial Network (Cyclic-GAN)

To further enhance the quality of synthetic MRI images, a Cyclic-GAN is implemented. The Cyclic GAN is trained using synthetic MRI images generated by the DCGAN to learn a mapping between the synthetic and real MRI domains. Through an iterative process of translation and reconstruction, the generated images are refined to improve realism and consistency.

18.4.3 Convolutional Neural Network (CNN) Implementation

A CNN model is developed utilizing frameworks like TensorFlow or PyTorch. The CNN model architecture is meticulously designed to perform tasks like brain tumor detection or other neuroimaging tasks. Training the model on both real and synthetic MRI datasets allows it to train from diverse data distributions and improve generalization.

18.4.4 Model Evaluation

The optimized CNN model is assessed using multiple indicators, including accuracy, precision, recall, and F1-score. Evaluation involves comparing the model's performance on real MRI images vs. synthetic MRI images. Statistical analyses may be performed to evaluate the relevance of any detected variations in performance.

18.4.5 Performance Analysis and Optimization

The effectiveness of the CNN model on artificial data is thoroughly analyzed to identify any discrepancies or limitations. Optimization techniques, like refining the GAN parameters or modifying the CNN architecture, can be implemented to further improve performance.

18.5 EXPERIMENTATION RESULT

18.5.1 Performance Analysis and Optimization

The effectiveness of the CNN model on generated data is meticulously analyzed to discern any disparities or potential limitations. To enhance performance, optimization techniques such as adjusting the hyperparameters of the GAN or the architecture of the CNN may be employed. In Figure 18.4 illustrates real images sourced from the dataset, while Figure 18.5 showcases the dataset comprising generated images from the DCGAN. Figure 18.6 displays images generated from the Cyclic GAN. Moreover, Figure 18.7 presents the accuracy comparison between real and synthetic images, elucidating the model's effectiveness. Notably, Figure 18.8 illustrates the pixel values of real vs. synthetic images, revealing a significant overlap and indicating the close resemblance between the two image types. Collectively, these figures offer a comprehensive depiction of our project's input and output, elucidating the efficacy of our approach.

After training the CNN model on our dataset, we evaluated its performance in terms of accuracy and loss. The model achieved a commendable accuracy of approximately 92.63% for real images and 93.33% for fake images. These accuracy values indicate the effectiveness of our model in correctly classifying the input images into their respective categories. Additionally, the loss function, which quantifies the the deviation between estimated and true values during training, was observed to be 0.2835 for real images and 0.2558 for fake images. These relatively low loss values suggest that the model has learned to minimize errors effectively during the training process. The accompanying graph illustrates the comparison of accuracy and loss between real and fake images, providing a graphical depiction of our model's effectiveness.

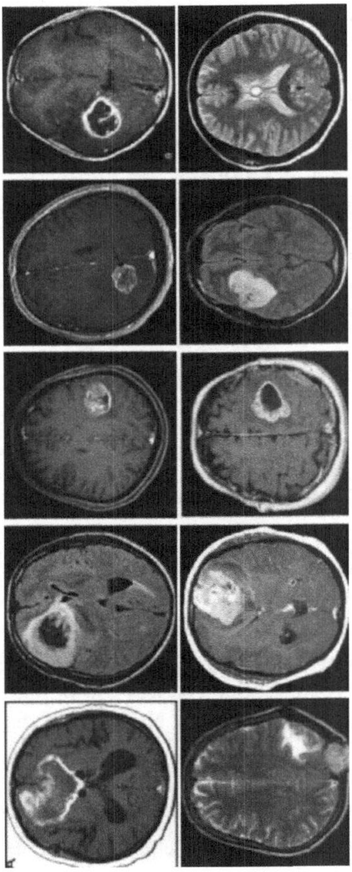

FIGURE 18.4 Real images.

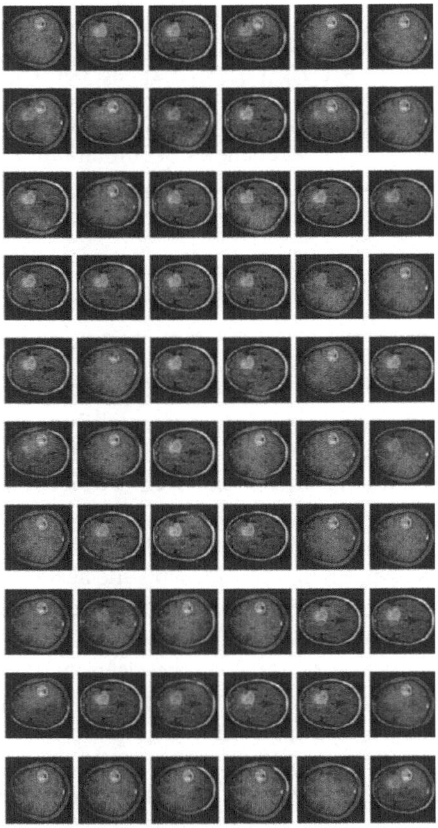

FIGURE 18.5 Generated images from DCGAN. .

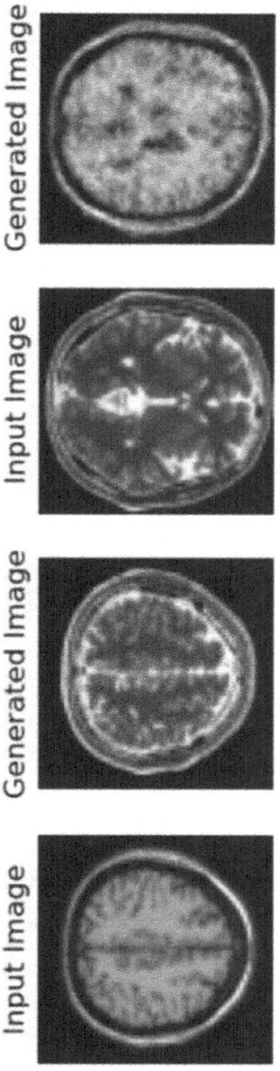

FIGURE 18.6 Generated image from Cyclic GAN.

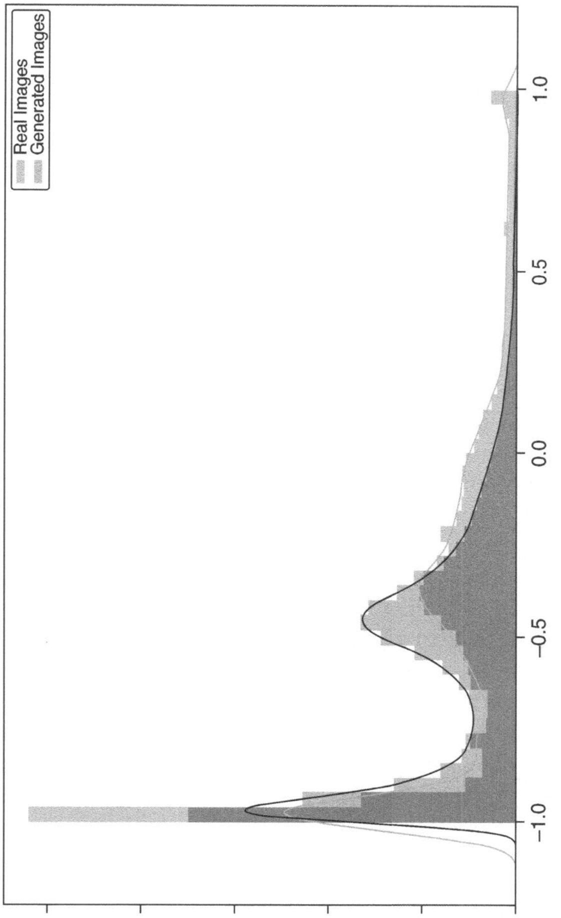

FIGURE 18.7 Analysis of real and fake images.

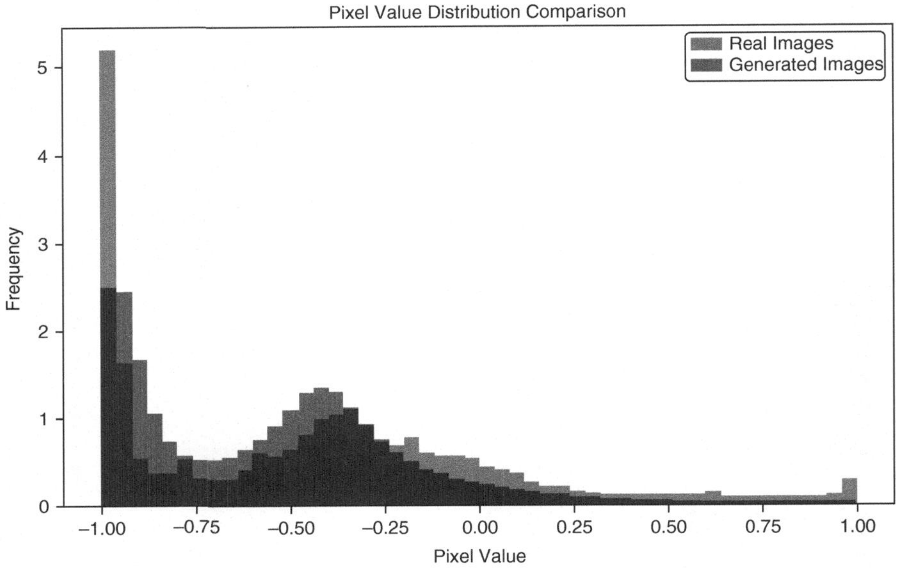

FIGURE 18.8 Analysis of real and fake images on the basis of pixel values.

The comparison graph in Figure 18.9 illustrates the performance of our model on real and fake images. The graph depicts the accuracy and loss metrics for both types of images, providing insights into how well our model distinguishes between real and fake data. Upon analysis, it is evident that the model achieved slightly higher accuracy for classifying fake images compared to real images as shown in Figure 18.10. This suggests that the model may have learned to identify patterns more effectively in the synthesized or generated data. However, it's essential to highlight that the loss values for both real and fake images are relatively low, indicating that the model has successfully minimized errors during training for both types of data. Overall, the comparison graph provides valuable insights into the performance of our model on different types of input images, highlighting its strengths and areas for further optimization (Figure 18.11).

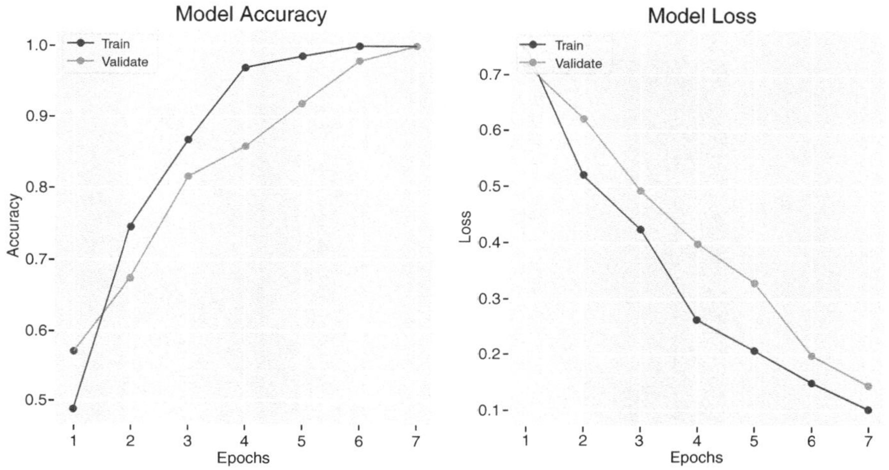

FIGURE 18.9 CNN model accuracy and loss.

18.5.2 CNN Model Performance Summary

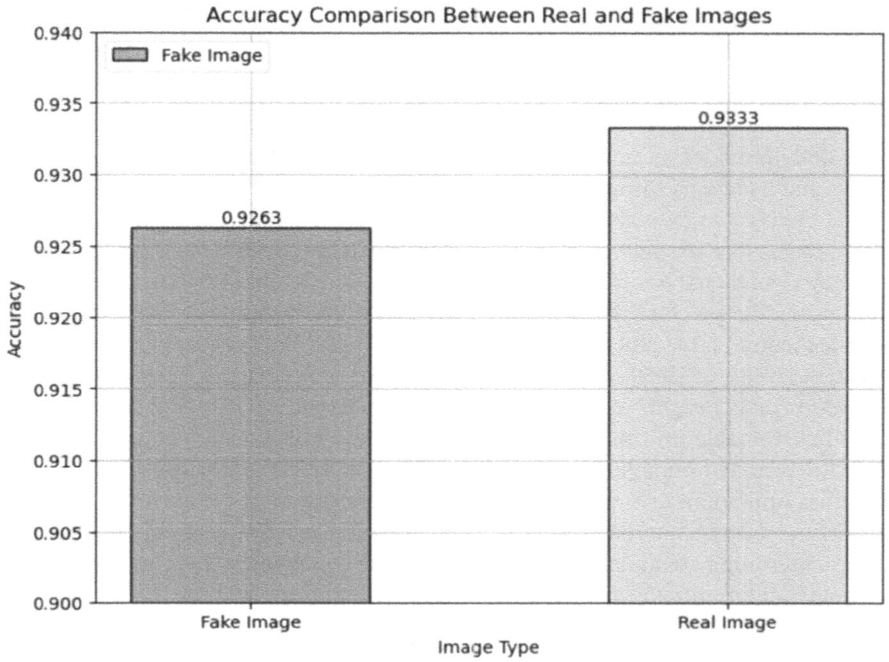

FIGURE 18.10 Accuracy comparison of real and fake images.

FIGURE 18.11 Loss between real and fake images.

18.6 CONCLUSION

This study provides an in-depth investigation of synthetic MRI data creation using GANs and its testing through CNNs. Through the implementation of DCGANs and Cyclic GANs, we demonstrate the feasibility of generating high-quality artificial MRI images that strongly mimic real scans. The iterative refinement process facilitated by Cyclic GANs enhances the fidelity of synthetic data over multiple epochs, showcasing the potential for continuous improvement. By training CNN models on both authentic and synthetic MRI datasets, we assess the performance and generalization ability of CNNs in tasks such as brain tumor detection. Our comparative analysis reveals insights into the similarity between real and synthetic MRI images, providing valuable information on the accuracy and effectiveness of synthetic data for AI-driven neuroimaging tasks. While this work yields promising results, it also identifies limitations and challenges, including computational complexity in training GANs and CNNs, interpretability of deep learning models, and ethical considerations regarding patient data privacy. Despite these challenges, the findings suggest the potential for utilizing synthetic MRI data in clinical applications, offering opportunities to augment limited real-world data and improve the robustness of AI-driven healthcare systems.

REFERENCES

1. Walia, M.S. 2020. *Synthetic data generation using* Wasserstein Conditional Gans with Gradient Penalty (WCGANS-GP). Dissertations. (Jan. 2020). https://doi.org/10.21427/js39-jq58.
2. Wang, Z. et al. 2024. *High-Fidelity Synthetic Data Applications for Data Augmentation*. IntechOpen.
3. Figueira, A. and Vaz, B. 2022. *Survey on Synthetic Data Generation, Evaluation Methods and GANs. Mathematics*. Multidisciplinary Digital Publishing Institute.
4. Bińkowski, M. et al. 2020. High fidelity speech synthesis with adversarial networks. *8th International Conference on Learning Representations, ICLR 2020* (Sep. 2020). https://arxiv.org/pdf/1909.11646
5. Pelikan, M. et al. 2002. A survey of optimization by building and using probabilistic models. *Computational Optimization and Applications*. 21, 1 (2002), 5–20. https://doi.org/10.1023/A:1013500812258.
6. Sarker, I.H. et al. 2024. Multi-aspect rule-based AI: Methods, taxonomy, challenges and directions towards automation, intelligence and transparent cybersecurity modeling for critical infrastructures. *Internet of Things*. 25, (Apr. 2024), 101110. https://doi.org/10.1016/J.IOT.2024.101110.
7. Murtaza, H. et al. 2023. Synthetic data generation: State of the art in health care domain. *Computer Science Review*. 48, (2023), 100546. https://doi.org/10.1016/j.cosrev.2023.100546.
8. Cheng, X. et al. 2021. Combating emerging financial risks in the big data era: A perspective review. *Fundamental Research*. 1, 5 (Sep. 2021), 595–606. https://doi.org/10.1016/J.FMRE.2021.08.017.
9. Buggineni, V. et al. 2024. Enhancing manufacturing operations with synthetic data: A systematic framework for data generation, accuracy, and utility. *Frontiers in Manufacturing Technology*. 4, (Feb. 2024), 1320166. https://doi.org/10.3389/FMTEC.2024.1320166.
10. Jordon, J. et al. 2022. Synthetic data—What, why and how? *arXiv preprint*. (May 2022).
11. Jia, Y. et al. 2019. CommunityGAN: Community detection with generative adversarial nets. *The World Wide Web Conference* (New York, NY, May 2019), 784–794.
12. Wetzel, S.J. 2017. Unsupervised learning of phase transitions: From principal component analysis to variational autoencoders. *Physical Review E*. 96, 2 (Aug. 2017), 022140. https://doi.org/10.1103/PHYSREVE.96.022140.
13. Varga, T. and Bunke, H. 2008. Perturbation models for generating synthetic training data in handwriting recognition. *Studies in Computational Intelligence*. 90, (2008), 333–360. https://doi.org/10.1007/978-3-540-76280-5_13.
14. Evsukoff, A.G. et al. 2009. Design of interpretable fuzzy rule-based classifiers using spectral analysis with structure and parameters optimization. *Fuzzy Sets and Systems*. 160, 7 (Apr. 2009), 857–881. https://doi.org/10.1016/j.fss.2008.08.010.
15. Bellovin, S.M. et al. 2018. Privacy and synthetic datasets. *SSRN Electronic Journal*. 22, (2018). https://doi.org/10.2139/ssrn.3255766.
16. Chan, C.Y. 2024. Data-driven innovation: The potential of synthetic data through generative AI. 39 pages. https://www.theseus.fi/bitstream/handle/10024/854321/Chan_ChungYin.pdf;jsessionid=94E04269881E4D1B441DB859843A1DBB?sequence=2
17. Bowles, C. et al. 2018. GAN Augmentation: Augmenting training data using generative adversarial networks. *arXiv preprint*. (Oct. 2018).
18. Majeed, A. 2023. Attribute-centric and synthetic data based privacy preserving methods: A systematic review. *Journal of Cybersecurity and Privacy*. 3, 3 (Sep. 2023), 638–661. https://doi.org/10.3390/JCP3030030.

19. El Emam, K., Mosquera, L., and Hoptroff, R. 2020. *Practical Synthetic Data Generation Balancing Privacy and the Broad Availability of Data*. Sebastopol, CA: O'Reilly Media.
20. Ravasio, C.S. 2022. Deep retinal optical flow: From synthetic dataset generation to framework creation and evaluation. Doctoral thesis, UCL (University College London) (Oct. 2022).
21. Han, X. 2017. MR-based synthetic CT generation using a deep convolutional neural network method. *Medical Physics*. 44, 4 (2017), 1408–1419. https://doi.org/10.1002/mp.12155.
22. Yamanakkanavar, N. et al. 2020. MRI segmentation and classification of human brain using deep learning for diagnosis of Alzheimer's disease: A survey. *Sensors*. 20, 11 (Jun. 2020), 3243. https://doi.org/10.3390/S20113243.
23. Wu, C. et al. 2021. Clinical applications of magnetic resonance imaging based functional and structural connectivity. *NeuroImage*. 244, (Dec. 2021), 118649. https://doi.org/10.1016/J.NEUROIMAGE.2021.118649.
24. Galgano, M. et al. 2017. Traumatic brain injury: Current treatment strategies and future endeavors. *Cell Transplantation*. 26 (Jun. 2017), 83–118. https://doi.org/10.1177/0963689717714102.
25. Chong, C.K. and Ho, E.T.W. 2021. Synthesis of 3D MRI brain images with shape and texture generative adversarial deep neural networks. *IEEE Access*. 9, (2021), 64747–64760. https://doi.org/10.1109/ACCESS.2021.3075608.
26. Ramadan, R.A. and Altamimi, A.B. 2024. Unraveling the potential of brain-computer interface technology in medical diagnostics and rehabilitation: A comprehensive literature review. *Health and Technology*. 14, 2 (Mar. 2024), 263–276. https://doi.org/10.1007/S12553-024-00822-1.
27. Hernandez, M. et al. 2022. Synthetic data generation for tabular health records: A systematic review. *Neurocomputing*. 493, (Jul. 2022), 28–45. https://doi.org/10.1016/J.NEUCOM.2022.04.053.
28. Dahmen, J. and Cook, D. 2019. SynSys: A synthetic data generation system for healthcare applications. *Sensors*. 19, 5 (Mar. 2019), 1181. https://doi.org/10.3390/s19051181.
29. Dankar, F.K. and Ibrahim, M. 2021. Fake it till you make it: Guidelines for effective synthetic data generation. *Applied Sciences*. 11, 5 (Feb. 2021), 1–18. https://doi.org/10.3390/app11052158.
30. Soltana, G. et al. 2017. Synthetic data generation for statistical testing. *ASE 2017—Proceedings of the 32nd IEEE/ACM International Conference on Automated Software Engineering*. (Nov. 2017), 872–882. https://doi.org/10.1109/ASE.2017.8115698.
31. Moya-Sáez, E. et al. 2021. A deep learning approach for synthetic MRI based on two routine sequences and training with synthetic data. *Computer Methods and Programs in Biomedicine*. 210, (Oct. 2021), 106371. https://doi.org/10.1016/j.cmpb.2021.106371.
32. Tariq, U. et al. 2023. *Brain Tumor Synthetic Data Generation with Adaptive StyleGANs*. Communications in Computer and Information Science. Cham, Switzerland: Springer.
33. Gonçalves, F.G. et al. 2018. *Synthetic Brain MRI: Review of Current Concepts and Future Firections*. Topics in Magnetic Resonance Imaging. Lippincott Williams and Wilkins.
34. Isola, P. et al. 2017. Image-to-image translation with conditional adversarial networks. In *Proceedings of the IEEE conference on computer vision and pattern recognition* (pp. 1125–1134).

19 Next-Generation Health Services

Integrating Mobile Chatbots with Health Monitoring Technologies

Megha Rathi and Jaya Srivastava

19.1 INTRODUCTION

A chatbot is an AI-powered, natural language-using interactive application. The artificially intelligent bot replicates human thought processes. Chatbot applications can help patients in any location in the world, including remote places and homes, and they can also make tasks for medical staff easier. We have made the decision to contribute to the field of health informatics by using this skill. Our research develops a text-to-text conversational bot that uses natural language to diagnose patients, identify potential diseases, and explain their conditions. To achieve high accuracy and a solid diagnosis, the system monitors the user's previous answers and utilizes those responses to formulate even more targeted queries. Furthermore, our project occasionally gathers information on specific critical tasks or statuses in order to monitor the patient's health. Our method consists of four main parts: (i) identifying and extracting symptoms from user interactions; (ii) precisely mapping the extracted symptoms to recorded symptoms; (iii) predicting the condition and, if required, directing to a suitable professional; and (iv) regularly checking on the patient's health in order to monitor their improvement. To get the mapping of symptoms with disease, clustering models are used to make the cluster of different symptoms based on similarity of symptoms and then a classification algorithm is applied to classify the symptoms to predict the disease.

The term Generative artificial intelligence (GAI) describes a class of artificial intelligence (AI) platform, which can create fresh content—text, images, music, or other types of data—that is frequently identical to what a human could create. GAI is intended to produce new, original data, in contrast to traditional AI, which is usually used for tasks like categorization or prediction based on existing data.

The most crucial aspect for a patient is tracking their health because, as we all know, prevention is always preferable to treatment. A patient can recognize and comprehend their own lifestyle as long as they are able to keep track of their routine task, including

DOI: 10.1201/9781003565703-19

exercise, medicine, and even appointments with doctors. They will be able to assess their development and become more driven as a consequence (Shankar et al. 2015).

Additionally, if necessary, a person may identify their symptoms without going to the doctor. This will be especially helpful for older folks or those who are hesitant about going to the doctor. Most of the time, a person cannot identify a common illness that simply requires certain home remedies to treat. It makes medical information more accessible to people and helps them make use of it.

With the help of this program, anyone can identify their ailment and, if necessary, find a cure—as long as it is not too serious or severe. Furthermore, since you won't need to see a doctor for each minor symptom, this is the cheapest method to identify the illness and even track your health. It is quite tough to pay for a doctor in today's world, particularly in a place like India where most of people avoid going to the doctor because they lack the funds to do so. Their needs will undoubtedly be greatly served by this application (Rahman et al. 2016).

As a result, the program will not only detect the ailment but also track the condition of the individual in relation to it. If there is something that is not going well, it is crucial to let the patient know so that they can take action. In any case, this will assist in averting any risk (Rahman et al. 2019). Additionally, the software can recognize certain motions when the scenario appears to be really serious and send a message to others who are close to the person. By doing this, he will warn those around him and avoid becoming disfigured. Furthermore, it would make it simple for the patient to locate a doctor by providing information about nearby ones (Veena and Gowrishankar 2021).

Because of this, patients will gain greatly from this app, which will minimize the doctor visit costs, the number of minor hospital visits, lower travel costs, early identification of illnesses, and low-cost medical treatments and tracking. Chatbot systems will also enable patients to get specialized medical care in neglected rural, semi-urban, and isolated regions.

We offer the following list of our system's aims as a summary of the goals: to ensure that everybody, even those residing in isolated and rural locations, has access to timely and effective healthcare, to lower the price of seeing a physician and a hospital stay, to advance analysis and handling diseases, to lessen the problems due to geographical separation and age losses, and to keep a watchful eye on individuals and track their change.

19.2 LITERATURE SURVEY

Every day, technological breakthroughs continue to increase exponentially, transforming every industry in their wake. Innovation is becoming more and more involved in practically every step of human health and environmental wellness procedures, from registering patients to data monitoring. For our understanding purpose, we have studied the following papers:

Recently, King (2023) analysed the chatbot's user logs for mental health. Chat Pal is a positive psychology-based chatbot. The persistence of this study is to observe the chatbot's log data in order to provide light on usage trends, user categories that utilize clustering, and correlations between how the app's features are used.

A novel method by Rathnayaka et al. (2022) gives a summary of the most recent developments and patterns in medical chatbot technology. The focus will be on chemotherapy for cancer, with in-depth discussions and examples of detection, therapy, surveillance, patient aid, workflow efficiency, and health promotion.

In yet another novel work, Bharti et al. (2020) propose that a chatbot setting is a more effective way for Behavioural Activation (BA) therapy and AI to materialize in order to offer ongoing emotional support, individualized help, and distant mental health nursing. IT will outline the creation and design of our AI chatbot based on BA, as well as its participation evaluation in a pilot research environment, which verified its efficacy in helping people with mental health concerns.

In another significant study, Aggarwal et al. (2023) conducted a systematic review to assess the viability, effectiveness, and intervention features of AI chatbots in order to encourage changes in health-related behaviour. AI chatbots have shown that interventions aimed at changing health-related behaviours are effective across a wide range of people.

The study by Rani et al. (2023) aimed to link patients with qualified professionals who can offer support and guidance, as well as to provide thorough information on various disorders and their treatments. The central component of saarthi is a chatbot that leverages cutting-edge AI algorithms to offer patients individualized, sympathetic care.

Another notable research by Duckki Lee et al. (2023) looks into the definition and design of chatbots that use AI. It also looks at numerous instances of AI chatbot applications in the medical field. Lastly, it lists the benefits and drawbacks of chatbots for healthcare that are powered by AI.

In their novel work, Barbara Onyekwelu et al. (2024) examined, via an analysis of 708 reviews, how people feel about four well-known chatbot apps linked to health. The findings verify that there is a discrepancy between the user experience that chatbots offer and what users require. There are recommendations for chatbot developers that may help close the gap between the functionalities that are offered and the needs of users.

Afsahi et al. (2024) summarized chatbots' application in healthcare. In the healthcare industry, chatbots—computer systems that mimic natural language conversations—have drawn interest. Advances in recent times have tackled problems such as dementia, obesity, sleeplessness, and oncology. For them to be widely adopted, their usefulness must be thoroughly evaluated.

In another significant study by Ingale and Zagade (2024), Chatbots evaluate patient complaints, offer early diagnosis, and suggest the best courses of action using complex algorithms. This facilitates prompt action, minimizes needless medical visits, and maximizes the use of available resources.

Balasubramanian et al. (2015) proposed a method of disease prediction based on symptoms using *K*-Means algorithm. A novel method by Shankar et al. (2015) suggested to use symptoms to identify illnesses and estimate how long they would take to heal. Another innovative study by Laskar et al. (2016) presented an automated illness prediction method (ADPS) that relies on guided user input is discussed. Dugane et al. (2017) extracted information from medical text, which forms the basis for other higher order analytics.

Harini and Natesh (2018) used Machine Learning (ML) algorithms for predicting the disease effectively. The research paper (Divya et al. 2018) discusses an idea to create an AI system, which has the ability to emulate human thought and behaviour in a computer that can give healthcare support to a patient based on inputs. Davis et al. (2008) gave insights about a Collaborative Assessment and Recommendation Engine (CARE), which trusts only on the medical history of a patient using ICD-9-CM codes to forecast upcoming diseases hazards. Patil et al. (2016) referred to a system that is grounded on a grouping of different data mining approaches like classification and clustering. In their novel work, Gomathi and Priyaa (2016) demonstrate the usage of data mining technique and how the number of tests required from the patient can be reduced.

In their article, Bharti et al. (2020) explain the Google Cloud Platform (GCP) conversational bot "AapkaChikitsak", which is being used to provide telehealth services in India. Its goal is to expand patient access to healthcare knowledge and use AI to close the gap between the supply and demand for human healthcare practitioners. By lowering obstacles to healthcare facilities and enabling intelligent remote consultations for prompt care and high-quality treatment, this conversational technology has significantly benefited society.

19.3 METHODOLOGY

One of the most significant challenges, both historically and in the present, is predicting illnesses. In the proposed research, a chatbot system has been developed that enables patients to determine their health condition solely based on their symptoms. Additionally, the application monitors the patient's vital signs associated with the identified illness, helping to prevent potential emergencies.

19.3.1 Dataset

The database we utilized is an information record that includes disease–symptom relationships derived from textual discharge summaries of patients admitted to New York Presbyterian Hospital in 2004. The dataset contains 133 types of symptoms and prognoses for 41 different diseases, as detailed in Table 19.1. Figure 19.1 illustrates the prognosis of diseases based on symptoms. The dataset comprises 4,920 samples.

TABLE 19.1
List of Different Symptoms and Respective Disease

S.No	Symptoms	Prognosis
1	Rashes, nodal_skin_eruptions	Fungal infection
2	Constant sneezing, shivering, chills	Allergy
3	Stomach pain, acidity, ulcers_on_tongue	GERD
4	abdominal_pain, vomiting	Chronic cholestasis
5	burning_micturition, spotting_ urination	Drug reaction
6	internal_itching, passage_of_gases	Peptic ulcer disease
7	muscle_wasting	AIDS

(Continued)

TABLE 19.1 (*Continued*)
List of Different Symptoms and Respective Disease

S.No	Symptoms	Prognosis
8	excessive_hunger, blurred_and_distorted_vision	Diabetes
9	diarrhoea, vomiting	Gastroenteritis
10	mucoid_sputum, family_history, High Fever, cough	Bronchial asthma
11	lack_of_concentration, loss_of_balance, dizziness	Hypertension
12	visual_disturbances, irritability, depression, excessive_ hunger, blurred_and_distorted_vision	Migraine
13	loss_of_balance, neck_pain, dizziness	Cervical spondylosis
14	weakness_of_one_body_side, headache	Paralysis (brain hemorrhage)
15	high_fever, weight_loss, itching, yellowish_skin	Jaundice
16	chills, vomiting, sweating, high fever	Malaria
17	headache sweating, high fever	Chicken pox
18	red_spots_over_body, muscle_pain	Dengue
19	belly_pain, toxic_look_(typhos), nausea	Typhoid
20	joint_pain, vomiting, yellowish_skin	Hepatitis A
21	fatigue, lethargy, dark_urine, yellowish_skin	Hepatitis B
22	family_history, nausea, loss_of_appetite	Hepatitis C
23	dark_urine, nausea, loss_of_appetite, abdominal_pain	Hepatitis D
24	dark_urine, nausea, loss_of_appetite, abdominal_pain	Hepatitis E
25	swelling_of_stomach, abdominal_pain	Alcoholic hepatitis
26	loss_of_appetite, breathlessness, sweating, fatigue, vomiting	Tuberculosis
27	fatigue, fever, cold, continuous_sneezing, chills	Common cold
28	chills, fatigue, cough, high fever, sweating, breathless	Pneumonia
29	pain_during_bowel_movements, pain_in_anal_region, bloody_stool, irritation_in_anus	Dimorphic haemmorhoids (piles)
30	chest_pain, breathlessness, sweating	Heart attack
31	fatigue, cramps, bruising, obesity	Varicose veins
32	dizziness, abnormal_menstruation, irritability, depression	Hypothyroidism
33	palpitations, irritability, slurred_speech, nausea, excessive_hunger	Hypoglycaemia
34	joint_pain, neck_pain, knee_pain, hip_joint_pain, painful_walking	Osteoarthritis
35	movement_stiffness, swelling_joints, stiff_neck, muscle_weakness	Arthritis
36	vomiting, spinning_movements, loss_of_balance, unsteadiness	(Vertigo) Paroxysmal positional vertigo
37	scurrying, blackheads, pus_filled_pimples, skin rash	Acne
38	bladder_discomfort, foul_smell_of urine, continuous_feel_of_urine	Urinary tract infection
39	skin_peeling, silver_like_dusting, small_dents_in_nails, inflammatory_nails	Psoriasis
40	blister, red_sore_around_nose, yellow_crust_ooze	Impetigo
41	dyschromic_patches, nodal_skin_eruptions, skin_rash, itching	Fungal infection

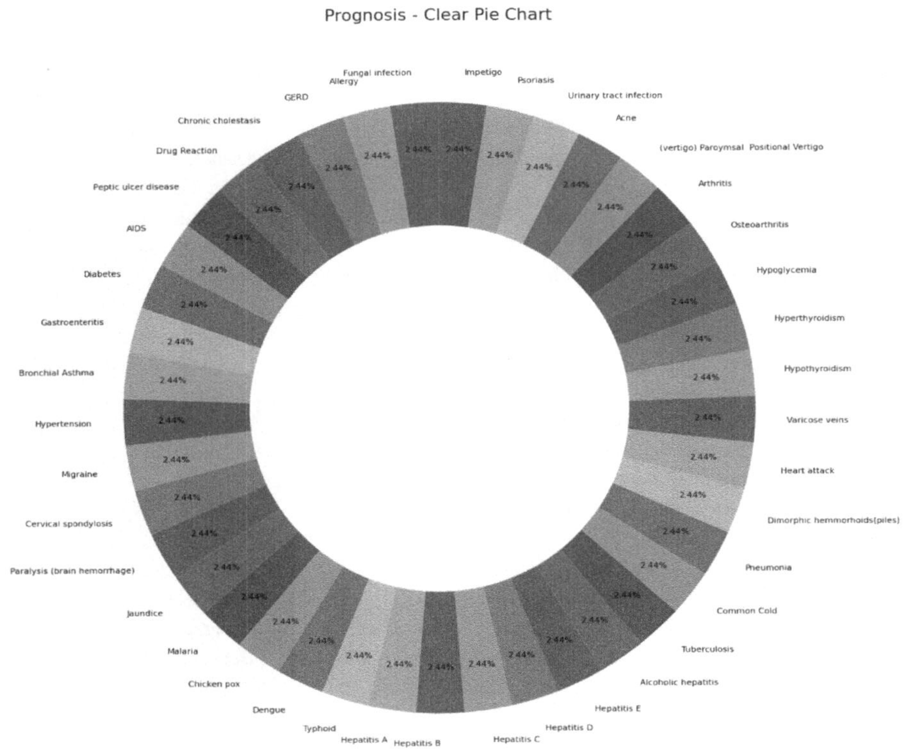

FIGURE 19.1 Prognosis of disease based on symptoms.

19.3.2 Preprocessing of Dataset

Real-life dataset is nearly certainly riddled through inaccuracies and is often conflicting, separated, and inaccurate. Arranging this data logically is one facet of data processing. The process, which also includes feature extraction, standardization, change, and data cleanup, culminates in a training sample. Preprocessing is necessary to increase the acceptability of data for data mining.

There are instances where data is distorted and worthless, or missing entirely. This leads to imprecise forecasting and may reduce the model's accuracy. Thus, it is imperative that this situation be handled. We have ensured that no parameter with a null value is processed when tuples with missing data ignored the null value. For every feature column in the dataset, calculate the Missing Value Ratio using the following equation:

$$\text{ratio} = \frac{\text{No. of missing values}}{\text{Total no. of values}} * 100 \qquad (19.1)$$

Transformation of information is the procedure of converting data from one format to another that is suited for use in training. Here, we have modified the data by choosing the specific features we desire and creating additional attributes based on our model. Then, we incorporate the min-max standardization technique for data standardization and cleaning using the following equation:

$$F(x) = \frac{X - \min(x)}{\text{Max}(x) - \text{Min}(x)} \tag{19.2}$$

Data reduction is the practice of reducing and simplifying data while maintaining essential information. The time and storage space needed for the dataset are decreased by this procedure. We have employed dimensionality reduction, which reduces the size of the dataset by identifying and eliminating redundant features.

19.4 PROPOSED INNOVATION

The proposed system is text-based and accepts commands in the form of written text. To store user data for future reference, the system first collects basic information such as name, age, and blood type. Following this, the system prompts the user to enter their symptoms. The system gathers one symptom at a time until a specific threshold is reached. Once this threshold is crossed, the system asks the user to either "Enter more symptoms" or select "No" if they have no additional symptoms. However, it is unreasonable to expect a user to input an excessive number of symptoms into a chatbot. A minimum of three symptoms is required for the system to classify an illness. The system encourages users to provide as many symptoms as possible, as a higher number of symptoms leads to more accurate disease prediction.

Chatbots are often heavily scripted, designed for structured interactions that simplify programming but restrict the range of topics a user can inquire about. However, advancements in AI techniques, such as ML, Natural Language Processing (NLP), and Deep Learning (DL) algorithms, have made chatbots significantly more sophisticated. These AI methods, though, require a large amount of data to generate accurate results. The more you interact with the bot, the more its accuracy will improve.

The chatbot program is managed by an administrator who has access to all user data and can manually add, remove, or update information on illnesses and symptoms. The chatbot is trained using data on symptoms and diseases. Based on the symptoms reported by the user, different clustering algorithms, such as *K*-means, DBSCAN, and Hierarchical clustering, are applied, followed by Random Forest classification to predict the illness using the dataset. Once the illness is identified, the system suggests the best treatment options.

One of the system's most critical tasks is the development of datasets for training and testing. To accurately categorize symptoms and diseases, a reliable information source was required. The most commonly used disease classification for training, along with independent test datasets, was utilized. The chatbot not only provides relevant diagnostics and treatment recommendations but also encourages users to discuss their medical concerns. In this context, NLP is used to process the text input by the user. The random forest ML algorithm is employed, where clustering helps to recognize symptoms based on user interaction. The clustering process then associates these symptoms with a specific ailment and suggests the most appropriate course of action to the user.

19.5 METHODOLOGY

Clustering is an unsupervised ML technique that groups similar items together within a large dataset based on a specific property known as a cluster. Each cluster is distinct and separate from others. This method allows the creation of a set number of categories without any prior knowledge of the number of clusters or the characteristics of each cluster. In this work, we apply various clustering algorithms to form different clusters associated with symptoms, which in turn help in predicting the accurate disease.

19.5.1 *K*-Means Algorithm

K-means clustering is a quick, reliable, and easy method that functions best when the datasets are well separated. By applying this method, the numbers of data are grouped together in a cluster in a way that reduces the squared distance among them and the centroid (Sinaga and Yang 2020). In order to make the centroids as minimal as feasible, the *K*-Means approach first counts the number of centroids before conveying every data point to the closest cluster. *K*-means approach of clustering is vulnerable to extreme values since they can readily modify a mean. *K*-Means allocates inputs to the closest centroid by calculating the squared distances among the inputs (also identified as input data points) and centroids using the following equation:

$$\sum_{k=1}^{K} \sum_{Xi \in Ck} \left\| xi - \mu k \right\| \tag{19.3}$$

where C_K represents the *K*th cluster, x_i represents the *i*th data point, and μk signifies the centroid of the *k*th cluster.

19.5.2 DBSCAN

It signifies density-based spatial clustering in noisy applications and is a widely utilized method in ML and data extraction. Density-based algorithms are used in nonparametric density-based clustering techniques to distinguish between higher- and lower-density clusters. The fundamental idea of DBSCAN states that a point is regarded as a component if it is next to other points inside a cluster. With noisy and outlier-filled data, it can find different-shaped clusters and dimensions. Like *k*-means, DBSCAN can identify clusters of any shape or size in the data and does not require a specific number of clusters to be present (Deng 2020). "Epsilon(ε)" is the range in which nearby points should be looked for and "MinPts" are the minimal number of points required to form a dense region or cluster.

$$|q \in D| d(p,q) \le \varepsilon | \ge \text{MinPts} \tag{19.4}$$

d(*p*,*q*) is the distance among points *p* and *q*, and *D* is the set of all points in the dataset.

19.5.3 Hierarchical Clustering

Hierarchical clustering is a method used to build a hierarchy of clusters, making it particularly useful for identifying nested clusters and illustrating the relationships between them. This approach typically begins with each observation in its own cluster and uses the agglomerative (bottom-up) method, where pairs of clusters are merged as one moves up the hierarchy (Roux 2018). Common linkage criteria include single linkage, complete linkage, and average linkage. In our implementation of agglomerative clustering, we employed single linkage, which defines the distance between two clusters as the shortest distance between any point in the first cluster and any point in the second cluster, as mathematically represented in the following equation:

$$d\min(A,B) = \min d(a,b) \quad a \in A, b \in B \tag{19.5}$$

19.5.4 Fuzzy C-Means Clustering

Fuzzy C-Means (FCM) clustering is a soft clustering technique in which a membership value determines the degree to which every point of data corresponds to a cluster. FCM permits partial participation in more than one cluster, in contrast to strict clustering techniques like *K*-means, which allocate every information point to precisely one cluster (Mohammadrezapour, Kisi, and Pourahmad 2020). Mathematical Expression is shown using the following equation:

$$Jm = \sum_{i=1}^{n} \sum_{j=1}^{c} U_{ij}^{m} . d\left(x_{i,} v_j\right) \tag{19.6}$$

19.5.5 Random Forest Classifier

The Random Forest approach is a powerful tree learning technique in ML. It produces several Decision Trees in the training phase. Every tree in the data collection is constructed using a random subset in order to assess a random group of characteristics in each division. When making predictions, the algorithm votes for classification tasks based on the output of each tree (Maylawati et al. 2021). This approach of group decision-making is assisted by the insights of several trees. Random forests are widely used for classification jobs because they can deal with complex data, reduce overfitting, and generate reliable estimates in a range of scenarios.

19.6 EVALUATION PARAMETER

The Confusion Matrix acts as a primary source of performance measurements for a classification model. The efficiency of the model is assessed using four parameters. Those are: True positive (TP): It is applied to predict the incident value extra accurately, and False Positive (FP): Essentially, it's used to determine an event's incorrect value.

True Negative (TN): This metric is used to anticipate the absence of an event value, and False Negative (FN): This is used to anticipate the no event value erroneously. It is necessary to choose performance metrics in order to calculate the confusion matrix.

The F1-score, accuracy, recall, and precision have all been calculated in order to analyse the performance. The fraction of properly classified examples comparative to all examples is identified as accuracy. The recall is described as the proportion of properly identified occurrences to correctly classed instances as positive. The fraction of precisely classified positive examples to positively classified examples is called the precision. The harmonic mean of recall and precision is identified as the F1-score. To evaluate the stability of the performance of the proposed classification model in the paper, the dataset was split using fivefold cross-validation into training and testing datasets.

$$\text{Precision} = \frac{\text{TP}}{\text{TP} + \text{FP}} \tag{19.7}$$

$$\text{Recall} = \frac{\text{TP}}{\text{TP} + \text{TN}} \tag{19.8}$$

$$\text{F1-Score} = \frac{2(\text{Precision*Recall})}{\text{Precision} + \text{Recall}} \tag{19.9}$$

$$\text{Accuracy} = \frac{\text{TP} + \text{TN}}{\text{TP} + \text{TN} + \text{FP} + \text{FN}} \tag{19.10}$$

$$\text{Silhouette Score } s(i) = \frac{b(i) - a(i)}{\max(a(i),\, b(i)} \tag{19.11}$$

19.7 CHATBOT TRAINING

Our chatbot functions as a virtual physician, using symptoms to make its own diagnosis. The patient and the bot can engage in a natural conversation about the troubling symptoms, which the bot identifies clearly using NLP. The chatbot is trained to learn from the symptoms using a pre-trained model. To ensure an accurate diagnosis, the bot asks relevant questions and requests pertinent information, such as a list of symptoms. The three crucial components of our bot are: Identifying and extracting potential symptoms that the user might be experiencing from their conversation with the chatbot, accurately mapping the extracted symptoms to those in the dataset, and determining the most likely illness based on the identified symptoms.

The ideal use case for our bot is as a primary diagnostic tool that individuals can use to evaluate their symptoms before consulting a physician. We trained the chatbot in Android Studio using Dialogflow, leveraging a previously trained dataset. Dialogflow is designed for natural language conversations and incorporates ML capabilities. It parses and analyses sentences to reveal their structure, along with information about individual words and their relationships. The platform supports over 14 languages, enabling the creation of conversational interfaces across multiple languages.

With the assistance of ML, an agent may convert natural language user inputs into structured data and identify pertinent parameters. ML techniques are used to match the user's purpose from the user request, as described in the Dialog flow language. In order to extract pertinent data from the user, the bot must comprehend their objective and ascertain what they would like to get out of the conversation. It does this by using entities, which are useful pieces of information such as symptoms. Since ML is the foundation of Dialog flow's NLP, one can input training data that the agent learns from and uses to improve its performance.

The created chatbot can serve as a medical assistant. When responding to inquiries from patients regarding their medical conditions, it ought to behave like a doctor. Before the user interacts using the application, the activity where the discussion is held begins. Using the patient's symptoms, the bot delivers the relevant information. The bot then gives the user the disease that the patient would be experiencing when they are in the diagnoses state. Lastly, the application's monitoring page with a list of disease-specific values is accessed by the user.

19.8 HEALTH MONITORING

The application gives users the ability to keep an eye on their health and make sure that a specific goal or set of goals is achieved. Graph representations allow the user to monitor their progress towards recovery from their ailment. The program offers the ability to input data based on an illness that has been identified or entered; this data is then kept in a database and displayed appropriately. Clinical recommendations and characteristics form the basis of the treatment plan's creation. The patient has to measure their health indicators on a frequent basis in order to obtain the high-quality care they are entitled to. Additionally, we used a validator, who is tasked to make sure that aberrant readings are entered into the application, to monitor the unusual readings of the patients. As of right now, no medical instrument automatically sends data to the application.

A few additional characteristics of our app are:

Reminder for Patients: Patients may find it difficult to enter their readings more than once due to their hectic lives. Periodic reminders to submit data and a record of their medicine intake are sent by this service.

Emergency SOS: This function will assist in promptly informing important others that the user requires immediate assistance, as the patient may be in a serious condition. The idea behind it is motion sensing, allowing the patient to alert the emergency contacts by shaking their phone.

Nearby Doctors: With the use of this tool, patients will be able to find expert physicians in their area based on their current place of residence.

The Firebase data houses all of the patient's data. It's a cloud-hosted NoSQL database that operates in real time. In order to synchronize the current diagnosis and give the patient more precise findings, we retain the patient's personal information in addition to previous records. Furthermore, because each user would receive tailored monitoring, the data will not be structured because it will vary from user to user.

NoSQL is the most appropriate in this scenario. Our program has been enhanced by using the Firebase database because it is incredibly quick and effective to get and synchronize data.

19.9 RESULTS

We have examined our system's overall performance as well as the output of various algorithms and system components. We have utilized both the separate test set and our own independent disease categorization dataset. The dataset is arbitrarily separated into ten similar sizes, jointly restricted subsets for fivefold cross-validation. The classification model has been trained and tested five times. The model is assessed on the final fold after being trained on fourfolds for each event. Figure 19.2 demonstrates the test accurateness result of the test graphically. In Table 19.2, there is comparison

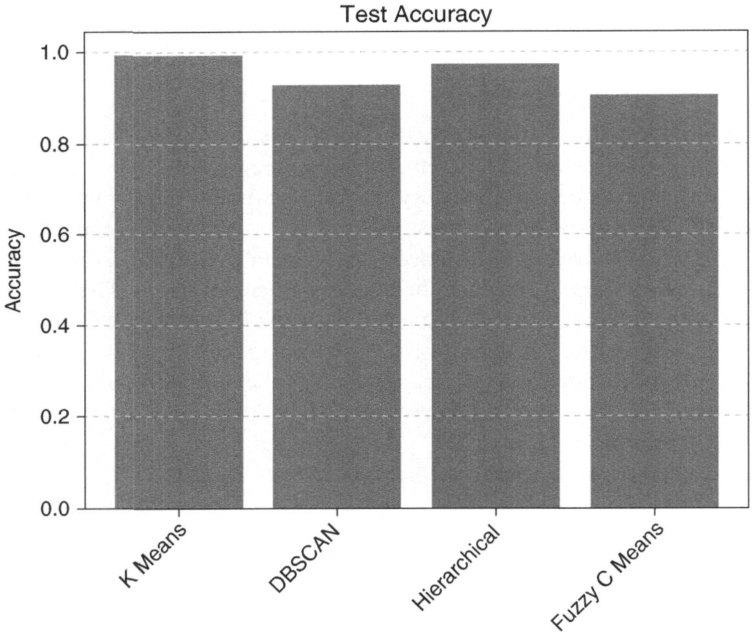

FIGURE 19.2 Comparison of accuracies of different clustering algorithms.

TABLE 19.2
Comprehensive Accuracy of Different Clustering Models

Model	Accuracy	Precision	Recall	F1-Score	Silhouette Score
K-Means	0.992	0.99	0.99	0.98	0.998
DBSCAN	0.9256	0.91	0.91	0.92	0.925
Hierarchical	0.9714	0.99	0.95	0.97	0.973
Fuzzy C-Means	0.903	0.92	0.88	0.90	0.891

of different clustering model followed by random forest classification algorithm. The *K*-Means clustering model has the best accuracy in predicting the disease that is 99.28%. In *K*-Means clustering method, the prognosis of disease is done by classifying the important symptoms that are classified by random forest classification algorithm.

The scattered plot of all clustering models is useful to describe that how the clustering algorithm grouped the data based on the similarity. In our dataset, there are different types of symptoms and the prognosis of disease is based on those symptoms. In this research paper, we applied *K*-Means Clustering, DBSCAN, Agglomerative clustering, and FCM clustering for grouping the symptoms based on similarity for predicting the disease. By applying different clustering algorithms, the two-dimensional visualization of different clusters is shown in Figure 19.3. The first two principal components found using Principal Component Analysis (PCA) are shown on these axes. PCA is frequently used to reduce the dimensionality of the data so that a 2D visualization may be created.

The clusters are distributed around the plot, with some clusters seeming more compact than others. The approximate centres of each coloured region would be the centroids of the clusters, which are not depicted in this plot. A distinct cluster found by the cluster technique is represented by each colour. Based on the features taken into account during the analysis, the points that are clustered together in the same colour indicate that they are comparable to one another.

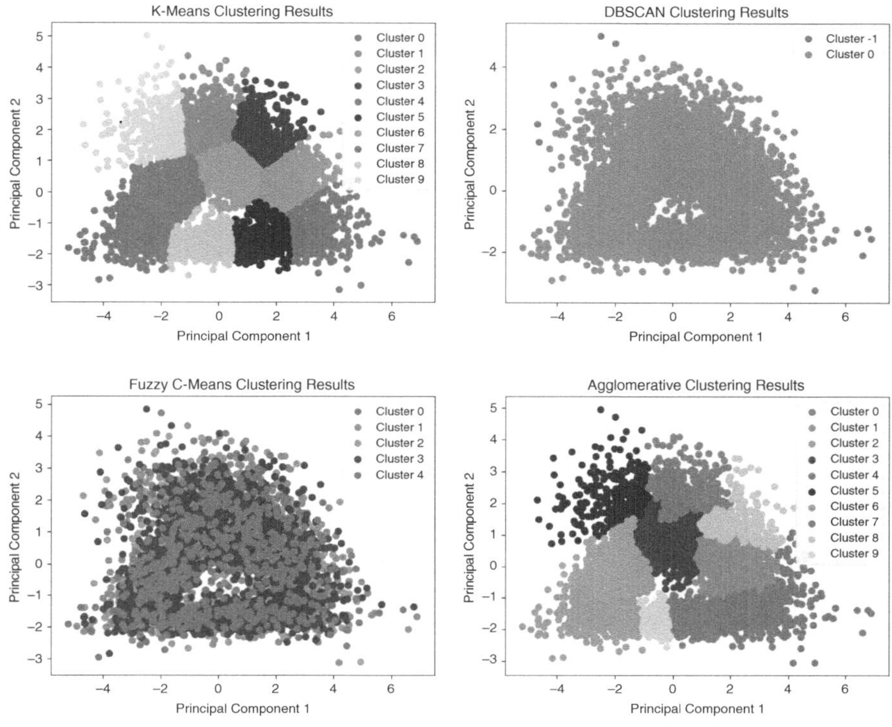

FIGURE 19.3 Scatter plot of all clustering models.

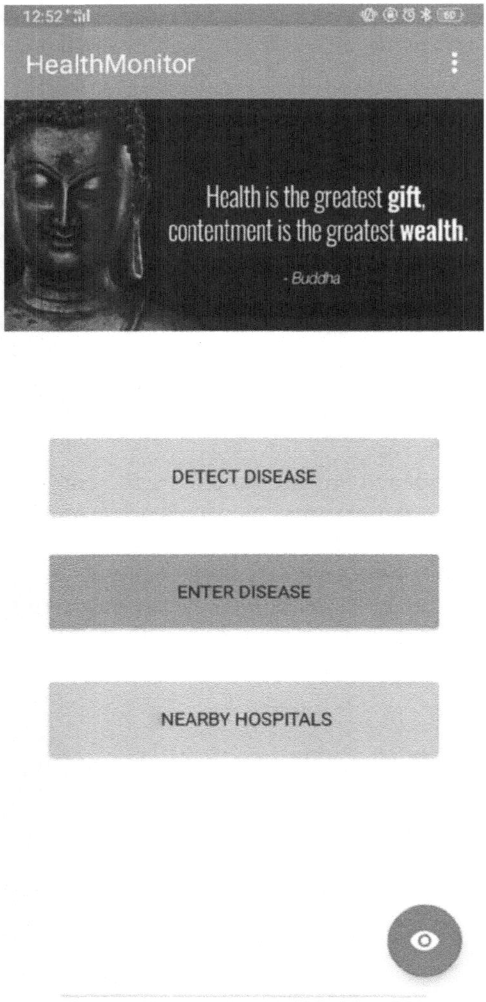

FIGURE 19.4 Main screen.

The user is led to the see progress screen after entering the disease on main screen, as seen in Figure 19.4, where the app requests certain disease-specific data regarding the patient's daily progress, as seen in Figure 19.5. The user input graphs for a depressed patient are displayed in Figure 19.6.

Figure 19.7 illustrates our app's "detect disease" feature, which uses a chatbot to determine a patient's illness based on their symptoms and responses to subsequent questions. Following detection, the program goes back to the user's home screen, where the ailment is shown and pertinent values can be entered.

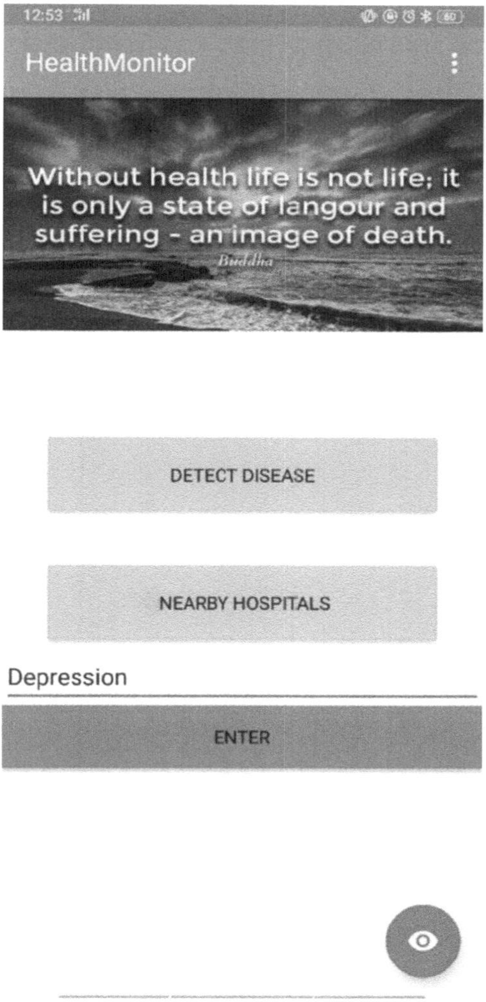

FIGURE 19.5 Enter disease.

Push notifications for medicine reminders and an SOS feature that instantly texts the user-specified contact in the event that the patient's phone shakes suddenly are two add-on features for our app shown in Figure 19.8. Additionally, the patient is prompted for confirmation before the message is sent (Figure 19.9).

Figure 19.10 shows an important feature where the patient can access the location of nearby hospitals through our app (Figure 19.11).

FIGURE 19.6 User input.

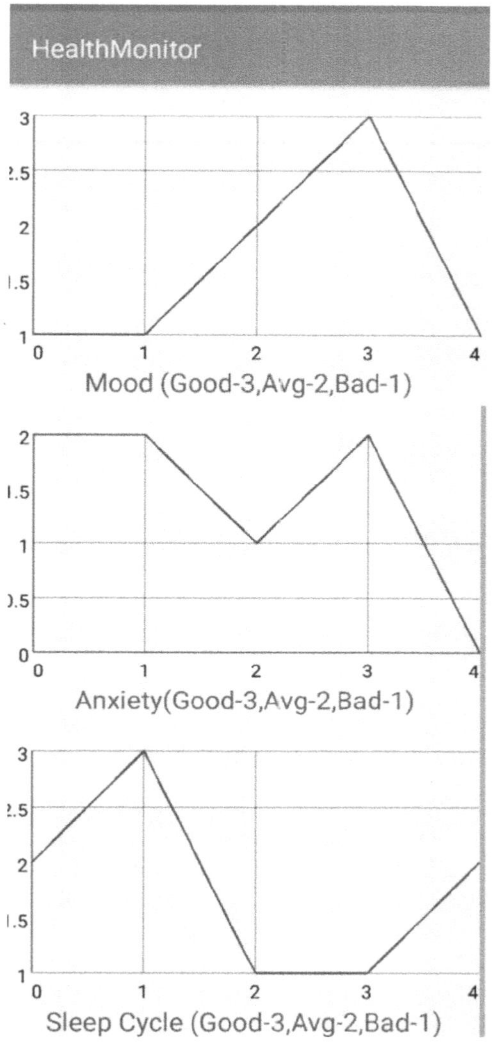

FIGURE 19.7 Monitoring graphs.

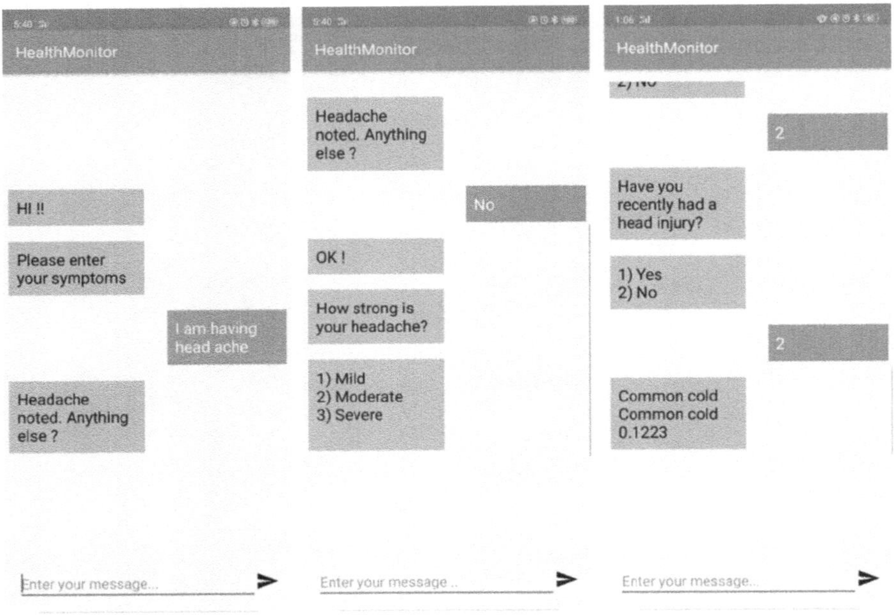

FIGURE 19.8 Chatbot.

19.10 CONCLUSION AND FUTURE SCOPE

This research offers a tool that will aid in the analysis and understanding of the health status of the human race. The application has a chatbot that uses symptoms to anticipate illnesses. Additionally, the patient may follow their progress and keep an eye on their health status thanks to graph visualizations. By using this, the patient can evaluate their own health and take preventative measures based on the outcomes. This will readily enable early disease diagnosis at no further cost. Furthermore, it is accessible from anywhere, regardless of geographic restrictions. The app has extra capabilities that will help the patient in unusual situations, such as an emergency SOS system, hospital searches nearby, and prescription reminders.

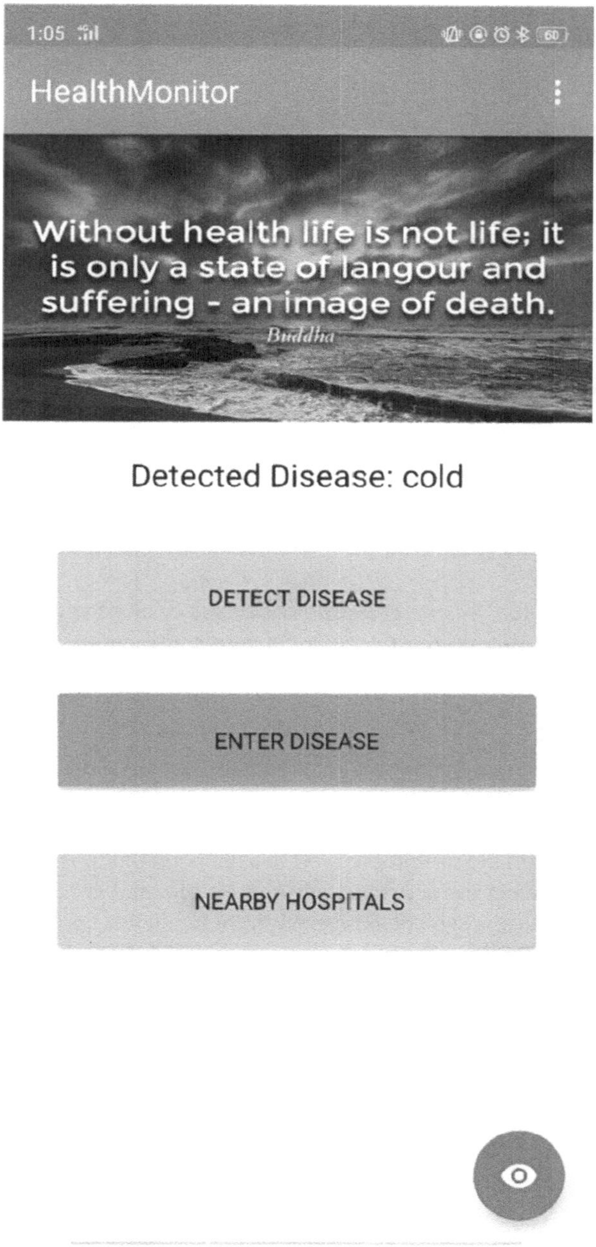

FIGURE 19.9 Predicted disease.

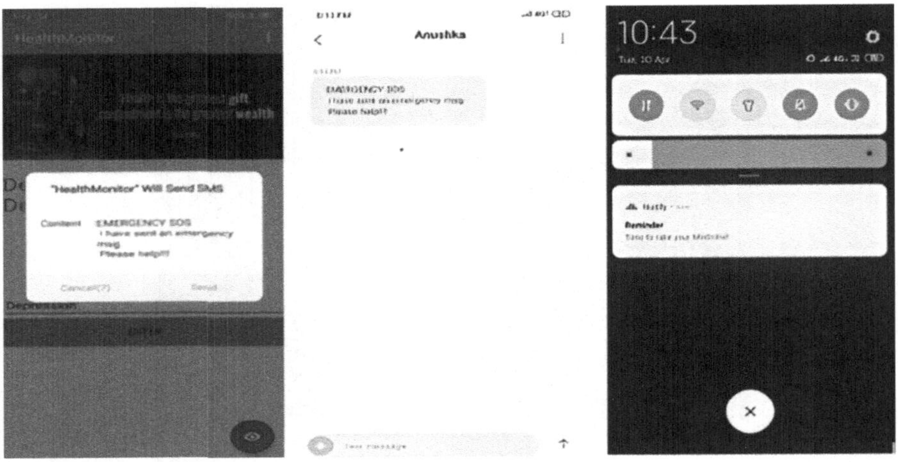

FIGURE 19.10 SOS and reminder.

This application has a very broad potential since, as scholars have previously noted, technology will rule the future. As a result, utilizing this program will be crucial to a patient's life because health is the most valuable resource and taking the right safeguards is crucial. By expanding the dataset or incorporating support for additional medical characteristics, such as geography and genetic history, the bot's ability to recognize symptoms may be substantially enhanced in the future. Additionally, IOT devices can be connected to the app to improve patient monitoring. Using GAI, a highly intelligent system can comprehend, analyse, and forecast possible diseases based on the symptoms. GAI has the potential to completely transform healthcare by improving the accuracy, personalization, and accessibility of medical advice. It can ease the strain on healthcare systems by facilitating early detection and precautionary care. To give the patient greater interaction with the application, audio systems can also be added.

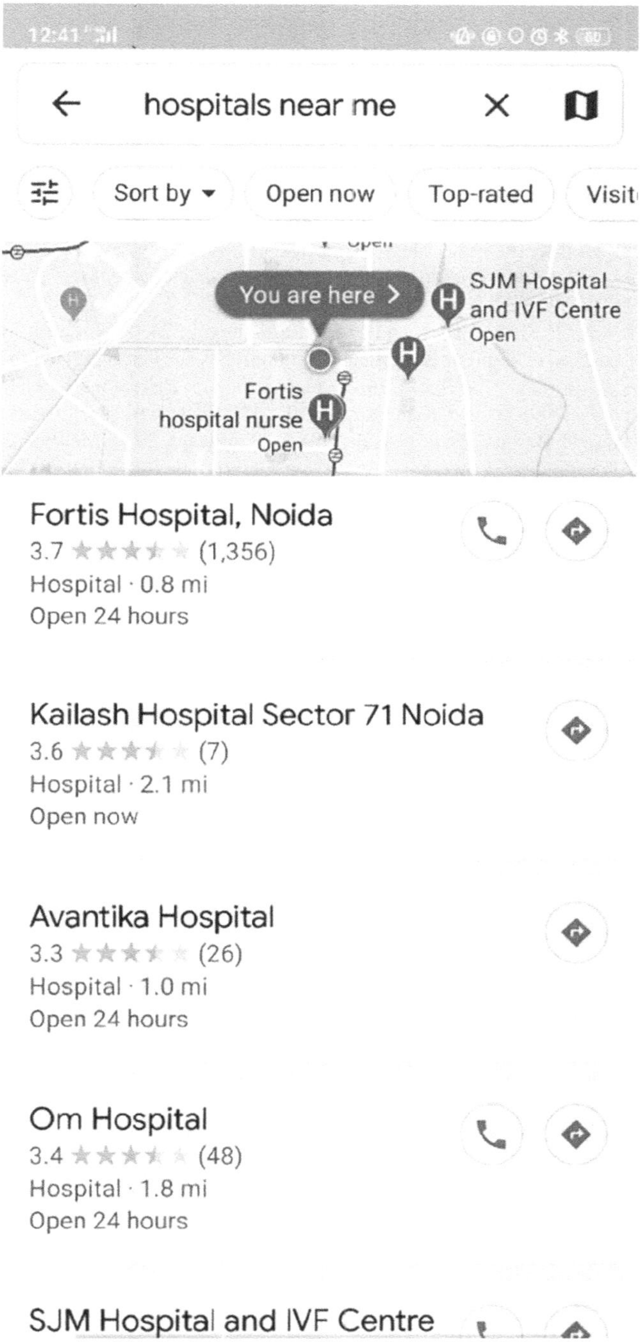

FIGURE 19.11 Location of hospitals.

REFERENCES

Afsahi, A.M., S.A. Seyed Alinaghi, A. Molla, P. Mirzapour, S. Jahani, A. Razi, P. Mojdeganlou, et al. 2024. Chatbots Utility in Healthcare Industry: An Umbrella Review. *Frontiers in Health Informatics* 13 (April 7): 200.

Aggarwal, A., C.C. Tam, D. Wu, X. Li, and S. Qiao. 2023. Artificial Intelligence–Based Chatbots for Promoting Health Behavioral Changes: Systematic Review. *Journal of Medical Internet Research* 25 (February 24): e40789.

Balasubramaniam, K., Pernille Ravn, Pia V. Larsen, Jens Søndergaard, and Dorte E. Jarbøl. Specific and unspecific gynecological alarm symptoms–prevalence estimates in different age groups: a population-based study. *Acta Obstetricia et Gynecologica Scandinavica* 94, no. 2 (2015): 191–197.

Barbara Onyekwelu, I., R. Shrestha, and F. Eika Sandnes. 2024. "Not Quite There Yet": On Users Perception of Popular Healthcare Chatbot Apps for Personal Health Management. In *Proceedings of the 17th International Conference on PErvasive Technologies Related to Assistive Environments*, Crete, Greece, 191–197. New York, NY: ACM.

Bharti, U., D. Bajaj, H. Batra, S. Lalit, S. Lalit, and A. Gangwani. 2020. Medbot: Conversational Artificial Intelligence Powered Chatbot for Delivering Tele-Health After COVID-19. In *2020 5th International Conference on Communication and Electronics Systems (ICCES)*, Coimbatore, India, 870–875. IEEE.

Davis, D.A., N.V. Chawla, N. Blumm, N. Christakis, and A.L. Barabasi. 2008, October. Predicting Individual Disease Risk Based on Medical History. In *Proceedings of the 17th ACM Conference on Information and Knowledge Management*, Napa Valley, CA, 769–778. ACM.

Deng, D. 2020. DBSCAN Clustering Algorithm Based on Density. In *2020 7th International Forum on Electrical Engineering and Automation (IFEEA)*, Hefei, China, 949–953. IEEE.

Divya, S., V. Indumathi, S. Ashwarya, M. Priyasankari, and S. Kalpana Devi. 2018. Survey on Medical Self-Diagnosis Chatbot for Accurate Analysis Using Artificial Intelligence. *International Journal of Trend in Research and Development* 5, no. 2: 1–7.

Dugane, S.Y., and K.G. Bagde. 2017, September. Framework for Disease Prediction from Symptoms and Health Related Data. *International Journal of Advanced Research in Computer Engineering & Technology (IJARCET)* 6, no. 9: 1–7.

Gomathi, K., and D.D.S. Priyaa. 2016. Multi Disease Prediction Using Data Mining Techniques. *International Journal of System and Software Engineering* 4: 12–14.

Harini, D.K., and M. Natesh. 2018, May. Prediction of Probability of Disease Based on Symptoms Using Machine Learning Algorithm. *International Research Journal of Engineering and Technology (IRJET)* 5, no. 05: 2395–0056.

Ingale, S., and T. Zagade. 2024. AI-Powered Healthcare Chatbots Designing Intelligent Virtual Assistants for Patient Education, Symptom Triage, and Remote Monitoring in Telemedicine Services. *African Journal of Biological Sciences*.

King, M.R. 2023. The Future of AI in Medicine: A Perspective from a Chatbot. *Annals of Biomedical Engineering* 51, no. 2 (February 26): 291–295.

Lee, D. 2023. AI-based healthcare chatbot. *International Research Journal of Engineering and Technology* 10: 563–568.

Laskar, M.T.R., M.T. Hossain, A.R.M. Kamal, and N. Rashid. 2016. Automated Disease Prediction System (ADPS): A User Input-Based Reliable Architecture for Disease Prediction. *International Journal of Computer Applications* 975: 8887.

Maylawati, D.S., A. Risqiati, C. Slamet, M.A. Ramdhani, N. Lukman, P. Dauni, and N.D. Arianti. 2021. Chatbot for Virtual Travel Assistant with Random Forest and Rapid Automatic Keyword Extraction. In *2021 IEEE 7th International Conference on Computing, Engineering and Design (ICCED)*, Sukabumi, Indonesia, 1–6. IEEE.

Mohammadrezapour, O., O. Kisi, and F. Pourahmad. 2020. Fuzzy C-Means and K-Means Clustering with Genetic Algorithm for Identification of Homogeneous Regions of Groundwater Quality. *Neural Computing and Applications* 32, no. 8 (April 19): 3763–3775.

Patil, R., P. Chopade, A. Mishra, B. Sane, and Y. Sargar. 2016. Disease Prediction System Using Data Mining Hybrid Approach. *Communications on Applied Electronics (CAE)* 4, no. 9: 48–51.

Rahman, Md. M., R. Amin, M.N. Khan Liton, and N. Hossain. 2019. Disha: An Implementation of Machine Learning Based Bangla Healthcare Chatbot. In *2019 22nd International Conference on Computer and Information Technology (ICCIT)*, Dhaka, Bangladesh, 1–6. IEEE.

Rahman, T., T. Hossain, A. Raihan, and N. Rashid. 2016. Automated Disease Prediction System (ADPS): A User Input-Based Reliable Architecture for Disease Prediction. *International Journal of Computer Applications* 133, no. 15 (January 15): 24–29.

Rani, K., H. Vishnoi, and M. Mishra. 2023. A Mental Health Chatbot Delivering Cognitive Behavior Therapy and Remote Health Monitoring Using NLP and AI. In *2023 International Conference on Disruptive Technologies (ICDT)*, Greater Noida, India, 313–317. IEEE.

Rathnayaka, P., N. Mills, D. Burnett, D. De Silva, D. Alahakoon, and R. Gray. 2022. A Mental Health Chatbot with Cognitive Skills for Personalised Behavioural Activation and Remote Health Monitoring. *Sensors* 22, no. 10 (May 11): 3653.

Roux, M. 2018. A Comparative Study of Divisive and Agglomerative Hierarchical Clustering Algorithms. *Journal of Classification* 35, no. 2 (July 7): 345–366.

Shankar, M., M. Pahadia, D. Srivastava, T.S. Ashwin, and G.R.M. Reddy. 2015. A Novel Method for Disease Recognition and Cure Time Prediction Based on Symptoms. In *2015 Second International Conference on Advances in Computing and Communication Engineering*, Dehradun, India, 679–682. IEEE.

Sinaga, K.P., and M.-S. Yang. 2020. Unsupervised K-Means Clustering Algorithm. *IEEE Access* 8: 80716–80727.

Veena, A., and S. Gowrishankar. 2021. Healthcare Analytics: Overcoming the Barriers to Health Information Using Machine Learning Algorithms. In Chen, J.I., J.M.R.S. Tavares, S. Shakya, and A.M. Iliyasu (eds.), *Image Processing and Capsule Networks*. ICIPCN 2020. Advances in Intelligent Systems and Computing, vol. 1200, 484–496. Springer, Cham.

20 From Pixels to Prognosis
The Role of Image Synthesis in Medical Diagnosis

Vijaykumar, Rachna Narula, Sarthak Khanna, Anil Kumar, Poonam Chaudhary, and Surbhi Bhatia Khan

20.1 INTRODUCTION TO MEDICAL IMAGE SYNTHESIS

Medical image synthesis is an uprising technology, which is radically changing how the medical visual data is created, manipulated, and interpreted in the field of medical imaging. Medical image synthesis has developed into an extremely valuable tool to tackle a diverse set of challenges and pull the potential of medical imaging technology by employing advanced computation techniques and machine learning algorithms [1].

However, Medical image synthesis is basically producing fake medical images that resemble real patient data. These fake images are very handy means for diverse applications in healthcare, research, and educational activities as they show features such as anatomic structures, illness-related anomalies, and imaging artefacts [2].

Medical image synthesis is very useful to us because it can overcome some of the limitations that are seen in traditional imaging medical techniques. Imaging rare or complex medical conditions is challenging, the availability of diverse and well-annotated images is scarce, and image quality varies across modalities among these limitations [3]. Medical image synthesis offers creative solutions to these challenges by generating synthetic images that enhance image resolution, mimic diseased conditions, augment existing datasets, and facilitate the creation and evaluation of imaging algorithms and protocols [4]. Also, medical image synthesis has strong potential to enable precision healthcare and personalized treatment. By making custom anatomical models and prediction simulations for each person, medical image synthesis aids doctors with prognosis of illnesses and planning treatments for improved therapeutic outcomes [5] (Table 20.1).

As this study advances, medical picture synthesis seems not only a tech advancement but also a game changer in the way we gather, process, and interpret visual medical information. By extending beyond current capabilities in medical imaging, medical image synthesis has the potential to upgrade the healthcare delivery system, enhance the accuracy of diagnoses, and ultimately improve patient care.

 DOI: 10.1201/9781003565703-20

TABLE 20.1
Impact of Medical Image Synthesis Across Applications

Sector	Primary Applications	Impact Level (1–5)	Key Challenges
Healthcare	Diagnostic imaging, Treatment planning	4.8	Ethical concerns, Data privacy
Medical Research	Algorithm training, Disease simulation	4.7	Lack of standardized validation protocols
Education	Teaching anatomical structures, Training practitioners	4.3	Dependence on synthetic accuracy
Technology	AI algorithm testing, Model Validation	4.6	Computational resource demands
Pharmaceuticals	Drug response prediction, Therapy optimization	4.5	Limited integration with existing workflows

20.2 IMPORTANCE AND APPLICATIONS OF MEDICAL IMAGE SYNTHESIS

Today, medical image synthesis has many applications and great importance in healthcare. Besides that, this technology greatly impacts medical research, diagnosis, treatment planning, and education. Moreover, there are numerous applications and importance of medical image synthesis.

20.2.1 Augmentation of Training Data

The generation of realistic images via medical image synthesis helps in enhancing training datasets and machine learning algorithms for classification and segmentation. It prevents the problem of scarcity of data by providing labelled medical data.

20.2.2 Simulation of Uncommon or Difficult Cases

With the help of artificial imagery, health professionals can learn how to recognize and treat rare diseases or anatomical anomalies. Also, they imitate complex processes like surgery, which ensures improving skills in a safe environment.

20.2.3 Assessment of Imaging Methods and Protocols

Computer-generated images assess imaging techniques without subjecting patients to needless radiation. In order to help evaluate the influence of artefacts on diagnostic accuracy, they also mimic them.

20.2.4 Treatment Planning and Personalized Medicine

For surgical simulation and personalized treatment planning, synthetic pictures are used to construct patient-specific anatomical models. Using predictive modelling helps ensure that every patient receives well-informed care [6].

20.2.5 Educational and Training Objectives

The purpose of synthetic images is to help medical practitioners learn anatomy, pathology, and diagnostic interpretation in virtual situations. To stay up to date with developments in medical practice, they are in favour of ongoing education.

20.2.6 Innovation and Research

In research, synthetic images facilitate the investigation of novel imaging modalities and computational strategies. By using computer models to facilitate virtual screening and molecular docking, they aid in the creation of new drugs [7].

As shown in Figure 20.1, the capabilities of medical image synthesis have grown exponentially over the past decade. Table 20.2 compares various applications of medical image synthesis, showcasing their key features, impact levels, and significance in improving healthcare, research, and training.

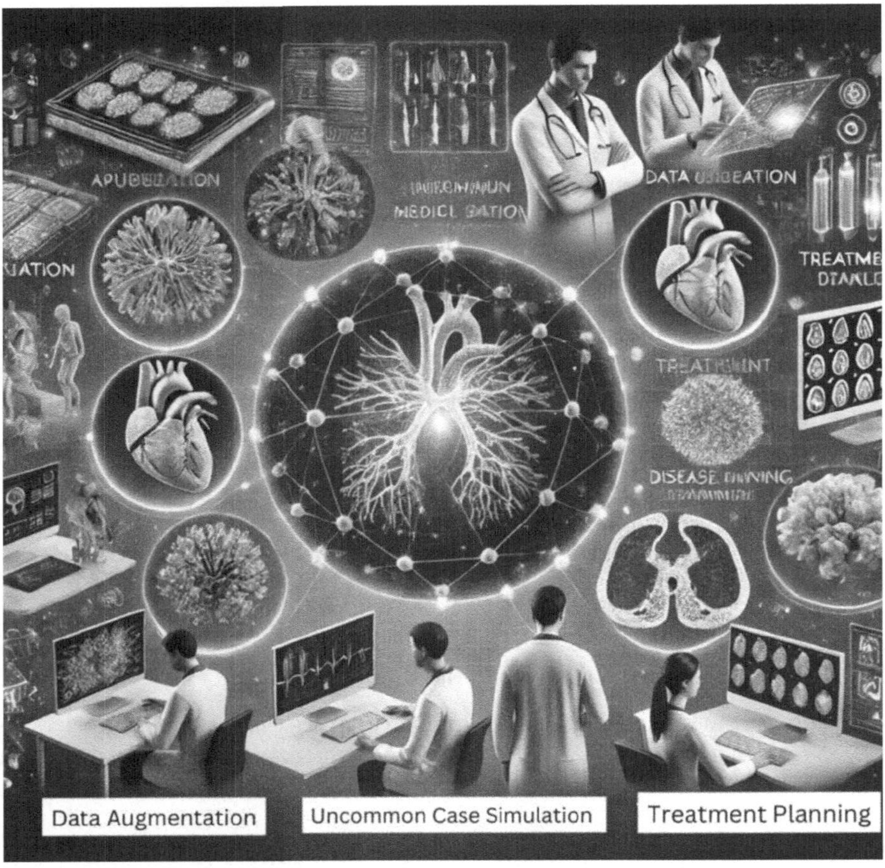

FIGURE 20.1 Applications of medical image synthesis in healthcare. (This image visually represents various applications such as data augmentation, simulation of rare cases, treatment planning, and education in a futuristic medical context.)

TABLE 20.2
Applications and Importance of Medical Image Synthesis

Application Area	Key Features	Significance	Impact Level
Augmentation of training data	Realistic image generation, Annotated datasets	Enhances machine learning models, addresses data scarcity	4.8
Simulation of uncommon/difficult cases	Artificial generation of rare conditions, Surgical simulations	Trains professionals in rare diseases and complex procedures	4.7
Assessment of imaging protocols	Synthetic evaluation, Artefact simulation	Reduces patient radiation, aids in diagnostic accuracy	4.5
Treatment planning and personalized medicine	Patient-specific models, Predictive modeling	Supports personalized treatment and surgical planning	4.9
Educational and training objectives	Anatomy/pathology simulation, Virtual learning environments	Improves diagnostics and supports continuous education	4.6
Innovation and research	Novel imaging modalities, Drug discovery	Facilitates new imaging techniques and drug development	4.9

Impact Level is rated on a scale of 1–5, where 5 represents a highly significant contribution to medical imaging and healthcare practices.

Medical image synthesis is an incredibly useful and important tool in today's healthcare system, providing creative answers to problems with medical imaging and enabling researchers, educators, and practitioners [8]. Its many uses highlight how crucial technology is to expanding medical knowledge, enhancing patient outcomes, and influencing how healthcare will be provided in the future.

20.3 FUNDAMENTALS OF MEDICAL IMAGE SYNTHESIS TECHNIQUES

The principles of medical image synthesis techniques comprise an extensive variety of approaches and algorithms designed to produce artificial medical images that have a striking resemblance to actual patient data. These techniques use state-of-the-art computational techniques including deep learning, machine learning, and image processing to replicate anatomical structures, disease abnormalities, and imaging features observed in clinical practice [9]. Some basic ideas of medical image synthesis techniques are as follows:

1. **Models that Generate**:
   Generative Adversarial Networks (GANs): GANs employ two neural networks, trained against each other, to make images like a Generator and a discriminator. In medical images, GANs are employed for creating synthetic image with disease and the synthetic anatomical structures to see [10].

Variational Autoencoders (VAEs): VAEs are probabilistic generative models that sample from the learned distribution to make new samples by learning latent representations of the input. In medical image synthesis, we apply VAEs to synthesize many types of images while still keeping data variance [11].

2. **Image-to-Image Translation**:

Conditional GANs: Images are generated with conditional GANs given some extra input, e.g., modality or segmentation masks. In the field of medical imaging, Conditional GANs are used for picture augmentation and cross-modality image synthesis [12].

CycleGANs: CycleGANs perform unsupervised image translation by training the mapping between image domains without any paired data. CycleGANs are used for methods including style transfer and domain adaptation in medical image synthesis [13].

3. **Geometric Transformations and Data Augmentation**:

Images of Medical Data are Geometrically Altered (translating, scaling, rotating, etc. to increase dataset robustness/resilience and variability). Intensity Transformations—Intensity transformations such as adding noise to the images, histogram equalization, and contrast enhancement are addressed for mimicking the artefacts that appear during image acquisition.

4. **Models for Specific Domains**:

Anatomical Models: An anatomical model, e.g., statistical normalization models (SSM) or deformable registration methods, is used to build high fidelity anatomical structures and shapes of organs [14].

Pathological Models: By simulating disease-specific properties and tissue parameter alterations, pathological models enable us to make a range of medical pathologies and anomalies in images [15].

5. **Pre-trained Models and Transfer Learning**:

Transfer of Learning: Transfer learning methods make it possible to train on tiny medical datasets with the help of pre-trained models that are previously learnt from large-scale image datasets for synthesis tasks [16].

Pre-trained Model: Convolutional Neural Networks (CNNs) and other pre-trained models like those on natural image datasets are retrained or adapted, for example, to shrink the training data needed in medical image synthesis tasks [17].

6. **Metrics for Evaluation**:

Perceptual Metrics Defines the structure, texture, and spatial coherence related perceptual metrics, rather than computing visual quality and similarity in artificial or real images from structures.

Numerical Metrics: These quantify objective criteria to judge synthesized images for the quality and condition: Optical flow, peak signal-to-noise ratio (PSNR), pixel-wise similarity, and structural similarity index [18] (Tables 20.3 and 20.4).

TABLE 20.3
Comparison of Performance of Medical Image Synthesis Techniques

Technique	Computational Overhead	Hardware Requirements	Training Time	Scalability	Notes
Generative Adversarial Networks (GANs)	High	GPUs/TPUs, High RAM	Long (days to weeks)	High	High-quality synthesis but very resource-intensive due to adversarial training loops
Variational Autoencoders (VAEs)	Moderate	GPUs recommended	Moderate (hours to days)	Moderate	Less computationally expensive than GANs; useful for diverse image generation
Conditional GANs (cGANs)	High	High-end GPUs, Large storage	Long (similar to GANs)	High	Conditional inputs increase complexity and training time
CycleGANs	Very High	Multiple GPUs, Distributed systems	Very Long (days to weeks)	High	Unsupervised learning adds overhead due to dual generator–discriminator pairs
Geometric/data augmentation	Low	Standard CPU/GPU	Very Short	Very High	Lightweight and easy to integrate; used as a baseline
Traditional morphological methods	Low	Minimal computational needs	Short	Low	Fast but lacks flexibility and realism of deep learning approaches
Pre-trained models + transfer learning	Moderate	GPU for fine-tuning	Short to Moderate	High	Efficient for small datasets, reduces training time
Anatomical/statistical models	Moderate	CPU/GPU for deformation registration	Moderate	Medium	Good for modelling organ shapes, but lacks high realism
Pathological models	Moderate to High	High memory/GPU for tissue simulation	Moderate	Medium	Computationally intensive depending on tissue complexity

In summary, we described the use of advanced computational algorithms in medical image synthesis approaches to create accurate and potentially physician-relevant images. By leveraging generative models, data augmentation, domain-specific models, image-to-image translation, and evaluation metrics, among others, these methods seek to stimulate creative avenues for enhancing medical imaging capabilities, providing support towards additional research and, ultimately, care.

TABLE 20.4
Computational Resource Analysis of Medical Image Synthesis Techniques

Technique	Runtime	Model Complexity	Scalability
GANs	High	High—Dual networks with millions of parameters	Scales with data parallelism; requires high GPU resources
VAEs	Moderate	Moderate—Probabilistic encoder-decoder model	Handles large datasets; quality may drop for complex domains
Conditional GANs (cGANs)	High	High—Extra conditional inputs add complexity	Excellent for tasks with labels; more GPU memory needed
CycleGANs	Very High	Very High—Four networks (2 Gen, 2 Disc)	Computationally demanding; good for unpaired data
Geometric/data augmentation	Very Low	Low—No networks involved	Extremely scalable; applicable in real time on large datasets
Traditional morphological methods	Low	Very Low—Mathematical operations only	Limited scalability; poor for modern high-dimensional image synthesis
Pre-trained models + transfer learning	Moderate	Moderate to High—Depends on base model	Highly scalable and efficient for both small and large datasets
Anatomical/statistical models	Moderate	Moderate—Shape priors and deformation fields	Moderately scalable; depends on quality of anatomical templates
Pathological models	Moderate to High	High—Simulated tissue and pathology models	Limited by diversity of training data and anatomical variability

20.4 DATA ACQUISITION AND PREPROCESSING FOR IMAGE SYNTHESIS

Preprocessing and Data Collection are important steps in getting ready for your supposed images. The raw data is taken from aggregated datasets (institutional, databases) and medical imaging modalities. Preprocessing is the process of noise removal, intensity normalization, artefact removal, and image registration. Ensuring training ground truth data is provided through annotation. Quality control makes sure that the data that is in your dataset is in fact what you intended; dataset augmentation adds diversity [19]. Without them, creating realistic-looking artificial medical images for healthcare and research applications is impossible.

20.5 MEDICAL IMAGE SYNTHESIS: TRADITIONAL VS. DEEP LEARNING APPROACHES

Two major ways of medical image synthesis are deep learning techniques and traditional approaches. In some ways, each one has its pros and cons.

20.5.1 Traditional Approaches

Model-Based: Traditional approaches often use mathematical models and algorithms to mimic anatomical components as well imaging characteristics. Techniques to create synthetic images are interpolation, morphological techniques, and picture warping [20].

Handcrafted Features: Conventional Approaches require a feature engineering. These are the features that are computed with respect to medical imaging tasks and use domain-specific information [21]. Examples of these are texture descriptors, shape attributes, intensity histograms, and so forth. Virtual case generation: Using this feature, the models synthesized are restricted and do not accurately capture the wide array of diseases and imaging abnormalities present.

Computational Complexity: Traditional methods are not as scalable and effective as deep learning techniques, especially for large-scale datasets since they commonly consist of intensive computational procedures and hyperparameter tuning.

20.5.2 Deep Learning

End-to-End Learning: Deep learning enables end-to-end learning, unsupervised from raw data, without unique functions or domain knowledge requirements [22]. Generative models and CNNs learn hierarchical representations of medical images to represent complex patterns and structures in a hierarchical manner [23].

Feature Extracting: To mine such patterns and features in medical image data automatically, deep learning models allow them to create hierarchical data representations. This results in efficient generation of images (synthesis) across different modalities and disorders jointly [24].

Scalability: Deep learning methods are highly scalable and robust to massive datasets as improvements in distributed computing and hardware accelerate several levels of the pipeline [25].

State-of-the-Art Performance: Data Augmentation, Super-resolution, and Image-to-image Translation have shown very promising outcomes with regard to how well they can generate high-quality synthetic images containing anatomical correctness and fidelity to disease [26].

To sum it up, deep learning-based approaches are very flexible, scalable, and performant for synthesizing medical picture, while traditional techniques have interpretability and simplicity. Deep learning-based methods have been extremely successful in detecting sophisticated patterns and producing fantastic synthetic images that have translated to phenomenal results in education, research, and healthcare. Nevertheless, deep learning vs. classical methods selection is based on the task itself, resources available at system, processing capacity, and proficiency interacting in the process.

20.6 CHALLENGES AND LIMITATIONS IN MEDICAL IMAGE SYNTHESIS

Some limitations and hurdles include the functionality and usability of medical image synthesis within research or clinical settings. To develop reliable and robust synthesis methods, it is important to have a good understanding of these barriers. Below are some of the major challenges and limitations in medical image synthesis:

Datasets Are Restricted and Not Balanced: Most datasets within the literature exhibit an unbalance in pathology cases making them challenging, and there are few image datasets in medical research.

Individuality of Patient Anatomy/Pathology: Representing human anatomy a million times over and the diversity of pathological situations makes it harder to perfectly capture all the anatomical features/illness symptoms in synthetic images.

Scalability of Resampling and Noise/Figletting in Imaging: Imaging figletting and noise make synthesis models challenging, resulting in inconsistency and lowered image quality. There are some artefacts and electrical interference cases.

Incomprehensibility of the Deep Learning: It is urgent to be interpretable, because some deep learning genetic synthesis models did not perform this. It becomes impossible for them to be used in places that need interpretability.

Legal and Ethical Implications: Strict adherence to regulations such as HIPAA and GDPR to vet client privacy issues, data security concerns, and the ethics of processing artificial pictures.

Challenge to Define Evaluation Measures and Benchmarks: There is a need of standard techniques and expert judgment for developing appropriate evaluation measures and benchmarks to measure the quality, realism, and clinical utility of synthesized images.

20.7 ETHICAL CONSIDERATIONS IN MEDICAL IMAGE SYNTHESIS

Before we can use medical image synthesis in healthcare, several important ethical issues need attention.

Protection of Patient's Data and Their Privacy: Synthetic images are created from authentic medical data, and so this data needs to be protected from unauthorized personnel as well as comply with regulations such as HIPAA and GDPR. This is essential for preserving public trust. Your AI tools must be designed to fend off any kind of reconstruction of real patient information in order to protect their privacy.

Correction of Bias in Data: Generative algorithms often tend to reflect bias if the dataset used in training is majorly based on a single demographic group, and, therefore, they will tend not to perform well on other groups. For example, training the system on faces from only one ethnic group will cause the solution to fail when it encounters people from other ethnic groups.

Regulatory Hurdles: The use of artificial medical images in this space will need regulatory clearance. Regulatory agencies will need to create rules and regulations to make the technology safe and understandable for all.

Mitigation of Risks of Misuse: AI can create images that look as good as the real thing, which can be misused. For example, fake medical records. We need tough security methods and checks so that these misuses do not happen.

20.8 EVALUATION METRICS FOR ASSESSING SYNTHESIZED MEDICAL IMAGES

It is important to analyse and evaluate the quality and usefulness of synthesized medical images. It combines the manager's quantitative analysis with their judgment [27].

Peak Signal-to-Noise Ratio and Structural Similarity Index can be used for several applications, including comparison of image quality and compression methods. However, PSNR suffers from some limitations. Therefore, PSNR and SSIM must be interpreted by professionals. PSNR refers to the Peak Signal-to-Noise Ratio that helps to measure the levels of noise found in the synthetic images with respect to the real images to find out the overall quality [28].

Mean Squared Error and Mean Absolute Error help to understand how accurate the images are. They do this through the average of difference in pixel values between the real and synthetic versions.

The Fréchet Inception Distance (FID) assesses how realistic and diverse the generated images are by comparing them to their real-world counterpart, with lower values being better.

Mean Structural Similarity checks the average SSIM in specified critical areas of synthetic images, which focuses on localized details.

Perceptual Hashing is the technique that helps in indexing and comparing images by generating compact representations. These representations are based on visual similarity [29].

Radiologists and physicians provide subjective evaluations regarding clinical relevance, diagnostic accuracy, and interpretability of the images.

All these measurements provide crucial feedback to improve design methods for imaging synthesis for specific healthcare applications.

20.9 RECENT ADVANCES AND FUTURE DIRECTIONS IN MEDICAL IMAGE SYNTHESIS

Recently, a lot has been done to produce medical images with advanced computer methods, which made them more robust and useful for medical use.

20.9.1 Recent Developments

New techniques, such as the Variational Autoencoder (VAE) and the GAN, have vastly improved the capacity to create clinically useful medical images [30].

Technical advances have made it much easier to transfer knowledge across various medical specialties and imaging modalities. As a result, systems are now able to manage diverse medical data. We are putting extra time and effort in bringing together images, like PET-MRI or CT-MRI, from different kinds of medical scans to enhance diagnostic accuracy [31].

As these systems become more useful in medical decision-making, evaluating the confidence of the results through a range of assurances has become a necessity.

New techniques are being developed to generate medical images without jeopardizing patient privacy. In future, this area will deal with the issues of rare data, ease of interpretation of models, and validation in a clinical setting.

20.9.2 Future Plans

Researchers are trying to create medical images that require less labelled data. They learn from small data, transfer knowledge across areas, and learn from data without labels.

Researchers want to create models that will help explain how images are created and which structures of the body are related to this.

To ensure that these images are acceptable to practitioners, further research will be undertaken to validate their value through thorough testing with expert practitioners such as physicians and radiologists.

Making medical images that are tailored to individual patients and situation scenarios is an increasing area of interest. Different models are created that adapt to specific body characteristics and medical requirements.

Researchers are attempting to incorporate these imaging processes into the doctor's workflow to support planning, decision-making during surgery, and medical education.

As these practices become more established in healthcare, research will be aiming to address ethical concerns, following regulations, and maintaining transparency in the development and application of these models.

In conclusion, recent developments are paving way for new opportunities in healthcare. The future of medical image creation is targeted on improving the effectiveness, clarity, and applicability of models, which will ultimately lead to better patient care (Figure 20.2).

20.10 CASE STUDIES

20.10.1 Case Study 1: Treatment Planning for Radiation Therapy

Background: Radiotherapy is an essential method employed in cancer treatment that initiates damage to cancer cells and preserves adjacent healthy tissues. The accuracy of this targeting is critical for successful treatment, and it is important to reduce radiation exposure to critical organs such as the heart, lungs, and spinal cord—commonly referred to as organs at risk (OARs).

Developing a detailed, personalized treatment plan is critical for delivering the right amount of radiation to the cancer while minimizing damage to healthy tissue. The traditional way for doctors is to use computed tomography (CT) and magnetic resonance imaging (MRI) scans to find the tumour as well as OARs. Nonetheless, this is a laborious process and dependent on the doctor's urine to define OARs and the radiation dose.

Synthetic medical images are revolutionizing this process to be more accurate and efficient.

These images help in ensuring that the correct dose is delivered to cancer while protecting healthy tissues.

a. To target the tumour and spare healthy tissues, dosing must be accurate.
b. Deep learning uses fake pictures to know radiation distribution.

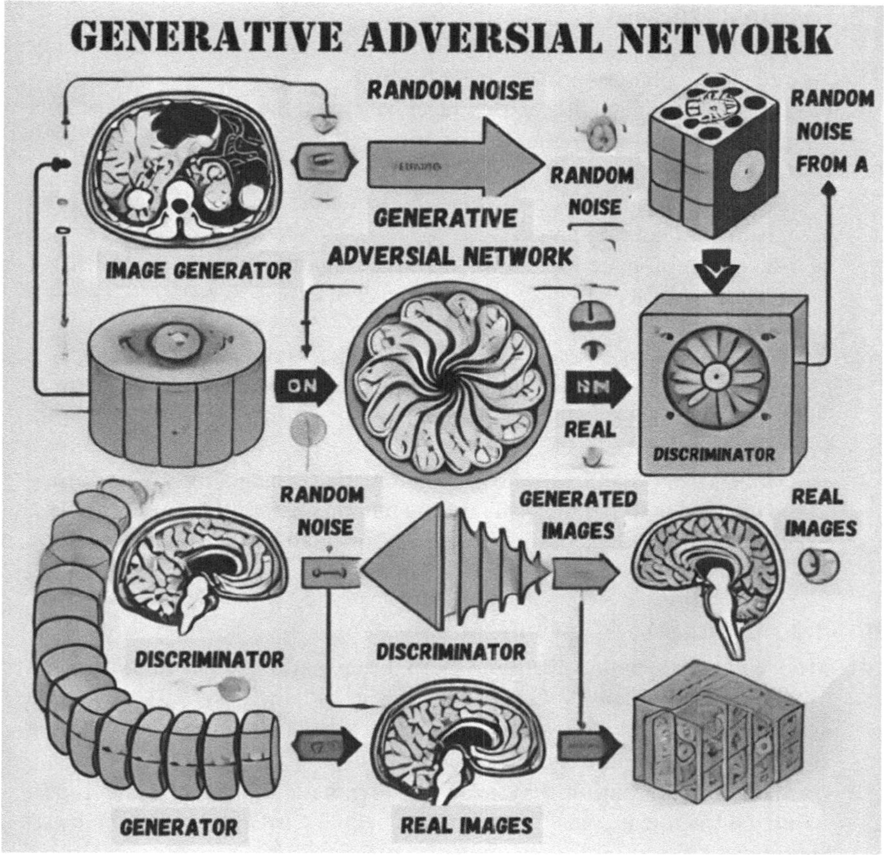

FIGURE 20.2 Schematic diagram of Generative Adversarial Network (GAN) architecture for medical image synthesis. (GAN operation schema. The generator generates replica medical images, while the discriminator assesses them by comparing them to real ones to verify their authenticity.)

c. Models that simulate the interaction of radiation dose with tissues help in optimizing tumour.
d. Dose prediction models are made more personalized to patients by incorporating synthetic images depicting diverse anatomical and clinical scenarios.

20.10.1.1 Benefits

1. **Improved Accuracy**:
   a. Putting together made images and real ones helps treatment planning to adapt to different bodies and diseases better.
   b. Improving the accuracy of organ segmentation makes tumour targeting more accurate and reduces damage to organs from radiation.

2. **Increased Efficiency**:
   a. Automating the detection of OARs and predicting radiation dose greatly reduce the planning time.
   b. Doctors can spend this extra time enhancing treatment plans of patients and engaging them rather than doing repetitive work.
3. **Better Consistency**:
   a. Using dummy images causes less human error, which standardizes the treatment planning process.
   b. Accordingly, patients receive the same quality of care regardless of what the facility has or the doctor's level of experience.
4. **Cost Savings**:
   a. Using machinery reduces cost by requiring less intensive manual effort.
   b. Automated workflows and synthetic image generation will help to deliver advanced radiation therapy in poorly resourced settings.
5. **Fewer Mistakes**:
   a. Models using synthetic data from automatic segmentation are less prone to errors due to fatigue or forgetting human anatomy.
   b. Overall, this enhances the safety and dependability of radiation therapy treatments.

20.10.1.2 Clinical Workflow Integration

1. **Model Training with Additional Data**: Deep learning algorithms are used to automatically identify OARs and predict radiation doses. Using these synthetic images allows fast and accurate treatment planning along with the ability to use these models.
2. **Immediate Application**: Physicians use synthetic images to develop personalized treatment plans. These images help in correcting radiation distribution instantly as per the anatomy of the patient.
3. **Ensuring Quality**: Synthetic images are made to pass through rigorous quality checks to meet medical standards. Radiologists' opinions are sought after comparing the images with the actual patient images.

20.10.1.3 Outcome

Synthetic medical images are contributing high to the radiation therapy given to the patient and the patient's response. Here's a breakdown of the same.

20.10.1.3.1 Better Tumour Control

a. With these images, doctors can pinpoint tumours with precision to annihilate with a full radiation dose.
b. They assist in keeping the radiation beam accurately lined up with the cancer as the shape and position of the tumour change.

20.10.1.3.2 Fewer Side Effects

a. The images help determine organs that may be affected by the radiation and where it will spread, minimizing doses to healthy areas.
b. Because of this, the risk of side effects such as damage to the spinal cord, heart, and lungs is lowered.

20.10.1.3.3 Personalized Treatment Plans

a. Patients receive treatment plans that are tailored to their unique body and illness conditions to improve their care.
Quicker Planning Time: Automated systems that identify different tissues and predict different doses can speed up the planning and help doctors.

20.10.1.4 Future Directions

1. **Advanced Imaging Techniques**: Using the combination of PET-CT or MRI-CT scans can improve synthetic medical images and assist in better targeting of the cancer cells.
2. **Federated Learning for Information Sharing**: Federated learning allows hospitals and research centres to work together on model training without sharing private patient data, which makes synthetic image-based models more robust.
3. **Flexible Treatment Plans**: Future systems might employ real-time synthetic images to modify treatment plans during therapy, adapting to changes in the patient's body or tumour size.

20.10.2 Case Study 2: Neuroimaging with Synthetic MRI

Background: MRI scanners use a powerful magnetic field and radio waves to create detailed images. MRI scanners help to observe the body tissues, diseases, etc. Generally, doctors request several scans, including T1-weighted, T2-weighted, and FLAIR, to investigate a certain area or condition. To get these scans, patients are required to lie down and remain still for a considerable amount of time. However, this can make the patient uncomfortable, and during the scan if there happens to be any motion, the images will get blurred. This is especially difficult for children, older people, or those who are very sick. Synthetic MRI solves these problems by speeding up scans and enhancing accuracy. It creates multiple types of images from one piece of data—enhancing tissue characterization and helping to make more accurate diagnoses.

Using Combined Medical Images: Synthetic MRI utilizes advanced computing to create images from the patient's data instead of multiple scans. This eliminates the need to scan each image type separately and speeds it up while still providing vital diagnostic information.

1. **Multi-Contrast Imaging**: Synthetic MRI uses a single dataset to create different types of images for medical analysis:

   T1-weighted images show a lot of important details like brain white matter.

   T2-Weighted Images show Fluid Accumulation and Swelling.

   FLAIR images are helpful to show lesions arising from multiple sclerosis by suppressing the visibility of CSF. It uses artificial intelligence to change data into pictures like MRI [32].

2. **Quantitative Analysis**: Synthetic MRI offers measurable insights into tissue characteristics:

   T1 Relaxation Time shows how fast excited protons return to equilibrium state, which helps in showing the tissue abnormality.

T2 Relaxation Time helps in differentiating between healthy and diseased tissues by specifying the duration of proton alignment loss.

3. **Clinical Diagnosis**: Radiologists and neurologists use synthetic MRI for brain-related conditions:

   People with **MS** (Multiple Sclerosis) find it easier if it is detected earlier so that a better treatment plan can be formulated.

   Synthetic MRI helps in monitoring the subtle change over time of brain tissue in neurodegenerative disorders like **Parkinson's and Alzheimer's**.

   Brain Tumours: Clearer images of tumour margins and oedema facilitate treatment planning.

   Stroke and Ischaemia: Improves the identification of injured brain tissue, helping in more accurate decisions in acute strokes.

20.10.2.1 Benefits

Synthetic MRI significantly improves brain imaging compared to traditional MRI methods.

1. **Reduced Scan Times**: Traditional MRI often requires 30–60 minutes and multiple scans, while synthetic MRI requires a single scan cutting the time down to 5–10 minutes. This time savings improves patient comfort, especially in children and those who have difficulty staying still or people who have a movement disorder.
2. **Increased Efficiency**: Synthetic MRI streamlines the process by eliminating the need for multiple scans, enabling hospitals to see more patients every day, which reduces the time people must wait to be seen.
3. **Improved Diagnostic Confidence**: The detailed images and data obtained through synthetic MRIs can be useful in monitoring the progression of diseases and enhancing diagnostic accuracy. The quantitative data available through this imaging technique enables clinical physicians to alter treatment strategies according to changes noted in tissue properties and evaluate the effectiveness of the treatments.
4. **Cost Savings**: Reducing the number of required scans reduces imaging costs for patients and healthcare providers alike.

20.10.2.2 Challenges and Limitations

Despite many positive aspects, some issues need to be solved before a broad application of synthetic MRI:

1. **Algorithm Accuracy**:
   a. Synthetic MRI is dependent on accurate machine learning algorithms. Errors in image reconstruction can lead to erroneous diagnoses.
2. **Data Dependency**:
   a. Synthetic images rely on training data with diversity and high quality. Having low datasets can create bias or poor performance.

3. **Clinical Integration**:
   a. If they were previously trained on conventional MRI techniques, it might take some time for the radiologists to get familiar with synthetic MRI technique to grade the synthetic MRI.
4. **Regulatory Approval**:
   a. The synthetic MRI systems need extensive validation and regulatory approval to ensure that they are safe and effective for clinical use.

20.10.2.3 Outcome

Multi-parametric imaging, facilitated by synthetic MRI in routine brain imaging, has allowed test performance to be more efficient and accurate, with the following salient benefits:

Faster Scans: Synthetic MRI cuts scan time, enabling doctors to see more patients in a day.

Improved Patient Care: Doctors can better plan and monitor treatments with reliable data emerging from scans. However, new synthetic MRI furnishes neurologist with the detailed information needed for the more efficient treatment of conditions like MS and brain tumour.

Enhanced Medical Decisions: Synthetic MRI helps the doctors to get a better understanding of diseases and assess treatment's effectiveness, steering to improved patient outcomes.

20.10.2.4 Future Directions

1. **Boosting Algorithm Precision**: Improve the accuracy and reliability of synthetic MRI algorithms across different patient groups [33].
2. **Broadening Synthetic MRI Applications**: Synthetic MRI is usually used for brain imaging but can also be used to look at muscles and bones and heart conditions.
3. **Integrating AI for Enhanced Efficiency**: By integrating synthetic MRI with AI, automated reports can be made, thus reducing workload.
4. **Performing Validation Studies**: By conducting necessary clinical trials, we can demonstrate the effectiveness of synthetic MRI.

20.10.3 Case Study 3: Cross-Modality Image Synthesis in Neuroimaging

Background: CT and MRI are specialized brain scans that look at the brain from different perspectives. CT excels at visualizing bone and acute haemorrhage; MRI excels at showing details of soft tissue. Having both scans done can also be costly, time-consuming, and occasionally impossible due to radiation or health concern. The approach taken to solve this is a new method called cross-modality image synthesis. It attempts to convert one type of scan to the other, providing information without performing both scans [34].

Cross-Modality Image Creation Process: The following steps are performed to transform images between different scan types:

1. **Data Gathering**: Get CT and MRI images from the same patient. Later, clean and align these images in order to denoise and enhance the quality.
2. **Model Training**:
   a. Train a GAN or CycleGAN to convert one type of image scan to another type of image scan.
   b. It consists of two parts—discriminator and generator. The discriminator measures the quality of generated images by contrasting them with actual CT images. The generator generates images that look like CT images from MRI images.
   c. A CycleGAN works better when CT and MRI cannot be perfectly matched.
3. **Image Creation**: After training, the model uses new MRI scans to generate artificial CT images that preserve the shape and brightness of actual CT scans.
4. **Evaluation and Testing**:
   a. To tell how well these artificial CT images compared with true images, experts employ several measures of evaluation. Mean Absolute Error (MAE) is the average difference between the fake and real images. The Peak Signal-to-Noise Ratio (PSNR) measures the quality of the image by comparing the desired signal to the background noise [35].
   b. Professional radiologists then evaluate the synthetic images' medical utility for particular diagnostic purposes, such as identifying fractures or haemorrhages.
5. **Clinical Workflow Integration**:
   a. For diagnostic purposes, the artificial CT pictures are used in conjunction with actual MRI scans.
   b. Without the need for further CT scans, they support thorough evaluations, surgery planning, and treatment monitoring.

20.10.3.1 Applications in Neuroimaging

1. **Tumour Diagnosis**: By integrating the advantages of many imaging modalities, cross-modality synthesis aids in the detection of brain tumours. Since calcifications and bone involvement are difficult to see in MRIs, synthetic CT images created from MRIs can shed light on these conditions.
2. **Stroke Assessment**: Although the detailed layers of an MRI make it more informative for ischaemic strokes, CT scans are preferable in acute stroke settings for the early detection of haemorrhages. This modality gap is bridged via the cross-modality synthesis, especially in the urgent condition where one is missing.
3. **Surgical Planning**: Doctors use various imaging methods to visualize both soft tissues and bones before surgery. This fusion of the MRI scans and synthetic CT images provides a holistic view of the patient anatomy.

4. **Radiotherapy**: For radiation treatment, synthetic CT scans play a very important role. They give exact radiation dose estimations and make sure to accurately target body areas, especially when standard CT scans are not available.

20.10.3.2 Benefits of Cross-Modality Image Synthesis

1. **Lower Radiation**: CT images made from MRI scans are also unlike the originals, and this makes it possible to reduce unnecessary radiation exposure—something that is especially important for children or those sensitive to radiation exposure.
2. **Cost Efficiency**: Synthetic imaging significantly reduces healthcare expenses by minimizing the need for multiple scans.
3. **Improved Access**: Reliable diagnosis potential is assured even in nations that do not have such equipment.
4. **Quick Turnaround**: Getting synthetic images fast helps to get all the decision-making right away. Whether it is stroke-like situations or major injuries from an accident.

20.10.3.3 Challenges and Limitations

1. **Accuracy**: While artificial images may look real, they often miss the details in the real things. Minor calcium deposits are important medical details.
2. **Data Quality**: The reliability of synthetic images depends on the diversity and quality of the training data. If the dataset is not big, the results are not generalizable and may introduce bias.
3. **Regulatory Approval**: Synthetic images must pass thorough testing and receive regulatory approval before being used in medical environments.
4. **Clinician Acceptance**: Some doctors might be reluctant to use synthetic images until they have proven accuracy and reliability in real-world applications.

20.10.3.4 Future Directions

1. **Using Multiple Scans**: Integrating scans like PET-MRI and SPECT-CT enhances the accuracy of medical diagnoses.
2. **Personalized Models**: Developing scan models tailored to individual anatomy boosts precision and clinical value.
3. **AI Verification**: Artificial intelligence tools ensure scan images meet medical standards, ensuring reliability.
4. **Federated Learning**: This approach allows hospitals to strengthen models with diverse data while safeguarding patient privacy.

Ultimately, a very large and unique step in brain imaging is using various scans to create images of the brain. Doctors are able to get full images without any radiation. So, it is safer and cheaper than combining different scans. This progress is driven by advanced AI technology. Using a particular tool, some problems still remain unsolved. With the technology research and development, the efficiency and vision of the tool are improving. Thus, it will be a major tool for the future (Figure 20.3).

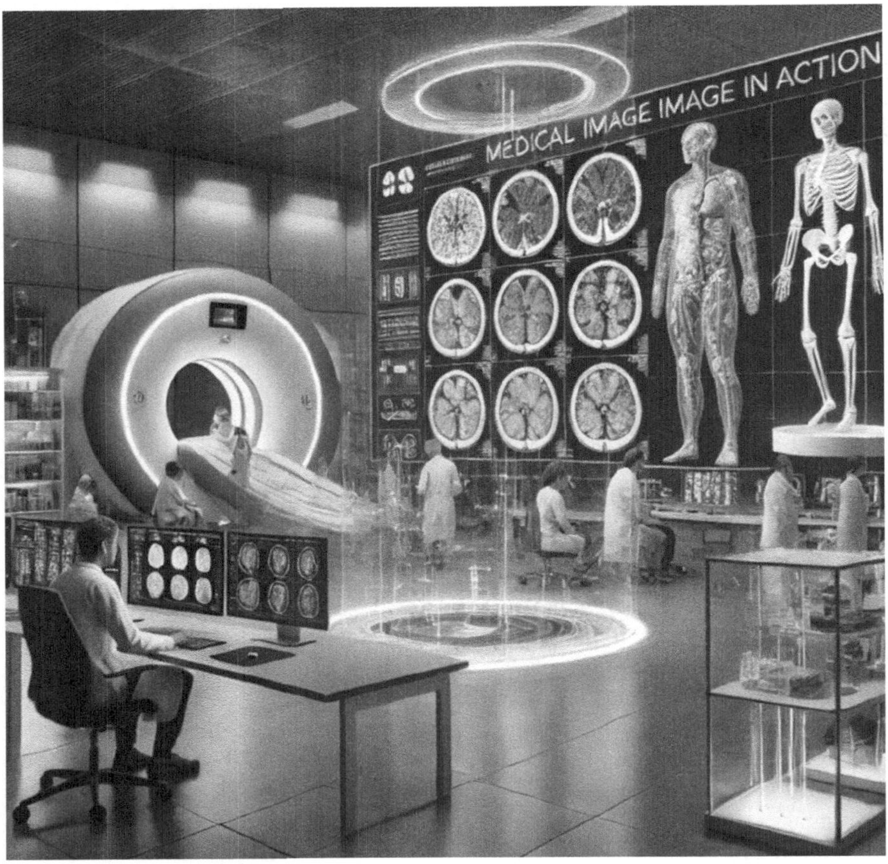

FIGURE 20.3 Realistic depiction of medical image synthesis in a modern lab. (The image shows a lab using artificial intelligence to create synthetic medical images including CTs and MRIs.)

20.11 CONCLUSION

Medical image synthesis is the future of medical imaging technology. Moreover, it can provide a cure for diseases, manage imaging of different sorts, and feature sets of various data. Using a supercomputer, AI, and this technology has been one of the most important of today's types of healthcare. It contributes significantly to increasing diagnostic accuracy, helping to tailor certain therapies, and providing extra training data for machine learning models.

20.11.1 Outlook

Synthetic medical images can enhance efficiency, accessibility, and effectiveness of healthcare delivery. Major research and development areas include:

1. **Precision Healthcare**: This refers to customizing techniques for every patient depending on their unique situation and medical conditions. It contributes to deciding treatment plans, predicting disease, organizing treatment plans, etc.
2. **Interpretability and Trust**: Improved lucidity of the produced model will help doctors confidently use it in clinical practice and decision-making.
3. **Ethical and Regulatory Considerations**: As these methods become more common, it's essential to address ethical concerns and meet regulatory standards, focusing on transparency, accountability, and protecting patient privacy.
4. **Prolonged Innovation**: Ongoing research will drive advancements in algorithms, learning systems, and measurement techniques, pushing the boundaries of medical imaging technology.

In conclusion, the technology behind the synthesis of medical images is highly innovative. It enhances the capacity for more accurate diagnostic tests and patient care. Also, it is virtually revolutionizing the techniques of tomorrow through the purpose of diagnosis and treatment as we strive towards a system that will enhance health and wellness.

REFERENCES

1. Chu, C., Zhmoginov, A., & Sandler, M. (2017). CycleGAN, a master of steganography. *arXiv:1712.02950.*
2. Mall, P. K., Singh, P. K., Srivastav, S., Narayan, V., Paprzycki, M., Jaworska, T., & Ganzha, M. (2023). A comprehensive review of deep neural networks for medical image processing: Recent developments and future opportunities. *Healthcare Analytics*, *4*, 100216. https://doi.org/10.1016/j.health.2023.100216.
3. Wang, T., Lei, Y., Fu, Y., Wynne, J. F., Curran, W. J., Liu, T., & Yang, X. (2020). A review on medical imaging synthesis using deep learning and its clinical applications. *Journal of Applied Clinical Medical Physics*, *22*(1), 11–36. https://pmc.ncbi.nlm.nih.gov/articles/PMC7856512/\.
4. Nie, D., Trullo, R., Lian, J., & Wang, L. (2018). Medical image synthesis with deep convolutional networks. *Medical Image Computing and Computer-Assisted Intervention (MICCAI)*, *65*(12), 2720–2730. https://pmc.ncbi.nlm.nih.gov/articles/PMC6398343/.
5. Niu, L., & Liu, Y. (2021). Recent advances in deep learning for medical image synthesis. *Medical Image Analysis*, *68*, 101876. https://doi.org/10.1016/j.media.2020.101876.
6. Guo, K., Chen, J., Qiu, T., Guo, S., Luo, T., Chen, T., & Ren, S. (2023). MedGAN: An adaptive GAN approach for medical image generation. *Computers in Biology and Medicine*, *163*, 107119.
7. Gu, X., Zhang, Y., Zeng, W., Zhong, S., Wang, H., Liang, D., Li, Z., & Hu, Z. (2023). Cross-modality image translation: CT image synthesis of MR brain images using multi generative network with perceptual supervision. *Computer Methods and Programs in Biomedicine*, *237*, 107571. https://doi.org/10.1016/j.cmpb.2023.107571.
8. Yin, X., Li, Y., Zhang, X., & Shin, B. -S. (2019). Medical image augmentation using image synthesis with contextual function. *2019 12th International Congress on Image and Signal Processing, BioMedical Engineering and Informatics (CISP-BMEI)*, Suzhou, China, pp. 1–6. https://doi.org/10.1109/CISP-BMEI48845.2019.8965817.
9. Goodfellow, I., Pouget-Abadie, J., Mirza, M., Xu, B., Warde-Farley, D., Ozair, S., & Bengio, Y. (2014). Generative adversarial nets. *Proceedings of the NeurIPS*. https://arxiv.org/abs/1406.2661.

10. Wang, T. -C., Liu, M. -Y., Zhu, J. -Y., Tao, A., Kautz, J., & Catanzaro, B. (2018). High-resolution image synthesis and semantic manipulation with conditional GANs. *2018 IEEE/CVF Conference on Computer Vision and Pattern Recognition*, Salt Lake City, UT, USA, pp. 8798–8807. https://doi.org/10.1109/CVPR.2018.00917.
11. Kingma, D. P., & Welling, M. (2013). Auto-encoding variational bayes. *Proceedings of the 2nd International Conference on Learning Representations (ICLR)*, May 2-4, Scottsdale, AZ, USA. https://arxiv.org/abs/1312.6114.
12. Meharban, M. S., Sabu, M. K.,& Krishnan, S. (2021). Introduction to medical image synthesis using deep learning: A review. *2021 7th International Conference on Advanced Computing and Communication Systems (ICACCS)*, Coimbatore, India, pp. 414–419. https://doi.org/10.1109/ICACCS51430.2021.9442041.
13. Goodfellow, A. (2020). Generative adversarial networks. *Communications of the ACM, 63*(11), 139–144.
14. Makhlouf, I., Maayah, M., Abughanam, N., & Catal, C. (2023). The use of generative adversarial networks in medical image augmentation. *Neural Computing and Applications, 35*(34), 24055–24068.
15. Fan, L., Bang, A., & Bonomi, L. (2024). Evaluating generative models in medical imaging. *Proceedings of the IEEE International Conference on Healthcare Informatics, 2024*, 553–555. https://doi.org/10.1109/ichi61247.2024.00084.
16. Zhou, T., Li, Q., Lu, H., Cheng, Q., & Zhang, X. (2023). GAN review: Models and medical image fusion applications. *Information Fusion, 91*, 134–148.
17. Suganthi, K. (2021). Review of medical image synthesis using GAN techniques. *Proceedings of the ITM Web of Conference, 37*, 01005.
18. Kalidindi, S., & Rathi, Y. (2019). Generative models for medical image augmentation. *IEEE Transactions on Biomedical Engineering, 66*(10), 2846–2856.
19. Rajpoot, N., & Zaidi, M. (2021). Advances in generative medical imaging. *Journal of Medical Imaging and Radiation Sciences, 52*(1), 25–34.
20. Miyato, T., & Koyama, M. (2018). CGANs with projection discriminator. *arXiv: 1802.05637.*
21. Mirza, M., & Osindero, S. (2014). Conditional generative adversarial nets. *arXiv: 1411.1784.*
22. Isola, P., Zhu, J. Y., Zhou, T., & Efros, A. A. (2017). Image-to-image translation with conditional adversarial networks. *IEEE Conference on Computer Vision and Pattern Recognition (CVPR)*, Honolulu, HI. https://ieeexplore.ieee.org/document/8100115.
23. Yi, Z., Walia, E., & Babyn, P. (2019). Generative adversarial network in medical imaging: A review. *Medical Image Analysis, 58*, 101552. https://www.sciencedirect.com/science/article/abs/pii/S1361841518308430.
24. Tajbakhsh, N., Shin. J. Y., Gurudu, S. R., Hurst, R. T., Kendall, C. B., Gotway, M. B., & Liang, J. (2016). Convolutional neural networks for medical image analysis: Full training or fine tuning? *IEEE Transactions on Medical Imaging, 35*(5), 1299–1312. https://doi.org/10.1109/TMI.2016.2535302.
25. Kebaili, A., Lapuyade-Lahorgue, J., & Ruan, S. (2023). Deep learning approaches for data augmentation in medical imaging: A review. *Journal of Imaging, 9*(4), 81. https://doi.org/10.3390/jimaging9040081.
26. Liu, J., Li, K., Dong, H., Han, Y., & Li, R. (2023). Medical image processing based on generative adversarial networks: A systematic review. *Current Medical Imaging, 20*, e15734056258198. https://doi.org/10.2174/0115734056258198230920042358.
27. Xun, S., Li, D., Zhu, H., Chen, M., Wang, J., Li, J., Chen, M., Wu, B., Zhang, H., & Chai, X. (2022). Generative adversarial networks in medical image segmentation: A review. *Computers in Biology and Medicine, 140*, 105063.

28. Zhang, Y., & Zheng, Y. (2020). Deep learning for medical image synthesis: From GANs to applications. *Journal of Healthcare Engineering, 2020*, 1–14.
29. Wang, S., Cao, G., Wang, Y., Liao, S., Wang, Q., Shi, J., Li, C., & Shen, D. (2021). Review and prospect: Artificial intelligence in advanced medical imaging. *Frontiers in Radiology, 1*, 781868. https://doi.org/10.3389/fradi.2021.781868.
30. Jeong, J. J., Tariq, A., Adejumo, T., Trivedi, H., Gichoya, J. W., & Banerjee, I. (2022). Systematic review of generative adversarial networks (GANs) for medical image classification and segmentation. *Journal of Digital Imaging, 35*(2), 137–152.
31. Islam, S., Aziz, T., Nabil, H. R., Jim, J. R., Mridha, M. F., & Kabir, M. (2024). Generative Adversarial Networks (GANs) in medical imaging: Advancements, applications, and challenges. *IEEE Access, 12*, 35728–35753. https://doi.org/10.1109/ACCESS.2024.3370848.
32. Liu, T., Peters, T., Staib, L., Essert, C., Zhou, S., Yap, P. –T., & Khan, A. (2019). Medical image computing and computer assisted intervention—MICCAI 2019. *Proceedings, Part II: 22nd International Conference*, Shenzhen, China, October 13–17, 2019. https://doi.org/10.1007/978-3-030-32245-8.
33. Anwar, S., Majid, M., Qayyum, A., Awais, M., Alnowami, M., & Khan, K. (2018). Medical image analysis using convolutional neural networks: A review. *Journal of Medical Systems, 42*, 226. https://doi.org/10.1007/s10916-018-1088-1.
34. Nguyen, H. V., Zhou, S. K., & Vemulapalli, R. (2015). Cross-domain synthesis of medical images using efficient location-sensitive deep network. *International Conference on Medical Image Computing and Computer-Assisted Intervention*, Munich, Germany, October 5–9, 2015.
35. Kazerouni, A., Aghdam, E. K., Heidari, M., Azad, R., Fayyaz, M., Hacihaliloglu, I., & Merhof, D. (2023). Diffusion models in medical imaging: A comprehensive survey. *Medical Image Analysis, 88*, 102846. https://doi.org/10.1016/j.media.2023.102846.

21 HybrBPF *Hybrid Bitcoin Price Prediction Framework Using Sentiment Analysis and Regression*

Abhinav Garg, Chhavi Dhiman, and Ajay Kaushik

21.1 INTRODUCTION

The rapid proliferation of digital technologies and the surge in online financial transactions have amplified the significance of predictive modeling in volatile domains such as cryptocurrency trading. Bitcoin (BTC) particularly being the most widely recognized and traded cryptocurrency has emerged as a focal point for investors and researchers alike. Accurate price prediction models are not only vital for informed investment decisions but also have broader applicability in domains like e-commerce and digital finance. For instance, sentiment-driven models can inform real-time product recommendation systems, enhance customer engagement strategies, or even contribute to fraud detection by identifying anomalous sentiment shifts. Furthermore, frameworks like the proposed HybrBPF have the potential to be adapted for real-time deployment, offering timely insights via mobile applications or edge devices. To enable such usage, optimization strategies such as model quantization, distillation, or lightweight transformer variants (e.g., DistilBERT) can be employed to reduce latency and computational overhead, making the model suitable for resource-constrained environments without significantly compromising performance.

Cryptocurrencies are digital assets that function as decentralized, intangible forms of virtual money, commonly represented as "coins." At present, the cryptocurrency market hosts over 20,000 distinct digital currencies. Public sentiment plays a crucial role in shaping investor interest, making it a valuable factor in predicting price movements. It is widely acknowledged that market sentiment influences asset prices. As highlighted by Baker & Wurgler (2007), the debate has shifted from questioning whether investor sentiment impacts stock prices to determining how to measure and analyze its effects. Investors are keen on market predictions, striving to establish a cause-and-effect relationship between available information and future trends. However, forecasting cryptocurrency prices is particularly complex due to the multitude of unknown factors influencing them. Unlike traditional assets, cryptocurrencies form a unique asset class that has garnered substantial attention from

 DOI: 10.1201/9781003565703-21

financial and investment communities. These digital currencies are traded on specialized crypto exchanges, which remain largely inaccessible to conventional investors.

BTC industry is challenging to analyze in a congenital way. Even the developers of the top digital currencies in the world have a tough time keeping an eye on all the latest coins, tokens, companies, and developments across the world that affect the BTC price. The industry is very young, largely untested, and unexplored meaning that there exist little theories, strategies, or prior proven models that can foresee and predict where the market is heading. Even when price prediction techniques use sophisticated analysis in an appropriate way, there are many factors that cryptocurrency community is unaware of or can control.

The COVID-19 pandemic has severely impacted communities, particularly those who are most vulnerable, including low-income individuals, the elderly, people with disabilities, young populations, and indigenous groups. It has disrupted various aspects of society, disproportionately affecting those in precarious social conditions. BTC experienced a steep decline in 2018, followed by only a slight recovery in 2019 (Peng, 2019). However, in the wake of the COVID-19 outbreak and the subsequent global implementation of quantitative easing, BTC's price surged beyond 40,000 dollars in the first half of 2020 (Peng, 2019). Given the significant risks posed by BTC's high price volatility, it is crucial for investors to adopt effective risk management strategies. Therefore, this study is valuable in providing a predictive support system to help investors minimize potential financial risks.

Stock market prediction techniques can generally be categorized into six main types: Statistical Methods, Pattern Recognition, Machine Learning, Deep Learning, Sentiment Analysis, and Hybrid Approaches (Shah, Isah, & Zulkernine, 2019). Among statistical models, the Autoregressive Moving Average (ARMA) is widely utilized for time-series forecasting. Pattern Recognition, a visually driven approach (Fu & Rosenfeld, 1976), has gained significant traction among traders, particularly with advancements in deep learning. Additionally, the development of machine learning-based recognition systems (Zhou & Hu, 2009) has further enhanced the importance of pattern recognition theory in market analysis. Another commonly used technique is Sentiment Analysis (Bollen, Mao, & Zeng, 2011), which operates on the principle of "the wisdom of crowds"—suggesting that collective opinions from multiple individuals can be as reliable as insights from a single expert. Social media, breaking news, press releases, and public events have become key sources for stock market forecasting through sentiment analysis. Hybrid methods, on the other hand, integrate multiple techniques from the aforementioned categories to enhance prediction accuracy.

The non-linearity in direction and magnitude of BTC price modulation makes it difficult to predict its prices in general. In many cases, tweets are duplicated or automated using bots, which makes the price prediction task more challenging. Tweets comprise features such as hashtags, profile mentions and URLs that cause a lot of noise. On top of this, use of sarcasm in tweets might cause erroneous or skewed predictions. Typically, predictive models focus on forecasting market movement in terms of direction, either upward or downward (Kilimci, 2020). While some research efforts have attempted to predict exact price values, their success has been limited (Pant, Neupane, Poudel, Pokhrel, & Lama, 2018).

Due to the challenges mentioned above, the analysis is not robust and powerful enough. Therefore, the proposed prediction framework introduces non-linearity to learn BTC price modulations using two-stage hybrid learning model. It learns sentiment features in stage 1 by performing classification and financial features in stage 2 using regression. This study aims to examine the impact of sentiment on predictive accuracy while exploring different approaches, including Machine Learning, Deep Learning, Sentiment Analysis, and Hybrid techniques (Shah, Isah, & Zulkernine, 2019) to predict BTC prices in near future.

The major contributions of the proposed work are threefold, which are given as follows:

1. An efficient BTC market price prediction framework is proposed that learns price modulations using BTC price data and user sentiments expressed as tweets during pre-COVID and post-COVID timelines in two stages. The first stage involves early BTC price prediction as a classification problem, which helps to understand the direction of BTC price modulation. The second stage involves early BTC price prediction as a regression problem, which helps to predict the magnitude of modulation in BTC price.
2. The Choquet fuzzy integral-based classifier and regressor is defined to combine the weighted integrals of multiple classification models—BERT and RoBERTa and regression models—Linear regressor, ARIMAX, XGBoost Regression, LSTM, and ensemble method.
3. The framework's performance is evaluated on a self-generated dataset of stock prices and tweets. An analysis is conducted to assess the effect of incorporating BTC price data and user tweets as hybrid data for confident decision-making over independent analysis. The experimental analysis supports the fact that sentiment analysis of tweets has reinforced the price prediction ability in understanding the direction of BTC price modulation.

The remainder of this chapter is structured as follows. Section 21.2 outlines the relevant literature. Section 21.3 details processes of data collection, data preprocessing, and an overview of the proposed framework. Section 21.4 presents the results derived from the proposed prediction model. Finally, Section 21.5 concludes the study and discusses potential directions for future research.

21.2 LITERATURE REVIEW

In this section, the recent literature about BTC price prediction is discussed in detail. It has been divided in two categories: (i) Regression-based methods and (ii) Sentiment analysis-based methods.

21.2.1 Regression-Based Methods

In recent years, regression models such as Autoregressive (AR) and Moving Average (MA), along with their combined forms—ARMA and ARIMA—have been widely employed for market forecasting. Since the 1970s, these models have gained popularity for time-series predictions, as introduced by Box, Jenkins, & Ayo (1970). Later, Ariyo, Adewumi, & Ayo (2014) utilized the ARIMA model to predict stock indices for the New York and Nigerian stock exchanges, demonstrating that ARIMA is not only effective for short-term forecasting but also a strong competitor against modern market prediction techniques. A review of previous research indicates that simpler and more linear models have been less frequently tested or evaluated over time. According to studies by Shah, Isah, & Zulkernine (2019) and Hiransha, Gopalkrishnan, Vijay, & Soman (2018), non-linear models tend to outperform linear ones, particularly in the presence of financial time-series challenges such as non-stationarity, seasonality, and high volatility. This motivated an examination of ARIMA, a well-established linear model. Deep Neural Network (DNN) architectures have emerged as a critical tool in forecasting applications. Researchers in cryptocurrency trading have leveraged state-of-the-art deep learning models such as CNN combined with RNN and Autoencoder-Decoder-based frameworks. Livieris, Kiriakidou, Stavroyiannis, & Pintela (2021) proposed a novel method for cryptocurrency price prediction using advanced deep learning techniques, including RNN and LSTM. The study incorporated multiple market factors, such as trading volume, market capitalization, circulating supply, and maximum supply, demonstrating promising results for specific dataset sizes. To improve the accuracy of cryptocurrency price predictions, researchers have explored ensemble learning techniques, integrating models such as bagging, stacking, and averaging with more complex deep learning architectures. The results highlighted that further optimization techniques and enhanced feature engineering could significantly refine predictive accuracy due to the sensitivity of hyperparameters in the CNN-LSTM model. Fang et al. (2021) applied an autoencoder-based augmented LSTM model to forecast the mid-price of eight cryptocurrency pairs. Meanwhile, Ali & Shatabda (2021) focused on data selection methodologies to train a linear regression model for BTC price prediction. Ma et al. (2023) utilized Support Vector Regression (SVR), Least Squares Support Vector Regression (LSSVR), and Twin Support Vector Regression (TWSVR) to forecast BTC prices. Their findings indicated that a hybrid model, XGBoost-WOA-TWSVR, delivered the most accurate predictions, with TWSVR excelling in both computational efficiency and predictive performance. Another study by Soni & Singh (2022) analyzed real-time BTC price data recorded at 24-hour intervals from 2009, concluding that the Bayesian Ridge Regressor outperformed all other regression-based models, followed by the Linear Regressor.

Additionally, a case study by Al-Haija (2023) introduced a non-linear autoregressive neural network with external input (NARX), identifying the regression coefficient that maximized prediction accuracy while minimizing normalized prediction error. More recently, Rungsrirat, Supratid, & Ariba (2023) examined multi-step BTC price forecasting using LSTM and GRU models, leveraging data from leading global cryptocurrency trading platforms.

21.2.2 Sentiment Analysis-Based Methods

Sentiment analysis refers to the process of examining and extracting emotions, opinions, subjectivity, and polarity from textual data. When applied to price prediction, sentiment analysis is often treated as a classification problem. It has various practical applications, such as automatically detecting positive, negative, or potentially harmful comments on websites and social media platforms. In recent years, sentiment analysis has been widely used to analyze data from Twitter. Several studies, including those by Kilimci (2020), Pant, Neupane, Poudel, Pokhrel, & Lama (2018), Stenqvist & Lönnö (2017), Kim et al. (2016), Kraaijeveld & Smedt (2020), Naeem, Mbarki, Suleman, Vo, & Shahzad (2020), Mohapatra, Ahmed, & Alencar (2019), and Hassan Kabir, Fahmi, & Rezzy (2021), have demonstrated a growing interest in predicting BTC prices using Twitter-based sentiment analysis. Since the introduction of sentiment analysis in the Natural Language Processing (NLP) field, various techniques have been developed to associate polarity with textual content (Marrese-Taylor, Rodriguez, Cristian, & Velasquez, 2012). Typically, sentiment classification categorizes text into three main groups: positive, negative, and neutral. One commonly used approach, the lexicon-based method, assigns sentiment scores by comparing text data with predefined sentiment lexicons. VADER (Valence Aware Dictionary and Sentiment Reasoner) Hutto & Gilbert (2015) is a well-known lexicon-based model that determines sentiment polarity using linguistic patterns. It incorporates tweet intensity rules derived from analyzing syntax, grammar, and valence scores of 800 tweets, integrating key features from three validated lexicons (Stenqvist & Lönnö, 2017). Kim et al. (2016) investigated the connection between user comments, reactions in online communities, and cryptocurrency price volatility across different platforms. Their findings revealed a mix of positive and negative discussions after performing web crawling on user-generated content. Additionally, Sinha, Kedas, & Kumar (2023) introduced entity-specific sentiment analysis for financial news interpretation. A recent study by Arratia, Avalos, Cabaña, López, & Mirambell (2021) emphasized how sentiment analysis contributes to investment and trading strategies by establishing correlations between financial news and market fluctuations (Wan et al., 2021). Some researchers have integrated multiple data sources for enhanced market forecasting. Dubey, Patil, Thakur, & Kanakia (2022) combined sentiment analysis from Yahoo Finance and Twitter polarity to predict stock market trends. Similarly, Fazlija & Harder (2022) examined the sentiment scores of news headlines to determine stock market index movements. Since stock prices are heavily influenced by investor sentiment, Lu, Ding, Shi, Fan, & Zhai (2021) developed a Vector Autoregression (VAR) model to assess how emotions impact stock price variations. Overall, these studies indicate that sentiment analysis plays a crucial role in refining stock market and BTC price predictions, helping investors make more informed decisions based on public sentiment trends.

Across the surveyed studies, several key gaps emerge, which motivated the design of our HybrBPF framework. First, many regression-based approaches—while effective at capturing linear dependencies (e.g., ARIMA, SVR)—struggle with the non-stationarity and high volatility inherent to BTC markets and often require manual hyperparameter tuning or extensive feature engineering (Shah et al., 2019; Hiransha et al., 2018). Second, sentiment-driven models typically treat textual signals and price series in isolation, limiting their ability to capture the joint influence of market psychology and trading behavior (Kilimci, 2020; Hassan et al., 2021). Third, deep learning and ensemble techniques, although powerful, frequently incur substantial computational cost and lack transparency, making them ill-suited for real-time deployment or resource-constrained environments (Livieris et al., 2021; Fang et al., 2021). Finally, existing hybrids seldom address both the direction and magnitude of price changes within a unified framework, thereby missing nuanced insights into short-term trend reversals vs. long-term volatility spikes (Ma et al., 2023; Rungsrirat et al., 2023). In contrast, our proposed two-stage HybrBPF addresses these limitations by first classifying directional shifts through transformer-based sentiment analysis on a combined pre- and post-COVID tweet-price corpus, then estimating fluctuation magnitude via an ensemble of regressors whose outputs are optimally fused using Choquet integral-based fuzzy logic. This design not only leverages complementary strengths of NLP and time-series models but also supports scalable adaptation for near-real-time inference, offering both interpretability and computational efficiency in handling BTC's complex dynamics.

21.3 RESEARCH METHODOLOGY

In this section, the proposed BTC price prediction framework is discussed in detail. It works in two stages. In the first stage, the direction of BTC price modulation in pre-COVID and post-COVID time duration is analyzed as classification problem. In the second stage, the magnitude of price modulation is evaluated using regression. It involves the collection of data samples and its preprocessing for the pre- and post-COVID sensitive time duration. The details are given in Sections 21.3.1 and 21.3.2. The proposed framework performs the BTC price modulation in terms of both direction and magnitude by treating respective tweets and BTC price data combined, instead of handling independently, as shown in Figure 21.1.

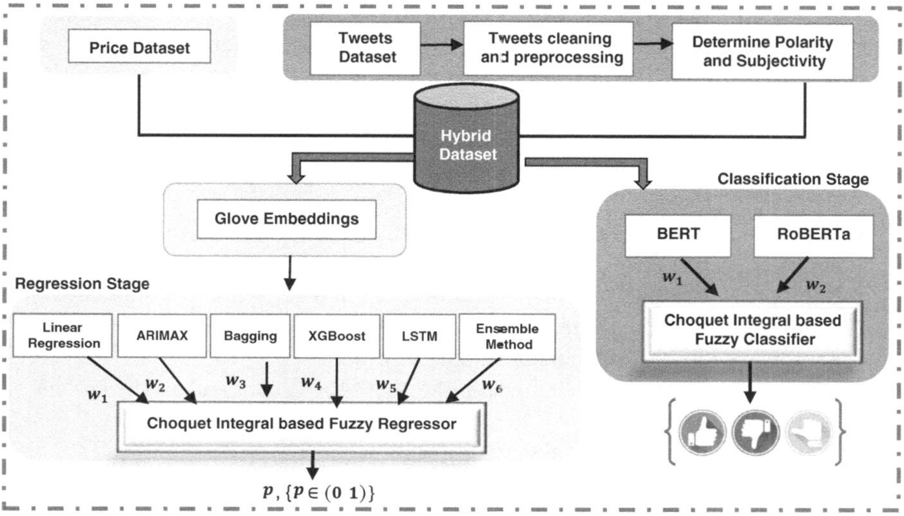

FIGURE 21.1 Proposed Bitcoin Price Prediction framework using combined price dataset and tweets dataset.

21.3.1 Data Collection and Preprocessing

The first and foremost step in predicting BTC price market trends involves searching for reliable and useful stock trading resources. The highly volatile nature of the BTC price market can't be decided using solely financial factors (Peng, 2019) or public opinions/sentiments (Pagolu, Reddy, Panda, & Majhi, 2016); so, we decided to adopt both financial (BTC/INR-Bitcoin Indian Rupee, 2020–2022) and sentiment features (Kaushik, 2022), (Badiola, 2019) as inputs.

For accurately determining the direction of BTC price modulation, tweets data is used for the sentiment analysis. For this purpose, we used twitter-based dataset (Kaushik, 2022; Badiola, 2019) for extracting tweets. The reasons for choosing this dataset are as follows. First, it embodies the "wisdom of crowds" concept (Goldberg, Nichols, Oki, & Terry, 1992), where collective opinions can be as valuable as expert insights. With approximately 500 million tweets published daily (Jethin, Higdon, Nelson, & Ibarra, 2018), Twitter serves as a vast, diverse, and dynamic platform. Its extensive user base and continuous flow of real-time data make it a valuable resource for gauging public sentiment on market trends, offering insights comparable to those of financial experts.

Second, the standardized structure of tweets sets them apart from other formats, such as newspaper articles. This distinct format implies that each tweet represents a specific combination of topic and sentiment, making it easier to analyze sentiment in a more granular and focused manner. Structure of self-generated dataset (pre-COVID and post-COVID) is shown in Figure 21.2c. Where stock price dataset includes the opening and closing price of the stock day wise along with highest and lowest stock values and the volume of stocks. On the other side, the twitter-based tweets dataset comprised of approximately 3.5 million tweets between 1 January 2018 to 21 January 2019 (pre-COVID) and 10 February 2021 to 25 May 2022 (post-COVID). Our dataset contained following fields: 'S.No', 'tweet', 'user_name', 'user_location', '*user_description*', 'user_created', 'user_followers', 'user_friends', 'user_favourites', 'user_verified', 'date', 'hashtags', 'source', 'is_retweet'. Except for 'S.No' and 'tweet', we dropped all the fields. For the experimentation, in the proposed work, hybrid dataset is utilized by combining date-wise stock variations with user sentiments during pre- and post-COVID. The preprocessing steps performed to handle tweet sample collection in the hybrid dataset are mentioned in Figure 21.2a.

i. Replaced symbols like '%' with 'percent', '₹' with 'rupee', '000' with 'k', etc.
ii. Replaced words like "didn't" with "did not," "it'd've" with "it would have," "y'all're" with "you all are," etc., i.e., all the words containing apostrophe are replaced by their expanded version of words.
iii. Removed all the special characters like '@,' '#', '*', etc., and all the emojis from the tweets.
iv. Removed hyphens and all the extra spaces between words.
v. Used TextBlob for Sentiment Analysis.

In the proposed model, it is ensured that the price dataset and Tweets dataset are combined to develop a hybrid dataset. It provides multi-dimensionality to the dataset.

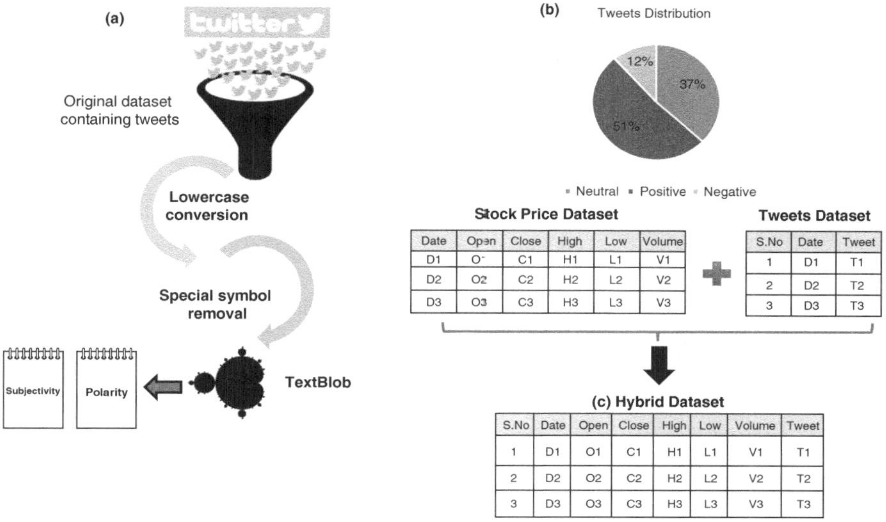

FIGURE 21.2 (a) Illustration of tweets preprocessing and cleaning steps. (b) Tweets distribution in twitter-based dataset. (c) Structure of self-generated dataset (pre-COVID and post-COVID).

And both regression and classification models are applied on hybrid dataset to observe both the magnitude and direction of the stock price values. For optimized performance, the best learning model is selected. Figure 21.2b represents the tweets distribution in twitter-based dataset used for training the classification models. We found Subjectivity and Polarity of each tweet and hence accordingly classified each tweet to positive, negative, and neutral.

21.3.2 Regression Stage

The proposed framework learns and predicts the magnitude of stock price variations by performing regression. The study includes four regression models to predict the BTC price data, i.e., Linear regression (Galton, 1886), ARIMA (Box, Jenkins, & Ayo, 1970), LSTM (Sepp & Jürgen, 1997), Bagging (Breiman, 1996), and XGBoost regression (Chen & Guestrin, 2016), and the Ensemble method. The Ensemble method utilizes multiple predictions generated by different decision trees. Each regression model generates independent predictions about stock variations. The proposed work utilizes Choquet integral-based fuzzy logic to make a final prediction.

21.3.3 Classification Stage

In the proposed work, two types of learning models are used. Regression models for learning BTC price prediction and classification models to learn the sentiments of the tweets.

21.3.4 Choquet Integral-Based Fuzzy Logic

The Choquet integral is a mathematical tool utilized in fuzzy logic to integrate information from various sources or rules, ultimately aiding in decision-making or classification. It is particularly effective in managing non-linear feature interactions and their respective contributions. This method consolidates individual scores into a single comprehensive score, commonly referred to as an aggregation operator. Other widely used aggregation operators include weighted averages and quasi-arithmetic means. Choquet Integral-based Fuzzy Logic is often favored over traditional aggregation methods because it accounts for the inherent uncertainty in decision scores—an aspect typically absent in standard ensemble techniques. As a result, it generalizes aggregation operators by incorporating confidence scores, which are represented as fuzzy measures. These fuzzy measures serve as weighted values assigned to each classifier. To apply Choquet Integral-based Fuzzy Logic, it is necessary to determine the fuzzy measure values for each classifier, denoted as s_i. These values not only define the strength of individual classifiers but also the collective strength of all possible classifier combinations. A schematic diagram illustrating the concept of Choquet Integral-based Fuzzy Logic is presented in Figure 21.3. In the case of the Choquet fuzzy integral, certain boundary conditions must be satisfied, which are as follows:

1. When all the classifiers are included in a combination, that combination possesses the highest possible strength.
2. If none of the classifiers are present in a combination, that combination does not hold any strength. The fuzzy measure captures three distinct types of interactions that can occur among the classifiers being analyzed. Suppose *X* and *Y* represent two mutually exclusive subsets of classifiers, meaning that they do not have any common element (i.e., $X \cap Y = \Phi$). Additionally, let $f(S)$ denote the strength of a given classifiers set *S* in the classification process. The three possible types of interactions are as follows:

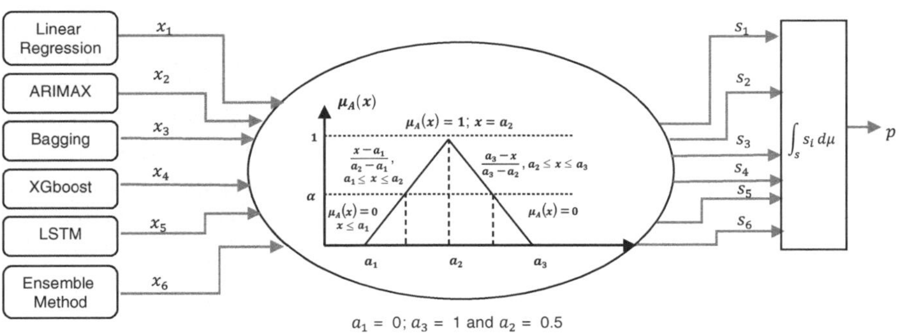

FIGURE 21.3 Schematic diagram of Choquet Integral-based Fuzzy Logic for triangle membership function. x_i: predictions generated by each learning model, s_i: fuzzy measure values mapped with fuzzy membership function μ; and p: final prediction.

i. **Fuzzy Rules Combined with Membership Functions**: In a fuzzy classifier, each feature or rule is associated with a membership function that assigns a membership value to each data point indicating the degree to which the data point belongs to a specific class. These membership values can be considered as a measure of evidence for each class.
ii. Capacities: Capacities are defined for each subset of features or rules, representing the importance or contribution of that subset in the classification process. Capacities are typically non-decreasing functions, and they contribute significantly to determine the relative importance of different subsets of features or rules. In our work, capacities have been learnt iteratively using Stochastic Gradient Descent. The hyperparameters chosen are epochs = 100 and Optimizer = Adam for $\text{learning}_{\text{rate}} = 0.01$
iii. **Combining Fuzzy Values**: The Choquet integral aggregates the fuzzy values (membership values) for each class based on the capacities. It calculates a weighted sum of the membership values for each class, where the weights are determined by the capacities.
iv. **Final Classification Decision**: After calculating the Choquet integral for each class, the class with the highest Choquet integral value is selected as the final classification decision.

Selecting the right fuzzy membership function is vital in fuzzy logic systems. It impacts data representation, interpretability, and overall system performance. The choice should match the data distribution and inherent uncertainty, enhancing model interpretability, robustness against noise, and computational efficiency. Matching domain knowledge can also guide the selection. Ultimately, the correct membership function plays a pivotal role in accurately modeling and processing data, ensuring improved system performance and generalization.

21.3.5 Experimental Results

The experimental analysis is divided into three parts. In stage 1, sentiment analysis of tweets on stock price modulation is performed using BERT and RoBERTa transformer Models, followed by classification. In stage 2, price prediction of the stock price data is carried out for various regression models like Linear regression (Galton, 1886), ARIMA (Box, Jenkins, & Ayo, 1970), and LSTM (Sepp & Jürgen, 1997) and ensemble methods—Bagging (Breiman, 1996) and XGBoost regression (Chen & Guestrin, 2016). Later, both sentiment features and financial features are learnt in a hybrid environment combinedly, and it is seen that the overall predictive strength of the proposed work improves over individual analysis. The structure of steps to perform Regression and Classification analysis on individual price dataset and tweets dataset, independently, are shown in Figure 21.4.

During training, for each regression and classification model, different parameters are utilized to optimize the learning. The experimental details of the regression models used in the proposed work are given as follows:

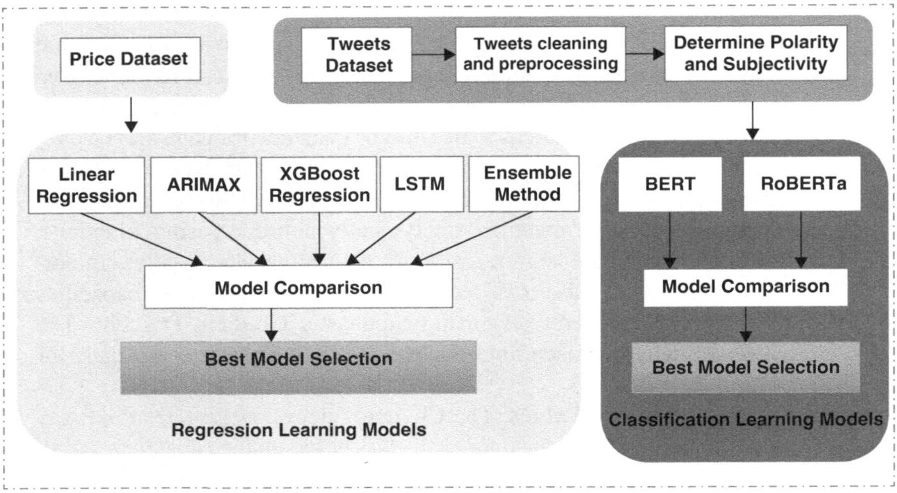

FIGURE 21.4 Regression and classification models built on individual price and tweets dataset.

i. **Linear Regression**: In our work, the model built is a simple linear regression (SLR) model, where one explanatory variable predicts a single dependent variable, and multiple linear regression (MLR) model, where various dependent variables predict a single dependent variable. A *train–test split* of 70–30 was followed on the price dataset. Different combinations of key features (independent variables) learn correlation between the price of BTC (dependent variable). Several independent variables are adding to the *Open* price feature and the observations of several features, including *Volume*, *High*, *Low*, and *Adj Close* of the current day. For this use case, we used a MLR model, mathematically represented as follows:

$$y = b_0 + \overrightarrow{b_1} \cdot \overrightarrow{x_1} + \cdots \overrightarrow{b_n} \cdot \overrightarrow{x_n} = b_0 + \sum_{i=1}^{n} \overrightarrow{b_i} \cdot \overrightarrow{X_i} \tag{21.1}$$

where the i refers to the ith independent variable in the dataset and n refers to the dimension of independent variables chosen.

ii. **Autoregressive Integrated Moving Average (ARIMA)**: A 70–30 train–test split was applied to the price dataset. To assess the stationarity of the time series, the Augmented Dickey–Fuller (ADF) test was conducted on both the original sequence and the first-order differenced sequence. A time series is considered non-stationary if the p-value exceeds 0.562 or if the Critical Value at a 1% significance level is greater than –3.44. Upon performing the ADF test, the results indicated that the original sequence was non-stationary, whereas the first-order differenced sequence was stationary. With $d = 1$ established, the next step was to determine $AR(p)$ and $MA(q)$ in ARIMA. This was achieved using Autocorrelation Figure (ACF) and Partial Autocorrelation Figure (PACF). The PACF indicated truncation at order = 2, suggesting that $AR\ (2)$ should be adopted. Meanwhile, the ACF displayed a long tail for all orders, indicating that $MA\ (0)$ should be selected. As a result, the final parameters chosen were $p = 2$ and $q = 0$.

iii. **Bagging**: This model is particularly effective where overfitting on the dataset is involved (high variance models). It reduces the variance without making the predictions biased. It reduces the absolute difference between training and testing accuracy smaller. A change in training accuracy might not be always visible, but test accuracy always improves in this case. We used the Bagging regression model of scikit learn. We tried training the regressor using different combinations of 'Open', 'High', 'Low', 'Adj Close', 'Volume' features and predicted the closing price of current day. A train–test split of 70–30 was followed on the price dataset. The hyperparameters chosen are $n\_estimators = 100$ and $max\_samples = 100$. A tenfold cross validation is adopted for $n\_repeats = 3$.

iv. **XGBoost**: In classification or regression problems involving unstructured data (like text, images, etc.), ANNs outperform all state-of-the-art algorithms because of their ability to capture non-linear relationship between input vectors. But while dealing with semi-structured or tabular data, decision tree-based algorithms are preferred. We used the Bagging regression model of scikit learn. We tried training the regressor using different combinations of *'Open'*, *'High'*, *'Low'*, *'Adj Close'*, *'Volume'* features and predicted the closing price of current day. A 70–30 train–test split was employed on the price dataset. The hyperparameters chosen are $n\_estimators = 100$, $max\_samples = 100$. A tenfold cross validation is adopted for $n\_repeats = 3$.

v. **LSTM**: As the time series evolves, the hidden state vector (h) updates frequently, whereas the memory cell vector (C) changes at a slower rate, maintaining long-term dependencies in Long Short-Term Memory (LSTM) networks. Each input vector x_t is influenced by the previous h_{t-1} and C_{t-1}, enabling LSTM to outperform standard Recurrent Neural Networks (RNNs) on longer sequences. To minimize computational complexity, we scaled stock prices between (0,1) as the first step of our implementation. Normalization is particularly beneficial when using RNNs with a Sigmoid activation function in the output layer. For forecasting the 61st price using 60-time steps, a specific data structure was required. After experimentation and testing, the number of historical timestamps was set to 60.

We utilized LSTM to build a Neural Network regressor for continuous value prediction, incorporating a Dropout layer for regularization. In each training cycle, 20% of the 50 neurons were randomly dropped to prevent overfitting. The second, third, and fourth LSTM layers were added following the same procedure.

vi. **Ensemble Model**: The learning model generates multiple subsets of data by randomly selecting samples from the training dataset with replacement. Each subset is then utilized to train its own decision tree. Consequently, this approach results in an ensemble of diverse models. To enhance reliability, the final prediction is derived by averaging the outputs of all decision trees, making it more robust than relying on a single decision tree. A train–test split of 70–30 was applied to the price dataset. The parameters used during training are as follows:

$$\text{'}base\_estimator\text{'}: [DecisionTreeRegressor()], \text{'}n\_estimators\text{'}: [200,500,1000],$$

$$\text{'}max\_samples\text{'}: [0.5,1.0, n\_samples / /2,], \text{'}max\_features\text{'}: [0.5,1.0, n\_features / /2,],$$

$$\text{'}bootstrap\text{'}: [True, False], \text{'}bootstrap\_features\text{'}: [True, False], \text{'}n\_iter\_search\text{'}: 100$$

For learning tweet's sentiments, the experimental details of the used classification models are as follows:

i. **BERT:** The selected hyperparameters include *padding* = *True*, *max length* = 512, which represent the maximum sequence length allowed for BERT. *Truncation* = True is applied to ensure that tokens exceeding this maximum length are truncated. A train–test–validation split of 60–20–20 is implemented. The model is trained for 1 *epoch* with a *batch_size* set to 8. BERT-base and BERT-large model specifications are listed in Table 21.1.

ii. **RoBERTa**: For the multi-class classification task, we implemented the RoBERTa base model. Hyperparameter tuning included *padding* = *True*, max *length* = 512, which defines the maximum allowed sequence length. Truncation was set to True to truncate tokens exceeding this maximum length. A split of 60–20–20 was used for train–test–validation; $batch_{size}$ of 8 was used for training, testing, and validation purpose; and the model was trained for 1 epoch.

Different evaluation metrics used for regression models are R^2 Score, Mean Squared Error (MSE), Root Mean Squared Error (RMSE), and Mean Absolute Percentage Error (MAPE). For classification, evaluation metrics such as model's precision, recall, F1, and support scores are used. Figure 21.5a and b represents confusion matrix obtained after training the BERT and RoBERTa models, respectively, on BTC

TABLE 21.1
BERT Model Specifications

Model	Layers	Attention Heads	Hidden Size	No. of Parameters (Millions)
BERT base (Devlin et al., 2019)	12	12	768	110
BERT large (Devlin et al., 2019)	24	16	1,024	340

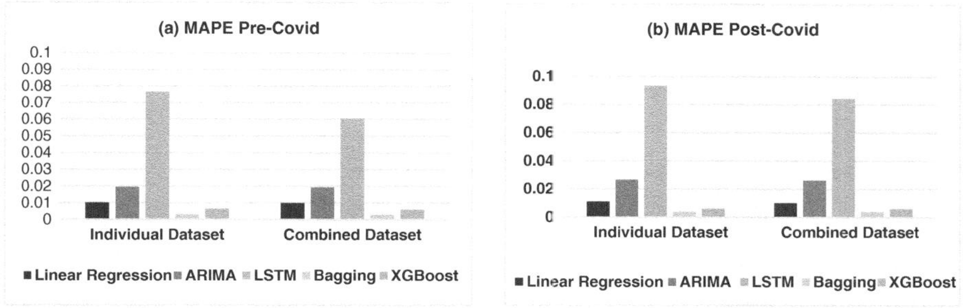

FIGURE 21.5 Confusion Matrix for (a) BERT on bitcoin price dataset, (b) RoBERTa on bitcoin price dataset, (c) BERT on hybrid dataset, and (d) RoBERTa on hybrid dataset.

TABLE 21.2
BERT and RoBERTa Classification Report on Tweets Dataset

Model Name	R^2 Score	MSE	RMSE	MAPE
Linear Regression	0.9964	1.8845	1.3728	1.0098×10^{-2}
ARIMA	0.9572	0.0001	0.0072	1.9437×10^{-2}
LSTM	0.9328	6.2430	2.4986	7.6381×10^{-2}
Bagging	**0.9982**	0.0001	0.0083	3.0179×10^{-3}
XG Boost	0.9951	0.0001	0.0125	6.4289×10^{-3}
Ensemble Model	0.9967	0.0012	0.00113	4.0021×10^{-3}

The bold value represents the best observed value of R^2 score

price dataset. Figure 21.5c and d denotes the confusion matrix obtained after training the BERT and RoBERTa models on hybrid dataset, respectively.

Tables 21.2 and 21.3 present the results obtained from two classification models, namely BERT and RoBERTa. As illustrated in these tables, RoBERTa outperforms BERT in the classification task for both the tweet dataset and the hybrid dataset by approximately 1%. This difference in performance can be attributed to significant variations in training methodology and dataset size between the two models. RoBERTa was trained on a massive 160 GB text corpus, which is over ten times larger than the dataset used for training BERT. Additionally, RoBERTa eliminates the Next Sentence Prediction (NSP) objective and employs a dynamic masking technique, enabling it to learn more robust and transferable word representations. These factors contribute to RoBERTa's superiority over similar models in NLP tasks such as text classification, language translation, and question answering. Its effectiveness has established it as a preferred model for various NLP applications in both research and industry.

The results presented in Tables 21.4 and 21.5 indicate that the Bagging ensemble method achieves the best performance among all models trained on the regression dataset. This can be attributed to Bagging's ability to mitigate overfitting in complex models and noisy data. By reducing the variance in predictions, it enhances

TABLE 21.3
BERT and RoBERTa Classification Report on Hybrid Dataset

	BERT				RoBERTa			
Classes/Metrics	**Precision**	**Recall**	**F1-Score**	**Support**	**Precision**	**Recall**	**F1-Score**	**Support**
Negative	0.93	0.92	0.93	2,446	0.94	0.93	0.94	2,446
Neutral	0.983	0.992	0.99	8,497	0.995	0.996	0.995	8,497
Positive	0.984	0.983	0.983	11,724	0.992	0.993	0.992	11,724
Micro Avg	0.98	0.98	0.98	22,667	0.99	0.99	0.99	22,667
Macro Avg	0.97	0.97	0.97	22,667	0.98	0.98	0.98	22,667
Weighted Avg	0.98	0.98	0.98	22,667	0.99	0.99	0.99	22,667
Samples Avg	0.98	0.98	0.98	22,667	**0.99**	**0.99**	0.99	22,667

Values in bold represent best observed value

TABLE 21.4
Performance of Various Regression Models on Bitcoin Price Dataset

	BERT				RoBERTa			
Classes/Metrics	**Precision**	**Recall**	**F1-Score**	**Support**	**Precision**	**Recall**	**F1-Score**	**Support**
Negative	0.92	0.91	0.91	2,446	0.93	0.92	0.92	2,446
Neutral	0.98	0.99	0.98	8,497	0.99	0.99	0.99	8,497
Positive	0.98	0.98	0.98	11,724	0.99	0.99	0.99	11,724
Micro Avg	0.97	0.97	0.97	22,667	0.98	0.98	0.98	22,667
Macro Avg	0.96	0.96	0.96	22,667	0.97	0.97	0.97	22,667
Weighted Avg	0.97	0.97	0.97	22,667	0.98	0.98	0.98	22,667
Samples Avg	0.97	0.97	0.97	22,667	0.98	0.98	0.98	22,667

TABLE 21.5
Performance of Various Regression Models on Hybrid Dataset

	Pre-COVID				Post-COVID			
Model Name	**R^2 Score**	**MSE**	**RMSE**	**MAPE**	**R^2 Score**	**MSE**	**RMSE**	**MAPE**
Linear Regression	0.9975	1.8289	1.3524	9.8991×10^{-3}	0.9992	1.4368	1.1987	9.9991×10^{-3}
ARIMA	0.9587	0.0001	0.0071	1.9176×10^{-2}	0.9395	0.0015	0.0396	2.6017×10^{-2}
LSTM	0.9429	2.8170	1.6784	6.0428×10^{-2}	0.9264	7.2215	2.6873	8.4283×10^{-2}
Bagging	**0.9993**	0.0001	0.0075	2.9726×10^{-3}	**0.9999**	0.0001	0.0073	3.8889×10^{-3}
XG Boost	0.9964	0.0001	0.0090	6.0955×10^{-3}	0.9967	0.0001	0.0084	6.0054×10^{-3}
Ensemble method	0.9985	0.0012	0.1001	4.0035×10^{-3}	0.9934	0.0025	0.02415	2.0001×10^{-3}

Values in bold represent best observed value of R^2 score for pre-covid and post covid test case.

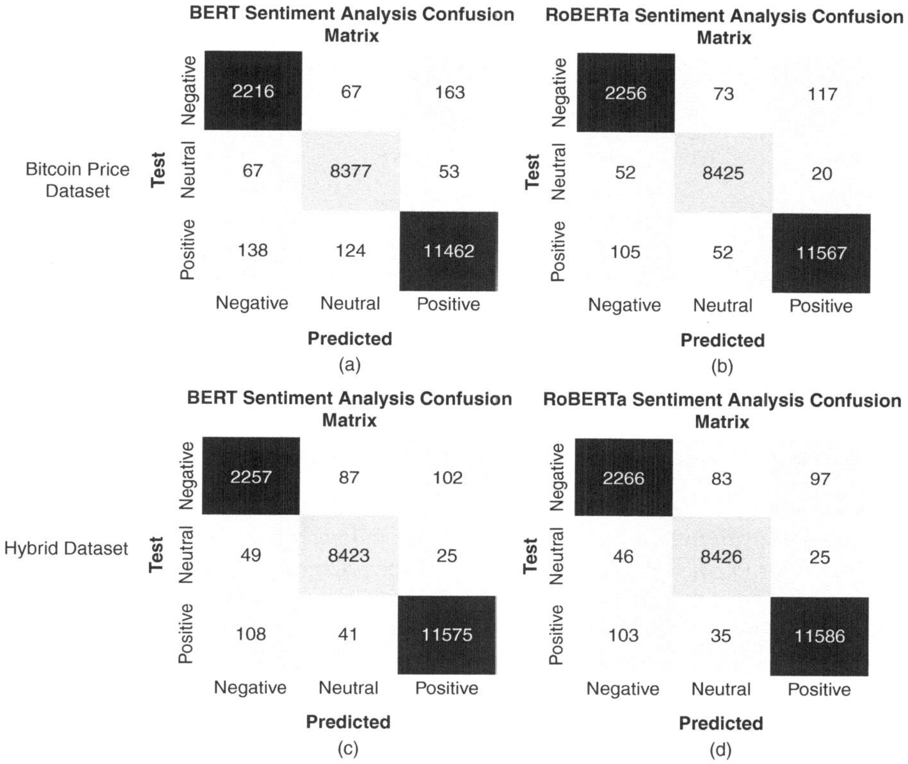

FIGURE 21.6 MAPE as reported on (a) pre-COVID dataset and (b) post-COVID dataset.

the model's stability. When comparing the results in Tables 21.4 and 21.2 with those in Tables 21.5 and 21.3, it becomes evident that the hybrid dataset produces more accurate predictions. Figure 21.5 further supports this finding, demonstrating a significant reduction in misclassifications when using a multi-dimensional hybrid dataset. Additionally, as shown in Figure 21.6a and b, the MAPE is lower for the hybrid dataset compared to individual datasets. This improvement is due to the integration of both price and tweet data, allowing for more informed and precise predictions. Consequently, the learning process on the multi-dimensional hybrid dataset surpasses that of individual regression models and classifiers. The overall effectiveness of the proposed framework stems from its ability to leverage a multi-dimensional hybrid dataset that integrates BTC price statistics and tweets rather than analyzing them separately. As evident from Table 21.6, utilizing this hybrid dataset leads to superior predictions when employing the RoBERTa-based classifier and the Bagging ensemble method. These approaches outperform independent analyses of tweets and BTC price data, particularly in capturing price trends during the COVID-19 and post-pandemic periods.

TABLE 21.6
Overall Observed Performance of the Proposed Framework

	Tweets Dataset	Bitcoin Price Dataset	Hybrid Dataset	
Method	(Precision/Recall)	R^2 Score	R^2 Score	(Precision/Recall)
RoBERTa	0.98/0.98	—	—	**0.99/0.99**
BERT	0.97/0.97	—	—	0.98/0.98
Bagging	—	0.9982	**0.9993**	—
Ensemble method	—	0.9967	0.9985	—
Linear Regression	—	0.9964	0.9975	

Values in bold represent best observed value.

21.4 CONCLUSION

In this work, a price prediction model is built using price statistics and tweets sentiment collected from twitter (BTC/INR-Bitcoin Indian Rupee, 2020–2022), respectively. The problem is divided into two parts for price prediction model. The price statistics are used to perform regression analysis, whereas sentiments of the tweets are analyzed by performing classification to comprehend the increase and decrease in price value. The contribution of our work is threefold. First, the tweet sentiments are analyzed using BERT and RoBERTa models to learn the direction of BTC price modulation. Second, various regression models such as LSTM and Bagging Methods are applied on price data to learn the amount of rise and fall of the BTC prices. Third, a hybrid model is defined, which merges both BTC price dataset and tweets dataset together and performs both regression and classification. This exhibits a relationship between the sentiments of the tweets and the predicted BTC prices to make a collective confident decision. The observed results showed that hybrid models outperformed the independently trained regression and classification models.

Our HybrBPF framework overcomes the inability of traditional linear models like ARIMA to capture BTC's non-stationary and non-linear volatility by integrating advanced non-linear regressors and transformer-based sentiment features to model complex market dynamics. Unlike resource-intensive deep-learning ensembles, HybrBPF leverages a Choquet integral-based fusion of lightweight regression outputs to reduce computational overhead and facilitate the deployment in resource-constrained environments. By jointly processing sentiment signals via fine-tuned transformer classifiers alongside price data within the same pipeline, our approach addresses the isolated treatment of textual and financial time series prevalent in lexicon-based sentiment models. Further, the two-stage design uniquely forecasts both the direction and magnitude of price movements in a unified framework—filling a gap identified in recent surveys of cryptocurrency forecasting methods. Finally, the adoption of fuzzy logic-based Choquet aggregation enables interpretability and robust handling of model interdependencies, enhancing prediction stability compared to simple averaging or majority-vote ensembles.

While our work emphasizes on using twitter sentiment analysis and prices dataset for BTC price determination, there are various trading strategies such as reversal trading, gap trading, pairs trading, and momentum trading. It would be felicitous to check upon them and draw a comparison with our methodology. Since other cryptocurrencies can also affect BTC prices, it would be worth investigating the correlation between various cryptocurrency prices. Also, investor attention and sentiment dispersion (Suardi, Rasel, & Liu, 2022) can be considered to further extend our work. Besides popular public figures (like Elon Musk) can also influence public opinions.

CONFLICTS OF INTEREST

The authors declare that they have no conflicts of interest to this work.

REFERENCES

Al-Haija, Q. A. (2023). Time-Series Analysis of Cryptocurrency Price: Bitcoin as a Case Study. *International Conference on Electrical Engineering, Computer and Information Technology (ICEECIT)*, Jember, Indonesia. https://doi.org/10.1109/ICEECIT55908.2022.10030536.

Ali, M., & Shatabda, S. (2021). A Data Selection Methodology to Train Linear Regression Model to Predict Bitcoin Price. *International Conference on Advanced Information and Communication Technology (ICAICT)*, Dhaka, Bangladesh. https://doi.org/10.1109/ICAICT51780.2020.9333525.

Ariyo, A., Adewumi, A. O., & Ayo, C. (2014). Stock Price Prediction Using the ARIMA Model. *16th International Conference on Computer Modelling and Simulation*, Cambridge, UK.

Arratia, A., Avalos, G., Cabaña, A., López, A. D., & Mirambell, M. R. (2021). Sentiment Analysis of Financial News: Mechanics and Statistics. In *Data Science for Economics and Finance* (pp. 195–216). Springer. https://doi.org/10.1007/978-3-030-66891-4_9.

Badiola, J. (2019). Bitcoin Tweets Sentiment Analysis. Retrieved from https://www.kaggle.com/datasets/jaimebadiola/bitcoin-tweets-and-price.

Baker, M., & Wurgler, J. (2007). Investor Sentiment in the Stock Market. *Journal of Economic Perspectives, American Economic Association, 21*(2), 129–152.

Bollen, J., Mao, H., & Zeng, X. (2011). Twitter Mood Predicts the Stock Market. *Journal of Computational Science, 2*, 1–8.

Box, G., Jenkins, G., & Ayo, C. (1970). *Time Series Analysis, Forecasting and Control*. Holden-Day Inc.

Breiman, L. (1996). Bagging Predictors. *Machine Learning, 24*(2), 123–140.

BTC/INR-Bitcoin Indian Rupee. (2020-2022). Retrieved from https://in.investing.com/crypto/bitcoin/btc-inr-historical-data.

Chen, T., & Guestrin, C. (2016). XGBoost: A Scalable Tree Boosting System. *KDD '16: Proceedings of the 22nd ACM SIGKDD International Conference on Knowledge Discovery and Data Mining*, San Francisco, CA.

Devlin, J., Chang, M., Lee, K., & Toutanova, K. (2019). BERT: Pre-training of Deep Bidirectional Transformers for Language Understanding. *North American Chapter of the Association for Computational Linguistics*.

Dubey, S., Patil, N., Thakur, P., & Kanakia, H. (2022). Stock Market Prediction Using Yahoo finance and Twitter Polarity. *IEEE 3rd Global Conference for Advancement in Technology (GCAT)*. Bangalore, India: IEEE, pp. 1–5, doi: 10.1109/GCAT55367.2022.9972013.

Fang, F., Ventre, C., Basios, M., Kanthan, L., Wu, F., & Li, L. (2021). Cryptocurrency Trading: A Comprehensive Survey. *Financial Innovation, 8*, 13.

Fazlija, B., & Harder, P. (2022). Using Financial News Sentiment for Stock Price Direction Prediction. *Mathematics, 10*(13), 2156. https://doi.org/10.3390/math10132156.

Fu, K., & Rosenfeld, A. (1976). Pattern Recognition and Image Processing. *IEEE Transactions on Computers, C-25*, 1336–1346.

Galton, F. (1886). Regression towards Mediocrity in Hereditary Stature. *Journal of the Anthropological Institute, 15*, 246–263.

Goldberg, D., Nichols, D., Oki, B. M., & Terry, D. B. (1992). Using Collaborative Filtering to Weave an Information Tapestry. *Communcation of the ACM, 35*, 61–70.

Hassan, M. K., Fahmi, H., & Rezzy, C. (2021). Mining Netizen's Opinion on Cryptocurrency: Sentiment Analysis of Twitter Data. *Studies in Economics and Finance, 39*, 365–385.

Hiransha, M., Gopalkrishnan, A., Vijay, K., & Soman, K. (2018). NSE Stock Market Prediction Using Deep-Learning Models. *Procedia Computer Science, 132*, 1351–1362.

Hutto, C., & Gilbert, E. (2015). VADER: A Parsimonious Rule-Based Model for Sentiment Analysis of Social Media text. *Proceedings of the 8th International Conference on Weblogs and Social Media, ICWSM 2014,* 8(1), 216–225. https://doi.org/10.1609/icwsm.v8i1.14550.

Jethin, A., Higdon, D., Nelson, J., & Ibarra, J. (2018). Cryptocurrency Price Prediction Using Tweet Volumes and Sentiment Analysis. *SMU Data Science Review, 1*(3), 1.

Kaushik, S. (2022). Bitcoin Tweets. Retrieved from https://www.kaggle.com/datasets/kaushiksuresh147/bitcoin-tweets.

Kilimci, Z. H. (2020). Sentiment Analysis Based Direction Prediction in Bitcoin Using Deep Learning Algorithms and Word Embedding Models. *International Journal of Intelligent Systems and Applications in Engineering, 8*, 60-65.

Kim, Y. B., Kim, J. G., Kim, W., Im, J. H., Kim, T. H., Kang, S. J., & Ki, C. H. (2016). Predicting Fluctuations in Cryptocurrency Transactions Based in User Comments and Replies. *Plos One, 11*, 161–197. https://doi.org/10.1371/journal.pone.0161197.

Kraaijeveld, O., & Smedt, J. D. (2020). The Predictive Power of Public Twitter Sentiment for Forecasting Cryptocurrency Prices. *Journal of International Financial Markets, Institutions and Money, 65*, 101188. https://doi.org/10.1016/j.intfin.2020.101188.

Livieris, I. E., Kiriakidou, N., Stavroyiannis, S., & Pintela, P. (2021). An Advanced CNN-LSTM Model for Cryptocurrency Forecasting. *Electronics, 10*(3), 287. https://doi.org/10.3390/electronics10030287.

Lu, Y., Ding, N., Shi, M., Fan, Z., & Zhai, Y. (2021). An Empirical Study on the Influence of Investor Sentiment and Stock Price——Based on VAR Model. *International Conference on Advanced Electronic Materials, Computers and Software Engineering (AEMCSE).* Changsha, China: IEEE.

Ma, J., Zhu, Y., Xu, J., Li, Y., Zhang, Y., & Wang, J. (2023). Research on Bitcoin Price Prediction Based on Support Vector Regression and Its Variant Combination Model. *International Conference on Computational Intelligence and Security (CIS),* Chengdu, China. https://doi.org/10.1109/CIS58238.2022.00046.

Marrese-Taylor, E., Rodriguez, C., & Velasquez, J. D. (2012). Web Opinion Mining and Sentiment Analysis. *Advanced Techniques in Web Intelligence, 2*, 105–126.

Mohapatra, S., Ahmed, N., & Alencar, P. (2019). A Real Time Cryptocurrency Price Prediction Platform Using Twitter Sentiments. *International Conference on Big Data (Big Data),* 5544–5551. IEEE. https://doi.org/10.1109/BigData47090.2019.9006554.

Naeem, M. A., Mbarki, I., Suleman, M. T., Vo, X. V., & Shahzad, S. J. (2020). Does Twitter Happiness Sentiment Predict Cryptocurrency. *International Review of Finance, 21*, 1529–1538. https://doi.org/10.1111/irfi.12339.

Pagolu, V. S., Reddy, K. N., Panda, G., & Majhi, B. (2016). Sentiment Analysis of Twitter Data for Predicting Stock Market Movements. *International Conference on Signal Processing, Communication, Power and Embedded System (SCOPES)*, Paralakhemundi, India, 1345–1350.

Pant, D., Neupane, P., Poudel, A., Pokhrel, A., & Lama, B. (2018). Recurrent Neural Network Based Bitcoin Price Prediction by Twitter Sentiment Analysis. *2018 IEEE 3rd International Conference on Computing, Communication and Security (ICCCS)*, 128–132, Kathmandu, Nepal.

Peng, D. (2019). Analysis of Investor Sentiment and Stock Market Volatility Trend Based on Big Data Strategy. *2019 International Conference on Robots & Intelligent System (ICRIS)*, Haikou, China, 269–272.

Rungsrirat, S., Supratid, S., & Ariba, T. (2023). An Investigation on Multi-Step Bitcoin Prediction Based on LSTM and GRU. *International Electrical Engineering Congress (iEECON),* Krabi, Thailand. https://doi.org/10.1109/iEECON56657.2023.10126509.

Sepp, H., & Jürgen, S. (1997). Long Short-Term Memory. *Neural Computation, 9*, 1735–1780.

Shah, D., Isah, H., & Zulkernine, F. (2019). Stock Market Analysis: A Review and Taxonomy of Prediction Techniques. *International Journal of Financial Studies, 7*(26), 1–22.

Sinha, A., Kedas, S., & Kumar, R. (2023). SEntFiN 1.0: Entity-Aware Sentiment Analysis for Financial News. *arXiv:2305.12257v1*, 1–32.

Soni, K., & Singh, S. (2022). Bitcoin Price Prediction—An Analysis of Various Regression Methods. *IEEE 12th Symposium on Computer Applications & Industrial Electronics (ISCAIE)*, Penang, Malaysia. https://doi.org/10.1109/ISCAIE54458.2022.9794532.

Stenqvist, E., & Lönnö, J. (2017). Predicting Bitcoin Price Fluctuation with Twitter Sentiment Analysis: Masters Project. *Master in Computer Science, School of Computer Science and Communication,* KTH, Stockholm, Sweden.

Suardi, S., Rasel, A. R., & Liu, B. (2022). On the Predictive Power of Tweet Sentiments and Attention on Bitcoin. *International Review of Economics and Finance, 79*, 289–301.

Wan, X., Yang, J., Marinov, S., Calliess, J. P., Zohren, S., & Dong, X. (2021). Sentiment Correlation in Financial News Networks and Associated Market Movements. *Scientific Reports, 11*, 3062. https://doi.org/10.1038/s41598-021-82338-6.

Zhou, B., & Hu, J. (2009). A Dynamic Pattern Recognition Approach Based on Neural Network for Stock Time-Series. *2009 World Congress on Nature & Biologically Inspired Computing (NaBIC)*, 1552–1555, Coimbatore, India.

22 A Framework Based on Generative AI for Stock Recommendation

Sonal Tuteja, Archit Malhotra, Aishwarya Shanker, and Yashika Agarwal

22.1 INTRODUCTION

The stock market is a complex and constantly changing ecology due to its inherent volatility and complexity. Financial analysis and stock recommendations have mostly depended on professional judgment and accepted statistical techniques. Despite their benefits, these strategies are limited in their capacity to grow and adjust to the volatile and fast-paced environment of contemporary financial markets. Financial decision-making has been revolutionized by the development of artificial intelligence (AI) and machine learning (ML), which have made it possible to develop advanced tools that can handle enormous volumes of data and produce incredibly accurate insights [1,2].

Long Short-Term Memory (LSTM) networks and Generative AI have recently emerged as key technologies in tackling the difficulties of contemporary stock recommendation systems. In order to provide individualized, real-time recommendations, Generative AI uses sophisticated Natural Language Processing (NLP) techniques to evaluate unstructured data sources, such as economic reports, social media conversations, and financial news [3]. However, by using historical information, LSTM networks are especially good at forecasting stock prices and trends due to their ability to analyze sequential data [4,5]. As a result of these advancements, hybrid models that combine the advantages of LSTM networks with Generative AI have been created, providing a thorough and reliable method of stock market analysis [6].

Despite these technological improvements, the employment of these tools in a user-friendly framework designed especially for the Indian stock market is still untapped. The gap between sophisticated financial instruments and the particular requirements of individual investors is growing as a result of many existing solutions' failure to address the crucial elements of personalization, transparency, and simplicity of access. Furthermore, the Indian stock market requires a system that can adjust to sudden changes in the market and provide fast, actionable insights due to its dynamic character, which is influenced by regional mood, global market trends, and local economic policies [7].

 DOI: 10.1201/9781003565703-22

This paper proposes an innovative framework designed to bridge these gaps by integrating Generative AI, LSTM networks, and technical analysis. The objective is to create an intuitive, scalable, and efficient stock recommendation platform that empowers both novice and experienced investors. The proposed system will provide tailored financial advice, accurate stock price predictions, and insightful visualisations through an interactive virtual assistant. By leveraging state-of-the-art technologies such as NLP, speech recognition, and predictive analytics, the framework aims to enhance financial literacy and enable informed decision-making for investors in the Indian market [8,9].

This paper is organized as follows. Section 22.2 summarizes the literature review. Section 22.3 represents the proposed framework. Section 22.4 presents the results. Section 22.5 concludes the paper with findings and future work.

22.2 LITERATURE REVIEW

22.2.1 Advancements in Stock Recommendation Systems: Generative AI and LSTM

Generative AI and LSTM networks have emerged as transformative technologies in stock recommendation systems. Generative AI utilizes advanced NLP and ML techniques to analyze vast amounts of unstructured financial data, such as news articles, social media trends, and financial reports. By extracting meaningful insights, these models generate real-time, dynamic stock recommendations tailored to individual investors' preferences and market conditions [3].

LSTM networks, a form of Recurrent Neural Network (RNN), excel in capturing temporal dependencies in time-series data, addressing the vanishing gradient problem, and retaining long-term information. These attributes make LSTM models particularly effective for stock price forecasting based on historical data. Studies by Qiao et al. [1] and Mehtab et al. [5] have highlighted the robustness of LSTM in predicting stock returns by capturing both short-term fluctuations and long-term trends. Integrating LSTM within stock recommendation systems enhances their predictive accuracy and adaptability to diverse market conditions [4].

22.2.2 Hybrid Models and Personalization in Financial Virtual Assistants

Hybrid models that combine Generative AI with LSTM networks represent a significant advancement in stock recommendation frameworks. These models leverage Generative AI's ability to process real-time, unstructured data (e.g., market news, social media sentiment) alongside LSTM's strength in analysing historical price data. This dual approach creates a comprehensive prediction system capable of adapting to volatile markets, such as India, where sudden economic or sentiment shifts can drastically influence stock prices [2].

Furthermore, Generative AI-driven frameworks are increasingly implemented as financial virtual assistants. These assistants provide real-time, personalized investment advice tailored to users' financial goals, risk tolerance, and preferences. By leveraging NLP, these systems make financial data accessible and actionable, enhancing user engagement, particularly for novice investors [7]. Such personalized systems cater to diverse investor profiles, making them indispensable tools in modern financial services [9].

22.2.3 Challenges and Future Directions

Despite the promising results, several challenges persist in implementing a Generative AI-based framework for stock recommendation. One major issue is the model's dependency on high-quality, real-time data for both NLP and LSTM components. Inaccurate or outdated data inputs can lead to misleading recommendations. Another challenge lies in the interpretability of Generative AI models, which can be perceived as a "black box" by users. The complexity of these models may hinder trust and transparency in financial decision-making. Future research must address these challenges by improving data preprocessing techniques, enhancing model interpretability, and refining hybrid models for real-time stock analysis [3]. Trivedi et al. [6] also emphasized the importance of transparency in AI-driven decision systems to build user trust and improve overall adoption.

22.3 PROPOSED WORK

This platform combines Generative AI, ML, deep learning (LSTM), and technical analysis to provide accurate, real-time stock and financial advice, along with an engaging learning experience. The aim is to improve financial and economic knowledge among users by giving them a complete investment guide. It predicts stock movements, records economic data, and gives insights into market behavior. Powered by Generative AI, the Fin-Bot solves financial literacy problems by answering user queries—from beginners to experts—making the platform more useful. With the help of advanced NLP, speech recognition, and prediction techniques, the system gives personalized stock suggestions, trend predictions, and detailed reports. All these are combined with Generative AI, as shown in Figure 22.1 [1,2].

The image explains the overall system architecture of the platform. It starts with data collection from sources like real-time stock APIs, news websites, and financial reports. This raw data is cleaned, transformed, and vectorized so that it can be used efficiently. Then, the processed data flows into the historical data analyzer and the LSTM-based predictor engine for analysis and forecasting. Simultaneously, the same data is stored as vectorized chunks for retrieval. When a user types or speaks a query, the query engine processes it using NLP, searches the relevant data chunk, and passes it to the Generative AI engine. This engine uses both a finance knowledge base (RAG approach) and a financial rule base to generate a smart and personalized response. The final results—stock predictions, analysis, and advice—are displayed on a user-friendly dashboard that supports learning and decision-making.

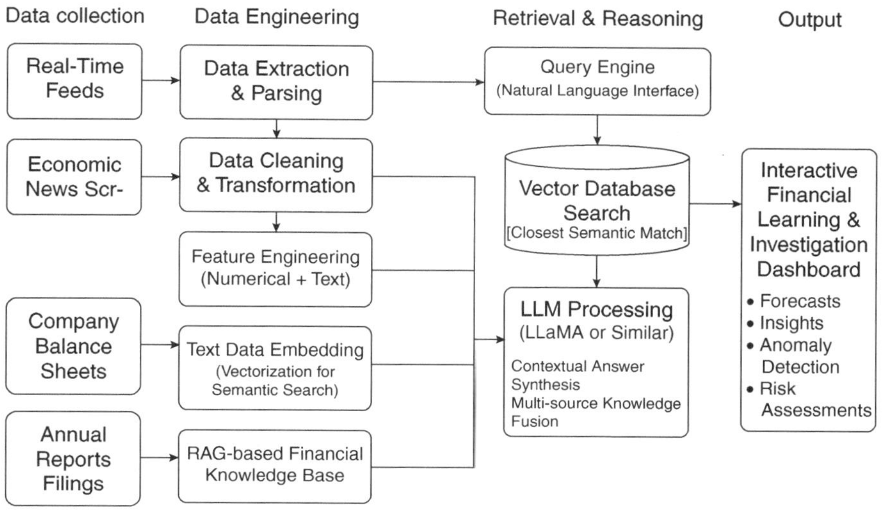

FIGURE 22.1 System architecture of the proposed work.

22.3.1 Data Collection and Preprocessing

To make sure the system gives accurate and trustworthy financial predictions, data is collected from multiple sources:

- Real-Time Stock Market APIs: These APIs give live data from the stock market, such as opening and closing prices, daily highs and lows, and the number of trades.
- Economic News from Websites: Financial news articles and updates that affect market trends are collected to understand wider economic situations [3].
- Companies' Balance Sheets and Financial Reports: These documents provide detailed company performance data to judge their financial health.

Once the data is collected, it goes through several preprocessing steps. The data extraction process takes out important information from the sources. Then, data cleaning and transformation removes mistakes, fills in missing values, and corrects data so that it's ready to use. Finally, data vectorization changes the cleaned data into a structured format so that advanced models and AI tools can understand and use it [4].

22.3.2 Historical Data Analysis Engine

The Historical Data Analyzer Engine is an important part of the system. It deeply studies past stock market trends by using the cleaned and processed data. Through yearly trend analysis, it finds long-term patterns in the stock market. These patterns help users understand how stocks behave over the years. This knowledge forms a solid base for users to make smart investment decisions [5].

22.3.3 AI Stock Predictor Engine

The AI Stock Predictor Engine leverages LSTM neural networks, a powerful deep learning model capable of processing sequential data. By analyzing both historical and real-time stock data, the engine predicts future stock prices and market trends.

The model architecture includes stacked LSTM layers that are fine-tuned to capture the temporal dependencies within time-series data. This enables the engine to make accurate forecasts. These predictions assist users in identifying promising investment opportunities and provide actionable insights to guide their decisions [1,6]. The predictive capability of this engine is a significant asset for both novice and experienced investors.

22.3.4 Query Engine

The system includes a Query Engine, which processes user inputs in natural language format. Users can type or speak queries such as *"What is the forecast for Tesla?"* or *"Which stocks are trending upward?"* The Query Engine utilizes NLP techniques to understand the context and intent of the queries [3,7].

To provide relevant responses, the engine searches for the closest matching data chunks from the processed data repository. This ensures that users receive precise and actionable information tailored to their queries.

22.3.5 Generative AI Engine (LLAMA)

The Generative AI Engine, powered by the LLAMA model, is responsible for generating highly personalized financial recommendations. This module combines data-driven insights with advanced AI capabilities to deliver tailored responses.

Two key components enhance the performance of this engine:

- Finance Knowledge Base (RAG Approach): This knowledge base retrieves relevant insights from a repository of financial data, enabling the engine to generate informed recommendations [8].
- Financial Rule Base: Predefined rules and heuristics are applied to interpret the data and provide actionable advice, such as buy/sell recommendations or portfolio diversification strategies [9].

The integration of Generative AI ensures that users receive comprehensive, transparent, and easy-to-understand financial guidance [2,8].

22.3.6 Financial Learning and Investigation Dashboard

The Financial Learning Dashboard serves as the primary user interface, consolidating the outputs of all system components. It provides users with a visually intuitive platform where they can access predictions, market analyses, and recommendations [10].

The dashboard is designed to enhance financial literacy by presenting information in a format that is easy to interpret. Users can view trend analyses, stock forecasts, and recommendations in real time. This module caters to a diverse audience, ranging from novice investors seeking basic guidance to financial experts requiring in-depth analysis [3].

22.4 EXPERIMENTAL EVALUATION

This framework integrates real-time stock market data, economic news, and financial reports to create a Financial Learning and Investigation Dashboard. Data extraction processes transform and clean input data, which is then vectorized and chunked [1]. Historical data is analyzed using trend analysis, while the AI Stock Predictor Engine (LSTM) forecasts stock performance [2]. A Query Engine identifies the closest data chunks for user queries, leveraging a GEN-AI Large Language Model (LLM) engine (LLAMA) integrated with a Finance Knowledge Base (RAG approach) and Financial Rules [3,4]. The system delivers insights and predictions efficiently, supporting financial decision-making and learning through an intuitive dashboard interface [5].

22.4.1 Experimental Setup

The proposed framework has been implemented on a system equipped with an Intel Core i7 processor, 16 GB of RAM, and an NVIDIA RTX 3060 GPU, running on Windows 11. The hardware configuration ensures efficient training and execution of ML models, while a stable Python environment is maintained for development [6].

Dataset: The framework uses datasets containing historical stock prices, financial indicators (e.g., RSI, MACD, and Bollinger Bands), and market news sentiment. These datasets are stored in PostgreSQL and processed during experimentation [7].

Tools and Technologies: The implementation leverages OpenAI's GPT-3.5 or GPT-4 for NLP tasks such as analyzing market sentiment and generating stock recommendations [8].

Evaluation Metrics: The system evaluates the accuracy of stock recommendations using Root Mean Square Error (RMSE) for price prediction [9].

This setup ensures a robust and scalable environment for generating reliable stock recommendations through advanced AI methodologies [10].

22.4.2 Evaluation Criteria

The study implements a comprehensive framework of Python-based tools for stock market analysis, utilizing machine learning, statistical models, and interactive dashboards [11]. Predictive modeling was conducted using ARIMA, SARIMA, and LSTM neural networks, which provided reliable stock price forecasts [12]. The LSTM model achieved the lowest RMSE, outperforming statistical models such as ARIMA and SARIMA [5]. Additionally, the LSTM model accurately captured temporal dependencies, achieving 68% accuracy in predicting uptrends and 32% accuracy in identifying downtrends [13].

Weekly performance metrics for Indian stocks, calculated using moving averages, P/E ratios, PEG ratios, and 52-week price ranges, identified top-performing stocks like Reliance and TCS. These insights were further augmented by a moving average crossover strategy that identified actionable buy and sell signals with an accuracy of 80% and 76%, respectively.

22.4.3 Results

The platform's performance was comprehensively evaluated using various methods and metrics to ensure that it delivers accurate stock price predictions, reliable trading signals, and an interactive user experience.

- Stock Price Prediction: Three different models were used to assess forecasting accuracy. The ARIMA model achieved an RMSE of 15.3, demonstrating its effectiveness in forecasting historical stock prices [12]. However, its inability to capture seasonal patterns was addressed by the SARIMA model, which reduced the RMSE to 12.7, offering improved accuracy by accounting for seasonality [5]. The LSTM model emerged as the most effective, with an RMSE of 8.5, showcasing its capability to learn and predict temporal trends with high precision [13]. Beyond error rates, the LSTM model achieved an uptrend prediction accuracy of 68% and a downtrend prediction accuracy of 32%, indicating its potential and room for improvement [13].
- Performance Metrics: Key financial indicators such as Moving Averages, P/E ratios, and PEG ratios were utilized to rank high-performing stocks. Based on these metrics, the platform successfully identified ten top-performing stocks, including *RELIANCE.NS*, *TCS.NS*, and *INFY.NS*, which consistently demonstrated strong financial performance and market stability. These insights validate the platform's ability to analyze financial data effectively and provide actionable investment recommendations.
- Trading Signals: The platform employed the Moving Average Crossover technique, which proved to be a reliable strategy for identifying buy and sell opportunities. The accuracy of buy signal predictions was 80%, while sell signals were identified with 76% accuracy, highlighting the system's capability to guide optimal trading decisions.
- User Experience and Interactivity: Interactivity was evaluated based on the performance of the Streamlit dashboard. The system was optimized for real-time analysis, achieving a response time of less than 2 seconds for user queries. This fast turnaround significantly enhanced the user experience by delivering timely insights. Additionally, the TF-IDF-powered news research tool achieved 85% query match accuracy, allowing users to retrieve highly relevant information from unstructured financial news sources (Figure 22.2).

The image depicted explains a bar chart titled "RMSE Comparison for Stock Price Prediction Models." It compares the RMSE values of three different models used for stock price prediction:

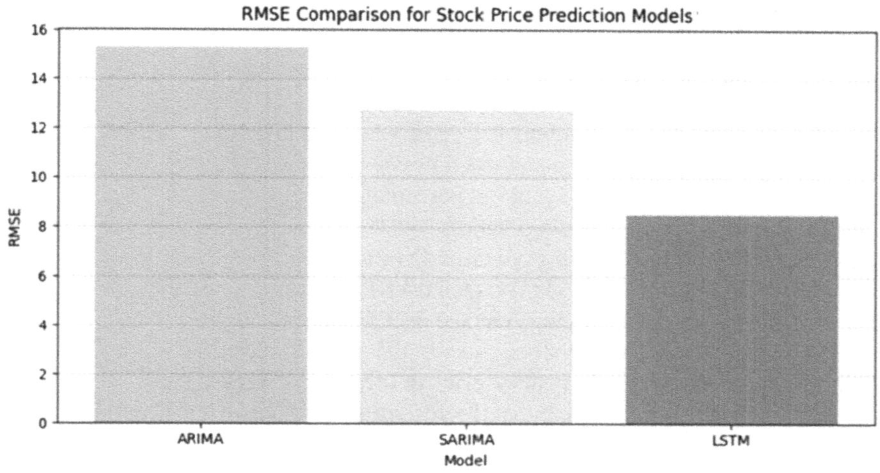

FIGURE 22.2 RMSE comparison for stock price prediction models.

S.No.	Ticker	Latest Price (₹)	Score (%)	PE Ratio	PEG Ratio	SMA 150 (₹)	SMA 200 (₹)	52 Weeks Low (₹)	52 Weeks High (₹)	Better By (%)
1	SBIN.NS	570.40	83.0	18.2	1.3	550.30	540.20	480.00	620.00	**12.63%**
2	ITC.NS	430.25	84.5	20.5	1.5	425.10	420.00	350.00	480.00	**11.58%**
3	HINDUNILVR.NS	2,550.80	85.0	31.2	2.2	2,520.20	2,500.60	2,100.00	3,000.00	**10.53%**
4	LT.NS	2,550.00	86.0	27.0	1.8	2,500.40	2,480.30	2,100.00	2,900.00	**9.47%**
5	KOTAKBANK.NS	1,900.15	87.5	28.4	2.1	1,870.50	1,860.40	1,600.00	2,200.00	**8.42%**
6	HDFCBANK.NS	1,670.25	88.0	24.8	1.7	1,630.10	1,610.50	1,400.00	1,900.00	**7.37%**
7	ICICIBANK.NS	940.30	89.0	22.3	1.6	920.40	910.20	800.00	1,050.00	**6.32%**
8	INFY.NS	1,525.60	90.5	26.7	1.9	1,500.50	1,490.80	1,200.00	1,750.00	**5.00%**
9	TCS.NS	3,450.75	92.0	30.2	2.0	3,400.10	3,350.45	2,850.00	3,950.00	**3.26%**
10	RELIANCE.NS	2,400.50	95.0	25.5	1.8	2,350.30	2,310.40	1,900.00	2,850.00	-

FIGURE 22.3 List of top 10 best performing stocks.

1. ARIMA—Has the highest RMSE value, slightly below 15.
2. SARIMA—Has a lower RMSE than ARIMA, approximately 13.
3. LSTM—Has the lowest RMSE, just below 9.

The *x*-axis represents different models (ARIMA, SARIMA, and LSTM), and the *y*-axis represents the RMSE values, which range from 0 to 16.

From this comparison, we can infer that LSTM outperforms both ARIMA and SARIMA in terms of prediction accuracy, as indicated by its lower RMSE (Figure 22.3).

The image explains the lists of the top 10 best performing stocks, and it provides detailed financial metrics for the top 10 stocks based on performance.

Here's a breakdown of the columns:

1. Ticker—Stock ticker symbols (e.g., SBIN.NS and ITC.NS).
2. Latest Price (₹)—Most recent stock price.
3. Score (%)—Performance score in percentage.
4. PE Ratio—Price-to-Earnings ratio.
5. PEG Ratio—Price/Earnings to Growth ratio.
6. SMA 150—Simple Moving Average over 150 days.
7. SMA 200—Simple Moving Average over 200 days.
8. 52 Weeks Low—Lowest stock price in the past 52 weeks.
9. 52 Weeks High—Highest stock price in the past 52 weeks.
10. Better By (%)—Percentage that the current price is above the 52-week high.

22.4.3.1 Notable Insights

- RELIANCE.NS has the highest score (95.0%) but no "Better By (%)" value listed.
- SBIN.NS leads in "Better By (%)" with 12.63%, indicating that it is performing significantly above its previous 52-week high.
- Stocks like TCS.NS, INFY.NS, and ICICI BANK.NS also show strong performances.

The table offers a snapshot useful for investors looking to evaluate top-performing Indian stocks based on a mix of technical and fundamental indicators.

22.5 CONCLUSION

This Generative AI and LSTM-Based Stock Recommendation Framework combines the power of deep learning, NLP, ML clustering, and Generative AI to deliver a comprehensive and user-friendly stock analysis tool [1,2]. The integration of real-time data, predictive models such as LSTM, and personalized recommendations using LLMs allows investors to make informed decisions with confidence [3,4].

The system leverages advanced AI techniques to conduct technical analysis, forecast stock trends, and extract sentiment insights from financial news, offering a multi-dimensional view of market behavior [5,6]. Its architecture incorporates vectorized and chunked data processing, Retrieval-Augmented Generation (RAG), and financial rules to support contextualized query responses [7,8].

The platform is scalable and efficient, and it provides reliable insights into stock trends, performance forecasts, and investment strategies [9]. It empowers both novice and professional investors with actionable intelligence through an intuitive and interactive dashboard that supports both data-driven analysis and conversational AI guidance [10,11].

Overall, this framework exemplifies how modern AI technologies can be synergistically applied to transform traditional stock analysis into a dynamic, real-time decision support system for financial markets [12].

REFERENCES

1. Qiao, R., Chen, W. and Qiao, Y., 2022. Prediction of stock return by LSTM neural network. *Applied Artificial Intelligence*, *36*(1), p.2151159.
2. Kumar, A., Hooda, S., Gill, R., Ahlawat, D., Srivastva, D. and Kumar, R., 2023, April. Stock price prediction using machine learning. In *2023 International Conference on Computational Intelligence and Sustainable Engineering Solutions (CISES)* (pp. 926–932). IEEE, Greater Noida, India.
3. Subhash, S., Srivatsa, P.N., Siddesh, S., Ullas, A. and Santhosh, B., 2020, July. Artificial intelligence-based voice assistant. In *2020 Fourth world conference on smart trends in systems, security and sustainability (WorldS4)* (pp. 593–596). IEEE.
4. Ding, G. and Qin, L., 2020. Study on the prediction of stock price based on the associated network model of LSTM. *International Journal of Machine Learning and Cybernetics*, *11*(6), pp. 1307–1317.
5. Mehtab, S., Sen, J. and Dutta, A., 2021. Stock price prediction using machine learning and LSTM-based deep learning models. In *Machine Learning and Metaheuristics Algorithms, and Applications: Second Symposium, SoMMA 2020*, Chennai, India, October 14–17, 2020, Revised Selected Papers 2 (pp. 88–106). Springer Singapore.
6. Trivedi, A., Pant, N., Shah, P., Sonik, S. and Agrawal, S., 2018. Speech to text and text to speech recognition systems—A review. *IOSR Journal of Computer Engineering*, *20*(2), pp. 36–43.
7. Roshini, M., Nandhika, S., Prathipa, S. and Velswamy, R., 2022, May. AI system with voice modulation using neural network, *International Journal for Research in Applied Science and Engineering Technology*, *10*(5), pp. 385–388. https://doi.org/10.22214/ijraset.2022.42193.
8. Manojkumar, P.K., Patil, A., Shinde, S., Patra, S. and Patil, S., 2023. AI-based virtual assistant using python: A systematic review. *International Journal for Research in Applied Science & Engineering Technology (IJRASET)*, *11*, pp. 814–818.
9. Mekni, M., 2021. An artificial intelligence based virtual assistant using conversational agents. *Journal of Software Engineering and Applications*, *14*(9), pp. 455–473.
10. Vijh, M., Chandola, D., Tikkiwal, V.A. and Kumar, A., 2020. Stock closing price prediction using machine learning techniques. *Procedia Computer Science*, *167*, pp. 599–606.
11. Ghosh, A., Bose, S., Maji, G., Debnath, N. and Sen, S., 2019, September. Stock price prediction using LSTM on the Indian share market. In *Proceedings of 32nd International Conference on Computer Applications in Industry and Engineering* (Vol. 63, pp. 101–110), San Diego, California, USA.
12. Roondiwala, M., Patel, H. and Varma, S., 2017. Predicting stock prices using LSTM. *International Journal of Science and Research (IJSR)*, *6*(4), pp. 1754–1756.
13. Mariani, M.M., Hashemi, N. and Wirtz, J., 2023. Artificial intelligence empowered conversational agents: A systematic literature review and research agenda. *Journal of Business Research*, *161*, p. 113838.

Index

B

C

D

E

F

G

J

K

L

M

S

T

X

Y

Z